Theoretical Evolutionary Ecology

Michael Bulmer
Rutgers University

Sinauer Associates
Publishers
Sunderland, Massachusetts

THEORETICAL EVOLUTIONARY ECOLOGY

 For information or to order address:

Sinauer Associates, Inc.
23 Plumtree Road
Sunderland, MA 01375-0407 U.S.A.

FAX: 413-549-1118
E-mail: publish@sinauer.com

Library of Congress Cataloging-in-Publication Data
Bulmer, Michael, 1931 –
Theoretical Evolutionary Ecology / Michael Bulmer.
p. cm.
Includes bibliographical references and index.
ISBN 0-87893-079-5 (cloth) ISBN 0-87893-078-7 (pbk.)
1. Evolution (Biology). 2. Ecology. 3. Population biology.
4. Evolution (Biology) — Mathematical models. I. Title.
QH366.2.B85 1994
575.01 — dc20 93–34424
CIP

Printed in U.S.A.
6 5 4 3 2 1

To Sally and Emma

Contents

Acknowledgments

I am grateful to the following colleagues who made valuable suggestions for improving the manuscript: Brian Charlesworth, Alan Grafen, Alasdair Houston, Peter Morin, Joel Peck, Andrew Pomiankowski, and Peter Taylor.

Andy Sinauer and his colleagues have been supportive throughout this project, and my copy editor David Baldwin has made a substantial contribution by his fine eye for detail.

I thank Rutgers University for granting me sabbatical leave to complete the manuscript and the Department of Genetics, Trinity College, Dublin for their hospitality during part of this leave.

Academic Press, Blackwell Scientific Publications, Cambridge University Press, Canada Communication Group, Evolution, Macmillan Magazines, Oxford University Press, Science, University of Chicago Press, and the respective authors, kindly gave permission for the use of previously published material.

CHAPTER 1

Introduction

"In considering the Origin of Species, it is quite conceivable that a naturalist, reflecting on the mutual affinities of organic beings, on their embryological relations, their geographical distribution, geological succession, and other such facts, might come to the conclusion that each species had not been independently created, but had descended, like varieties, from other species. Nevertheless, such a conclusion, even if well founded, would be unsatisfactory, until it could be shown how the innumerable species inhabiting this world have been modified, so as to acquire that perfection of structure and coadaptation that most justly excites our admiration....

In the next chapter the Struggle for Existence amongst all organic beings throughout the world, which inevitably follows from their high geometrical powers of increase, will be treated of. This is the doctrine of Malthus, applied to the whole animal and vegetable kingdoms. As many more individuals of each species are born than can possibly survive; and as, consequently, there is a frequently recurring struggle for existence, it follows that any being, if it vary however slightly in any manner profitable to itself, will have a better chance of surviving, and thus be *naturally selected*. From the strong principle of inheritance, any selected variety will tend to propagate its new and modified form."

—Charles Darwin, *On the Origin of Species* (1859)

Ecology is the study of plants and animals at home, that is to say, in their natural environment (from the Greek word *oikos*, a house). Evolutionary ecology is the branch of this subject that considers how organisms have evolved to become adapted to their environment, including in this term their interactions with members of their own and other species (the biotic environment) as well as the physical environment; it examines the selective pressures imposed by the environment and the evolutionary response to these pressures.

Darwin (1859) proposed natural selection as a unifying principle to explain two things: the transmutation of species and the adaptation of organisms to their environment. Evolutionary ecology takes the second of these as its field of study. Its aim is to explain, in the light of current knowledge, "how the innumerable species inhabiting this world have been modified, so as to acquire that perfection of structure and coadaptation that most justly excites our admiration."

THE EVOLUTION OF CLUTCH SIZE

To illustrate the main ideas of the subject I shall consider the evolution of clutch size in birds, a problem first considered in a classic paper by Lack (1947). Why do different species of birds lay very different numbers of eggs? Lack began by pointing out that this question can be answered at two different levels:

> There has been much confusion in discussions on clutch-size through failure to distinguish between the ultimate factors affecting survival value and the proximate factors affecting physiological control. ... To illustrate this point from another problem, that of breeding seasons, it may be said that a songbird breeds in spring (a) because there is then enough food to raise a brood, and (b) because daylength stimulates the growth of the gonads. Food and daylength both influence the breeding season, but at quite different biological levels.

The distinction between a proximate, physiological explanation and an ultimate, evolutionary explanation in biology is fundamental. They are not alternative explanations between which one must choose, but explanations at different levels that are equally important.

Lack was interested in the ultimate factors determining clutch size, those concerned with its survival value. He first destroyed an erroneous view that was popular at the time, and that still misleads some people today, that clutch size has been adjusted by natural selection to fit the mortality of the species. It had been suggested, for example, that northern races tend to have larger clutches than tropical races to balance the higher mortality experienced by the former, either in surviving a cold winter or in migratory journeys. Lack argued that this interpretation is wrong for two reasons. He wrote:

> It is a truism that if a species is neither increasing nor decreasing in numbers, then clutch-size and total mortality from egg to adult must be balanced. [Some authors] have taken this to mean that clutch-size has been adjusted by natural selection to fit the mortality of the species. ...
>
> However, the fundamental assumption is wrong. While reproductive rate closely affects the potential rate of increase of a species, it is not the primary cause of the steady density that a species eventually reaches. If a population is approximately steady, this is probably due to the operation of one or more mortality factors whose effectiveness increases with the density of the species, so that when the species is scarce a smaller proportion, and when it is abundant a larger proportion, of the individuals are killed. ...
>
> Hence the existence of balance does not really imply that clutch-size is adjusted to mortality. It is, of course, still true that clutch-size and mortality equate together. But the correct inference to draw from this is that, if more eggs are laid, the annual mortality has to be greater. ...
>
> Another strong argument against [this] view is the extreme difficulty of seeing how natural selection could act on clutch-size so as to bring about its alleged adjustment to total mortality. Natural selection acts on the individual and its family rather than on the species as a whole. If some individuals of a species lay four eggs and others five, then, given that clutch-size is inherited, the five-egg birds are bound to predominate rapidly in the population, as they have more descendants, even if this leads to "over-population"—unless the laying of five eggs instead of four is a disadvantage in itself, either to the brood, or to the parents, or to both.

In this passage Lack lucidly explained two important principles of modern population biology: the significance of density-dependent factors in regulating population size and the primacy of individual selection over group selection. He then proposed his own theory of the evolution of clutch size. Most birds are indeterminate layers; if eggs are removed during laying, the female continues to lay further eggs until the clutch is made up to the normal number typical of the species. Thus clutch size is not limited by the number of eggs a female can lay. Why then does she limit the clutch to the size typically observed? Lack proposed that, at least in nidicolous species that feed their offspring until they fledge, the female lays a clutch that will give the maximum number of surviving young. He wrote (the italics are his):

> How might the laying of too large a clutch be a disadvantage *per se* to a bird? ... It is easy to see why a species which normally lays four eggs and raises four young should not, instead, lay only three eggs. The difficult problem is to discover why such a species should not normally lay five eggs. I believe that, *in nidicolous species, the average clutch-size is ultimately determined by the average maximum number of young which the parents can successfully raise in the region and at the season in question,* i.e. that natural selection eliminates a disproportionately large number of young in those clutches which are higher than the average, through the inability of the parents to get enough food for their young, so that some or all of the brood die before or soon after fledging, with the result that few or no descendants are left with their parent's propensity to lay a larger clutch. This view has much to recommend it *a priori,* and in addition there is much circumstantial evidence in its favour."

In support of this view Lack (1954) gave the data shown in Table 1.1 on the relation between brood size and reproductive success for Common Swifts in England. Nestling mortality was so much higher in broods of 3, presumably because the parents were unable to bring them all enough food, that the average number of young fledged was no higher than in broods of 2. Since they may also have suffered a higher postfledging mortality, it seems likely that a brood size of 2 is optimal.

However, several subsequent studies have found that the most productive clutch size is larger than the commonest one. Several factors may account for this apparent failure of Lack's theory, of which I shall consider two. First, laying more eggs may be detrimental to the parents' subsequent fecundity or survival. Reid (1987) showed that adult survival over the following winter in Glaucous-winged Gulls was 83% in birds with natural broods (having 1–3 eggs) while it was only 74% in birds with experi-

Table 1.1. The relation between brood size and fledging success in Common Swifts.

Brood size	Number of broods	Percentage young fledged	Young fledged per brood
1	36	83	0.8
2	204	84	1.7
3	96	58	1.7

After Lack, 1954.

Table 1.2. The "silver spoon" effect; clutch size in bold type, number fledged in roman type.

	Genotype		
Nutritional state	*aa*	*Aa*	*AA*
Low	**4** 3.00	**5** 3.50	**6** 3.00
Medium	**6** 3.75	**7** 4.25	**8** 3.75
High	**8** 4.50	**9** 5.00	**10** 4.50

mentally enlarged broods (having 4–7 eggs), due to deterioration in body condition. This means that the expected reproductive lifespan was reduced from 1/0.17 = 6 years to 1/0.26 = 4 years. This reduction in lifespan would need a 50% increase in productivity in enlarged broods to offset it; it seems likely that the observed clutch size is optimal in the sense of maximizing the parents' lifetime reproductive success. (Data on offspring survival in broods of different sizes are unfortunately not available.)

Another explanation for the apparent failure of Lack's theory in some cases is the "silver spoon" effect. Some birds find themselves, for accidental reasons, in better physical condition than others. These birds are predisposed to lay larger clutches, which is adaptive since they are better able to care for more young. This situation is illustrated in Table 1.2 (adapted from Price and Liou, 1989). Clutch size (in bold type) is determined by a combination of genotype and nutritional state; the number of fledglings (in roman type) is determined by a combination of clutch size and nutritional state. Suppose that the three nutritional states are equally likely, and that the genotypes *aa*, *Aa*, and *AA* have frequencies $\frac{1}{4}$, $\frac{1}{2}$, and $\frac{1}{4}$, respectively. A plot of the number fledged as a function of clutch size for all the data (ignoring nutritional state and genotype) shows a strong upward trend, with an optimal clutch size of 9, whereas the average clutch size is 7, and one might predict that selection should operate to increase clutch size. But in fact the fitnesses of the three genotypes (in terms of numbers of surviving offspring) are 3.75, 4.25, and 3.75, respectively; the optimal genotype is *Aa* and a balanced polymorphism is achieved at the observed genotype frequencies with no directional selection on clutch size.

It seems likely that this effect is operating in the Great Tit. In this species the modal brood size is about 8, while the most productive brood size is in most years substantially higher, averaging about 11 (Perrins and Moss, 1975). Pettifor et al. (1988) have determined the reason for this discrepancy by a thorough statistical analysis of survival in natural broods and in broods that had been manipulated by adding or removing young just after hatching. This manipulation had no effect on survival of the parents or on their fecundity in subsequent years, so that there is no evidence of a trade-off between current brood size and future reproductive success in this species. Among birds with a given initial clutch size, any manipulation in the brood size (either up or down) lowered the average number of offspring recruited into the breeding population; the actual clutch size is the optimal clutch size. However, birds laying more eggs did consistently better in terms of subsequent recruitment, presumably reflecting some environmental advantage; the clutch size laid by an individual is an accurate reflection of its ability to raise young that survive to breed. An analysis restricted to natural broods will therefore show the most productive brood size to be substantially higher than the modal brood size. The true situation can only be revealed by experimental manipulation. (An alternative theory invokes the idea of bet-hedging, discussed in Chapter 5, to balance the effects of good and bad years [Boyce and Perrins, 1987; Yoshimura and Clark, 1991].)

This account of the evolution of clutch size illustrates some of the recurring themes in evolutionary ecology: the importance of distinguishing ultimate from proximate causes, the importance of ecological factors like density-dependence, the primacy of individual selection over group selection, the maximization of lifetime reproductive success (or some other appropriate measure of fitness) by natural selection, the importance of trade-offs between components of fitness like fecundity and longevity, and complications introduced by environmental factors that can only be resolved by experimental manipulation.

MATHEMATICAL MODELS

I chose the evolution of clutch size as an illustration because it could be explained by largely verbal reasoning, together with the numerical model in Table 1.2. In many cases it is helpful to construct a more formal mathematical model of the process. This clarifies one's thoughts by forcing the underlying assumptions to be explicitly stated, as well as ensuring that the correct conclusions are drawn from those assumptions by mathematical reasoning.

Two types of model are commonly used. The first is a *population genetic model*, in which different genotypes determine different traits (for example, clutch sizes); the model can be investigated, if necessary by simulation, to determine which behavior predominates. This type of model mimics the behavior of natural selection directly and can therefore be guaranteed to give the right prediction, provided the correct assumptions have been incorporated into it! However, the study of population genetic models is often difficult, and it may be simpler and more illuminating to consider instead an *optimization model* in which it is assumed that selection maximizes some quantity closely related to fitness, like lifetime reproductive success in life-history models, without bothering about the genetic details whereby this is done. The success of this strategy depends on the accuracy of the assumption that the quantity maximized in the model is in fact maximized by selection, which requires separate justification.

A second distinction is between tactical and strategic models (Levins, 1968). Any model is a compromise between generality and simplicity on the one hand and biological realism on the other. The more biological details are included in specifying a model, the more complicated and specialized it becomes. A good model is like a caricature that includes all the biology relevant to the current purpose while excluding inessential details. How much detail is relevant depends on what the current purpose is. If one wants, for example, to be able to predict the detailed behavior of an ecological system for management purposes, many biological details about the behavior of its components must be incorporated into the model in order to have confidence that its predictions will be accurate. This is called a *tactical model*; it will usually be purpose-built for each application, with little generality, and is likely to be so complex that it must be analyzed by computer simulation rather than mathematical analysis. On the other hand, if one is more interested in obtaining insight into the way in which the system will behave qualitatively, rather than in precise quantitative predictions, a much simpler *strategic model* will suffice, with the advantage that general qualitative conclusions about its behavior can be obtained by mathematical analysis.

Much progress has been made in developing theoretical models in evolutionary ecology, and there is a fruitful interaction between theoretical and empirical work. This book provides a reasonably elementary, systematic account of these theoretical models from first principles. Chapters 2 and 3 discuss classical models of population dynamics, which form the ecological background to many evolutionary models, and use them to describe the properties of dynamical systems and the technique of local stability analysis. Age-structured populations are introduced in Chapter 4 as a tool for the study of life-history evolution in Chapter 5. The remaining chapters cover the other theoretical topics of major interest in contemporary evolutionary ecology—optimal foraging theory, evolutionary game theory, kin selection and inclusive fitness, the sex ratio, sexual selection, and the evolution of sex. Most of the models considered are at the strategic rather than the tactical end of the spectrum. Maynard Smith (1977), in developing a game-theoretic model of parental care, writes:

> I propose some simple mathematical models of sexual strategies. Such models have an obvious air of unreality when compared to ... qualitative and verbal models .... They have the corresponding advantage of forcing one to make one's assumptions clearer. The purpose of mathematical formulation in this case is almost entirely to clarify the assumptions made.

In some cases, however, modeling can make precise, numerical predictions about the evolutionary process; sex-ratio theory provides the most successful example.

I have aimed at a biological rather than a mathematical readership, and I presuppose only some familiarity with elementary calculus. A set of mathematical appendixes at the end of the book describes the other mathematical tools required, in particular matrix algebra and its application to local stability analysis. Some more specialized material is included in appendixes at the ends of chapters. The exercises are an important component of the book, since one can only understand mathematics by doing it. Some of them require only pencil and paper, but others will be facilitated by access to some computing facility, for example for manipulating matrices or doing simulations. *Mathematica* and *Maple* are the most powerful programs for computer modeling; they have built-in functions for matrix manipulation, solution of differential equations, etc., they can be used for symbolic as well as numerical calculations, and they are also programming languages. Many other possibilities are open. I have used *Mathematica* in teaching a course from which this book arose, and a set of *Mathematica* Notebooks giving model solutions to the exercises is available.[1]

FURTHER READING

The biological background of evolutionary ecology is covered by Cockburn (1991) and Krebs and Davies (1993). Williams (1966, 1992) provides a thorough and influential treatment of the ideas of natural selection. (You should of course also read Darwin, 1859.) Gould and Lewontin (1979) make a vigorous attack on the adaptationist program, dismissing much of the work as unscientific and unsubstantiated storytelling; Williams (1985) has made an equally vigorous, and to my mind convincing, defense. Maynard Smith (1978a) discusses the use of optimization the-

---

[1] These Notebooks have been deposited in *MathSource*, an electronic library of *Mathematica* material. The best way to access this library is by electronic mail (mathsource@wri.com). For further information about using *MathSource*, type "Help Intro"; for the Notebooks themselves type "Send 0207–201". *MathSource* can also be accessed by ftp (mathsource.wri.com) or by dial-up (217-398-1898).

ory in evolution. The evolution of clutch size has been reviewed by Godfray et al. (1991).

EXERCISE

1. From Table 1.2 and the genotype and nutritional state frequencies in the text calculate the average number of young fledged from clutches of size 4, 5, ... , 10 and plot the results. Find the average number of young fledged for the three genotypes. Comment.

CHAPTER 2

Population Dynamics

Single-Species Models

In this chapter and the next I shall discuss the classical models of population dynamics, which seek to describe and explain the factors determining the abundance of natural populations of plants and animals. These models are important in their own right and also illustrate the most interesting features of the behavior of dynamical systems, which have applications in many areas of population biology and will be used extensively in later chapters.

Suppose n is the number of individuals of a particular plant or animal species in a particular area at time t. A population dynamic model tries to describe how n varies with t. The model may be formulated as a differential equation in continuous time or as a difference equation in discrete time. Continuous-time models are suitable for populations growing continuously in a nonseasonal environment (for example, microorganisms in batch culture), or for populations with overlapping generations censused at the same time each year (for example, a perennial plant). On the other hand, populations with nonoverlapping generations, such as an insect population with one generation each year or an annual plant, are better represented by a difference-equation model in discrete time.

Another distinction is between models that consider only the numbers of a single species and models that represent explicitly the interaction between two or more species (for example, they might be competing species or predator and prey). Single-species models are considered in this chapter; models for two or more species will be considered in Chapter 3.

CONTINUOUS-TIME MODELS

The rate at which the population size, n, is changing will depend on its current value. This can be expressed in continuous time by the differential equation

$$dn/dt = f(n) \tag{1}$$

where $f(n)$ defines the population growth rate as a function of n. It is sometimes easier to think about the model in terms of the per capita growth rate, defined as

$$c(n) = f(n)/n \tag{2}$$

which represents the difference between the birth rate and the death rate per individual. (If $b(n)$ is the per capita birth rate [the average number of offspring each individual produces in unit time] and $d(n)$ is the death rate [the average mortality per unit time], then $c(n) = b(n) - d(n)$. Density-dependent factors may cause the birth rate to decline or the mortality to increase with population size.) Equation 1 can be written

$$dn/dt = c(n)\, n \tag{3}$$

This model assumes the population to be homogeneous and ignores age or sex differences between individuals. We shall now consider models based on this formulation by postulating different functional forms for the per capita growth rate.

Exponential growth

The simplest model of population growth in continuous time assumes that the per capita growth rate is a constant, independent of population size: $c(n) = r$. This leads to the differential equation

$$dn/dt = rn \tag{4}$$

whose solution is

$$n(t) = n_0 e^{rt} \tag{5}$$

where n_0 is the population density at time zero. If $r < 0$ (death rate exceeds birth rate), the population size decreases exponentially to zero; if $r > 0$ (birth rate exceeds death rate), it increases exponentially without limit.

This is a good description of the initial phase of growth of a population growing in optimal conditions, such as when a species is first introduced in a vacant niche. For example, Lack (1954) records that in 1937 two cock and six hen pheasants were set down on a small island off the coast of the state of

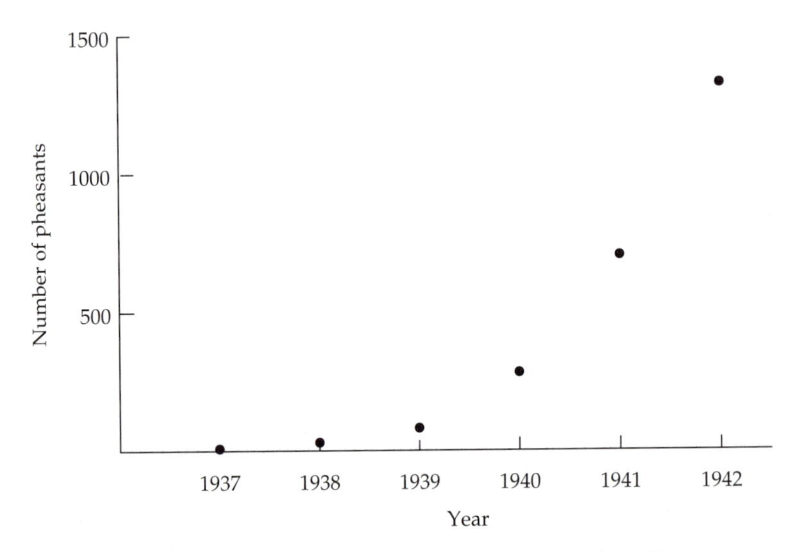

Figure 2.1. Exponential growth of pheasants. After Lack, 1954.

Washington where the species was previously unknown. The population size for six years is shown in Figure 2.1. There is a spectacular and approximately exponential increase in numbers. Lack remarks: "The figures suggest that the increase was slowing down and was about to cease, but at this point the island was occupied by the military and many of the birds were shot."

The logistic model

The exponential-growth model is unrealistic because it does not take into account density-dependent factors, such as resource depletion and competition for territory, which are expected to ensure that the per capita growth rate is not constant but is a decreasing function of population density. The simplest way of representing this in the model is to make the per capita growth rate decline linearly with population size in the form

$$c(n) = r\,(1 - n/K) \tag{6}$$

In this representation, r is the intrinsic rate of increase, the maximum growth rate when the population density is small and resources are abundant, and K is the carrying capacity, the critical population density the environment can support, below which the growth rate is positive and above which it is negative.

The equation for population growth becomes

$$dn/dt = r\,(1 - n/K)n \tag{7}$$

This is the logistic equation, whose solution starting from size n_0 at time zero is

$$n(t) = K / [1 + (K/n_0 - 1)e^{-rt}] \tag{8}$$

Because of the exponential term, which tends to zero as t becomes large, n increases or decreases smoothly with time to the equilibrium value K according to whether its initial value is below or above K. Thus K is a *globally stable* equilibrium value; the population will eventually tend to K whatever its initial value (except zero).

Gause (1934) did pioneering work on the growth of microorganisms, both in single culture and in competition. Figure 2.2 shows some of his data (together with the best-fitting logistic curve) on the growth of a population of *Paramecium caudatum* in single culture, starting with two individuals on day zero and grown on a growth medium with a fresh supply of bacteria renewed daily for 25 days. Gause concluded that in these growth conditions population growth was limited, and K was determined, by the amount of bacteria supplied as food. There is a slight tendency for oscillations about the carrying capacity, which may be due to the food being supplied daily rather than continuously.

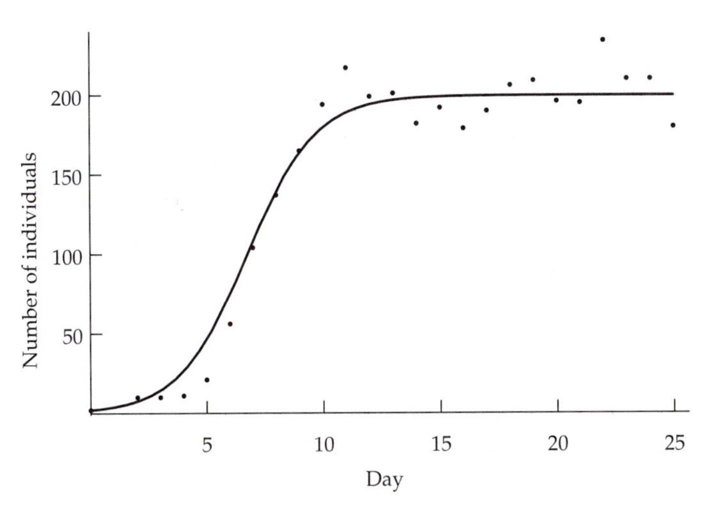

Figure 2.2. Logistic growth of *Paramecium caudatum*. Source: Gause, 1934.

Stability analysis

It is clear from the form of Equation 8 that K is the unique globally stable equilibrium, but in more complex models it may not be possible, or illuminating, to find an explicit solution for the growth curve. In this more usual situation, useful information about the stability of an equilibrium can be obtained either graphically or analytically.

The graphical approach is illustrated in Figure 2.3 that plots the population growth rate for the logistic model, $f(n) = r(1 - n/K)n$, against n with parameter values $r = 0.5$ and $K = 100$. The arrows are vectors showing the rate of growth at different population sizes. (If the arrows are turned 90 degrees counterclockwise they touch the $f(n)$ curve.) They point right or left according to whether the population size is increasing, with $f(n) > 0$, or decreasing, with $f(n) < 0$, at n.

The equilibrium values are the values of n for which $f(n) = 0$, where the function crosses the n axis and the arrows vanish; they are 0 and K. If the population size is at one of these stationary points it will stay there forever. But what happens if the population is near one of these points? Will it converge to it (stability) or diverge from it (instability)? This question is important because an unstable equilibrium is unlikely to be observed in nature except for fortuitous reasons. The question can be answered by observing the direction of the arrows near the equilibrium. It is clear from Figure 2.3 that zero is an unstable equilibrium since the arrows near it point away from it, while K is a stable equilibrium since all arrows point toward it. Any nonzero population will ultimately get closer and closer to K (in this case 100).

The logistic model is a simple descriptive model that should not be taken too literally; the assumption in Equation 6 that the per capita growth rate declines linearly with population density was chosen as the simplest way of representing density dependence and is unlikely to be exactly true. However, the qualitative conclusions from the graphical stability analysis above remain valid for any per capita growth-rate function $c(n)$ that is positive for all values of n less than some critical value K and negative when n

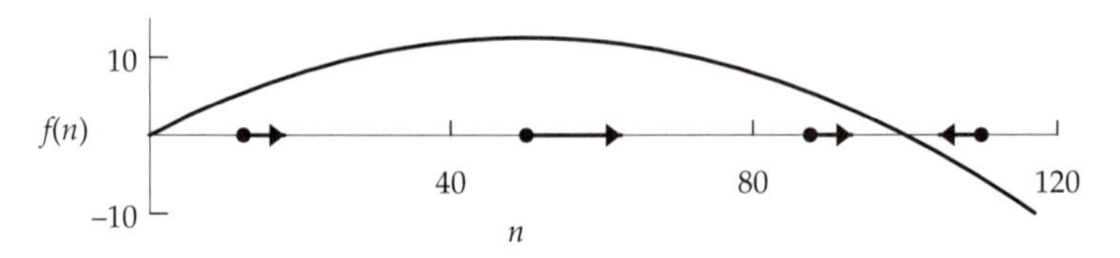

Figure 2.3. Stability analysis of the logistic model with $r = 0.5$ and $K = 100$.

exceeds K. In this case the population growth rate $f(n) = c(n)n$ will have the same form as in Figure 2.3, being positive for $0 < n < K$ and negative for $n > K$, and in consequence K will be the globally stable equilibrium value. For example, Maynard Smith (1989) discusses a model in which the per capita growth rate is

$$c(n) = \frac{b}{n+1} - d \tag{9}$$

This represents a situation in which the death rate d is constant but density-dependent factors reduce the birth rate $b(n)$ from b at low densities to zero at high densities. The behavior of this model is similar to that of the logistic model with $r = b - d$, $K = (b - d)/d$.

It was suggested by Allee (see Allee et al., 1949) that the per capita growth rate might decline at very low population densities and even become negative due to factors such as the difficulty of finding a mate. A simple model is shown in Figure 2.4a in which we suppose that the per capita growth rate is

$$\begin{aligned} c(n) &= -0.5 + 0.15n \qquad n \leq 50 \\ c(n) &= 0.5 - 0.005n \qquad n \geq 50 \end{aligned} \tag{10}$$

For population sizes below 50, the per capita growth rate increases with increasing n because of factors such as better mating opportunities, but for population sizes above 50 it decreases with increasing n because of factors such as overcrowding. The graphical stability analysis is shown in Figure 2.4b. There are three equilibria: 0, 33 and 100. The first and third are stable equilibria while the second is unstable; if the initial population size is less than 33, the population will inevitably go extinct; while if the initial size exceeds 33, it will end up at 100. In contrast to the logistic model, there is no globally stable equilibrium that will attract the population from any starting value except zero; the two stable equilibria are only locally stable in the sense that they will attract the population if it starts near enough to them. This may be expressed in terms of their *domains of attraction*, which are all points less than 33 for 0 and all points greater than 33 for 100. The possibility of an Allee effect has obvious relevance for conservation biology.

The results of this graphical stability analysis can be summarized as follows. Write the model as

$$\mathrm{d}n/\mathrm{d}t = f(n) \tag{11}$$

The equilibrium values $\hat{n}$ at which the population size is stationary satisfy

$$f(\hat{n}) = 0 \tag{12}$$

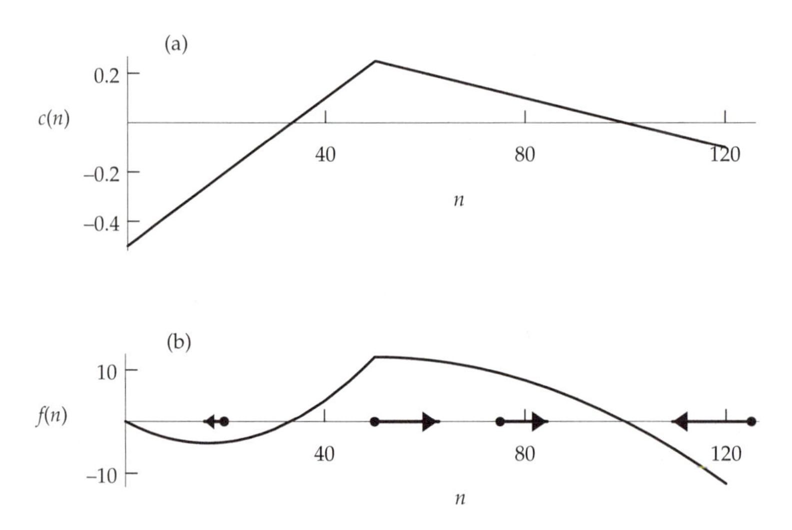

Figure 2.4. Stability analysis under the Allee effect.

An equilibrium value $\hat{n}$ is stable or unstable according to whether $f(n)$ is a decreasing or an increasing function of n at that point. In other words, if we write $f'(\hat{n})$ for the derivative of $f(n)$ evaluated at $n = \hat{n}$, then

$$\begin{aligned} &\hat{n} \text{ is locally stable if } f'(\hat{n}) < 0 \\ &\hat{n} \text{ is locally unstable if } f'(\hat{n}) > 0 \end{aligned} \qquad (13)$$

If there are several equilibria, then stable and unstable equilibria must alternate; the domain of attraction of a stable equilibrium is between the unstable equilibria on either side (see Figure 2.4).

The local stability criterion in Equation 13 follows from the following analytic argument that, though less intuitively appealing, is easier to generalize to models with two or more species (Appendix H).[1] Write p for a small perturbation from the equilibrium value, so that $n = \hat{n} + p$. If we linearize $f(n)$ about $n = \hat{n}$, we find the approximate differential equation for p

$$dp/dt \cong pf'(\hat{n}) \qquad (14)$$

whose solution is

$$p = p_0 e^{f'(\hat{n})t} \qquad (15)$$

[1] Appendix H refers to an appendix at the end of the book, while Appendix 5.1 would refer to the first appendix at the end of the fifth chapter.

Thus the perturbation will converge exponentially to zero or diverge exponentially from it, giving stability or instability, according to whether $f'(\hat{n})$ is negative or positive, in confirmation of Equation 13. The magnitude of $f'(\hat{n})$ determines the rate of convergence or divergence. Equation 14 is only valid for a perturbation small enough that the curvature in $f(n)$ is negligible, but it is sufficient to determine the local stability of the equilibrium under small perturbations. In contrast, the graphical analysis is valid under perturbations of any size.

For the logistic model, the equilibria are $\hat{n} = K$ and $\hat{n} = 0$, and the derivatives at these values are $-r$ and r respectively. Thus the equilibrium at K is locally stable, while the equilibrium at 0 is unstable, in confirmation of the global stability analysis in Figure 2.3.

DISCRETE-TIME MODELS

The exponential and logistic growth laws are expressed by differential equations in continuous time. They are suitable for populations growing continuously in a nonseasonal environment, or for populations with overlapping generations censused at the same time each year. On the other hand, populations with nonoverlapping generations, such as an insect population with one generation each year, in which next year's adults are the offspring of this year's, are better represented by a difference-equation model in discrete time. This can be expressed as

$$n_{t+1} = F(n_t) \tag{16}$$

where n_t is the population size in year (or generation) t and $F(n)$ is the growth function relating population sizes in successive years. The per capita growth function, defined as

$$C(n) = F(n)/n \tag{17}$$

represents the average number of offspring that an individual in a population of size n will have in the following year. Then, Equation 16 can be written as

$$n_{t+1} = C(n_t)n_t \tag{18}$$

The simplest model of population growth in discrete time assumes that the per capita growth function is a constant, independent of the population size, $C(n) = R$, leading to the difference equation

$$n_{t+1} = Rn_t \tag{19}$$

This equation can be solved iteratively starting with n_0 individuals at time 0:

$$\begin{aligned} n_1 &= Rn_0 \\ n_2 &= Rn_1 = R^2 n_0 \\ n_3 &= Rn_2 = R^3 n_0 \end{aligned} \tag{20}$$

and so on. The general solution at time t is clearly

$$n_t = n_0 R^t \tag{21}$$

This is the same as Equation 5 if we equate R with e^r. This is the discrete-time analogue of exponential growth in continuous time.

From the analogy that $C(n)$ in discrete time is equivalent to $e^{c(n)}$ in continuous time, the discrete-time analogue of logistic growth is

$$n_{t+1} = \exp[r(1 - n_t/K)]n_t \tag{22}$$

The behavior of this model is most easily studied by simulation, starting with some initial value and iterating the process. The results of such simulations with $K = 1000$ and with an initial population size of 750 are shown in Figure 2.5 for different values of r. The carrying capacity K is a scale parameter that will not affect the qualitative behavior of the result; exactly the same results would be obtained with $K = 100$ and with an initial value of 75, apart from a factor of 10. However, changes in the intrinsic rate of increase r have a big qualitative effect for reasons that I shall now try to explain.

The equilibrium values $\hat{n}$ at which the population size is stationary from one generation to the next for any discrete-time model specified by Equation 16 satisfy

$$F(\hat{n}) = \hat{n} \tag{23}$$

That is to say, they are the points at which the graph of $F(n)$ plotted against n intersects the 45° line in Figure 2.6. The stability of these equilibrium values is most easily investigated by the analytic approach (see Appendix J). Write p_t for a small perturbation from the equilibrium value, so that $n_t = \hat{n} + p_t$. If we linearize $F(n)$ about $n = \hat{n}$, we find the approximate difference equation for p_t:

$$p_{t+1} \cong F'(\hat{n})p_t \tag{24}$$

whose solution is

$$p_t = [F'(\hat{n})]^t p_0 \tag{25}$$

The perturbation decreases to zero geometrically if $0 < F'(\hat{n}) < 1$; it decreases to zero but with alternating sign each generation if $-1 < F'(\hat{n}) < 0$; and it diverges from zero otherwise (with alternating sign if it is negative).

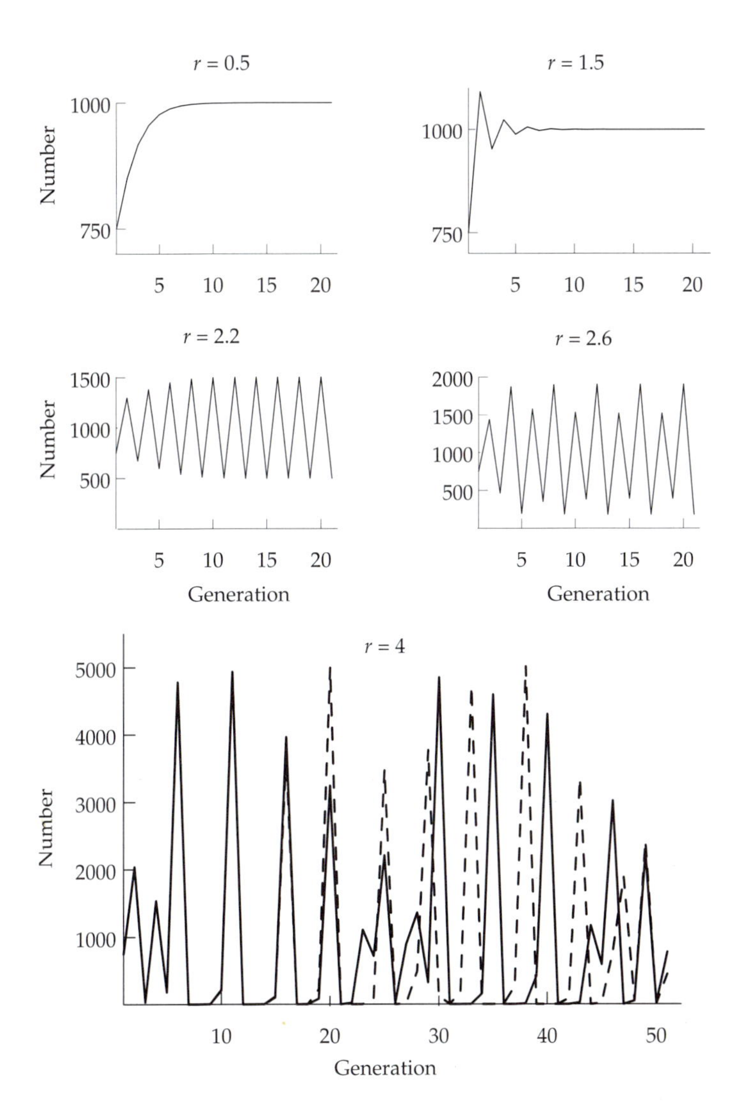

Figure 2.5. Behavior of the discrete-time logistic model with $K = 1000$ and an initial population size of 750 for different values of r. For $r = 4$, simulations starting at 750.1 are also shown (dashed lines).

Hence

$$\begin{aligned} &\hat{n} \text{ is locally stable if } |F'(\hat{n})| < 1 \\ &\hat{n} \text{ is locally unstable if } |F'(\hat{n})| > 1 \end{aligned} \tag{26}$$

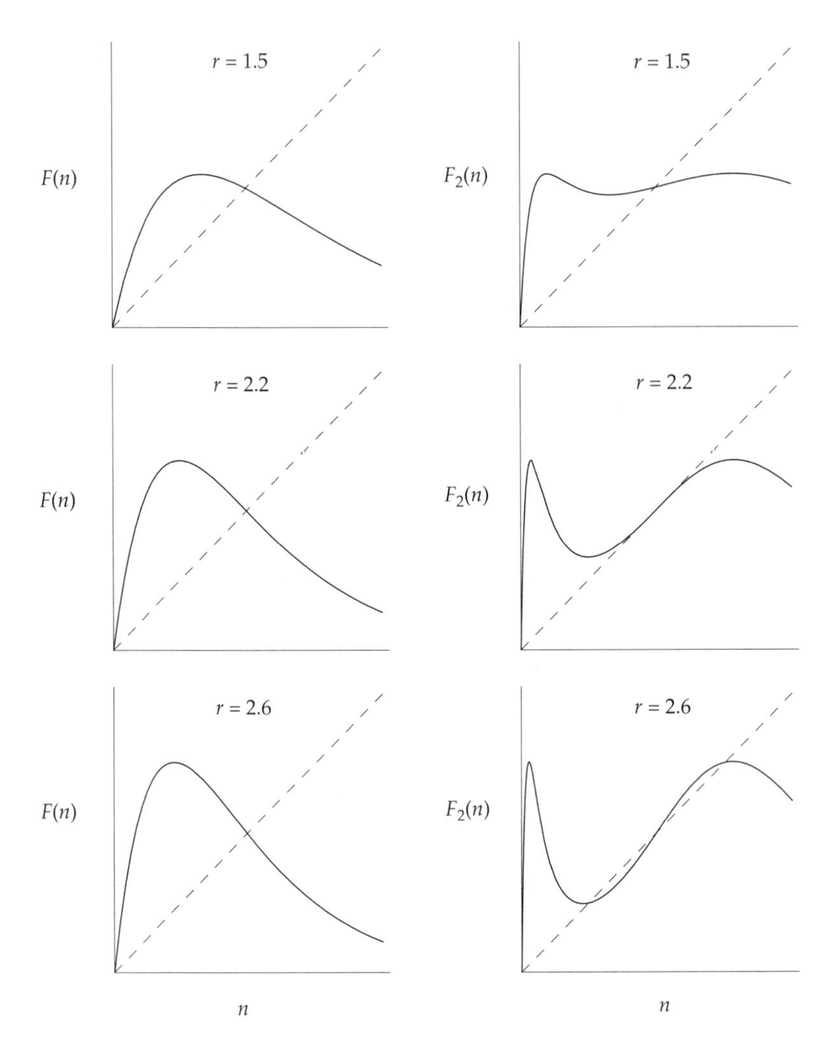

Figure 2.6. The first and second iterates of the discrete logistic function (solid curves) in relation to the 45° line (dashed).

The discrete-time logistic model (Equation 22) has equilibria satisfying Equation 23 at 0 and K. The derivative of F at 0 is $\exp r$, which is greater than 1, so that this equilibrium is locally unstable as expected. The derivative at K is $1 - r$. When $0 < r < 1$, the perturbations decline geometrically to zero, so that the system is stable in the same way as the logistic model in continuous time (Figure 2.5 with $r = 0.5$). When $1 < r < 2$, then $1 - r$ is negative; the perturbations still decline geometrically to zero in absolute value, since $|1 - r| < 1$, but they alternate in sign each generation (Figure 2.5 with $r = 1.5$). When $r > 2$, the system becomes unstable since $|1 - r| > 1$; the perturbations increase geometrically in absolute value as well as alternating in sign (Figure 2.5 with $r = 2.2$, 2.6, and 4).

The reason for instability with large values of r is that increasing r makes the function $F(n)$ rise more steeply near the origin, thus generating a greater tendency to overshoot. There is a general tendency for discrete-time models to be less stable than their continuous-time analogues because of the inherent time lag in a discrete-time model; the system responds to the population size in the previous generation in a discrete-time model, while it responds to the instantaneous population size in a continuous-time model.

When $r > 2$, Equation 25 shows that the system diverges locally from the equilibrium value K, but more detailed study is required to see what happens then. More complex behavior is observed than in the continuous-time model. The dynamical behavior of the system is summarized in Table 2.1.

When r exceeds 2, the stable equilibrium at $\hat{n} = K$ becomes unstable but is replaced by a pair of points between which n ultimately alternates from one generation to the next (a stable two-point cycle); this is called a *bifurcation*. As r grows above 2.526, each of these points bifurcates, giving rise to a four-point cycle. This process continues until r reaches 2.692. Beyond this, the system exhibits behavior that is called *chaotic*. A mathematical description of this behavior is difficult, but in practical terms a realization shows the following features: (1) It is aperiodic, so that the same pattern is never repeated no matter how long the simulation runs. (2) It is very sensitive to initial conditions, so that two realizations starting from very similar values will ultimately diverge. This is often called the *butterfly effect*: the flapping of a single butterfly's wing today produces a tiny change in the state of the atmosphere, which will be magnified over time, making long-term weather prediction impossible.

Table 2.1. Dynamical behavior of the discrete-time logistic model.

Range of r	Dynamical behavior
$0.000 < r < 2.000$	Stable equilibrium value
$2.000 < r < 2.526$	Stable 2-point cycle
$2.526 < r < 2.656$	Stable 4-point cycle
$2.656 < r < 2.685$	Stable 8-point cycle
$2.685 < r < 2.692$	Stable 2^n-point cycle, $n > 3$
$r > 2.692$	Chaotic behavior (for nearly all values of r)

From May, 1981a.

These features are illustrated in Figure 2.5 in simulations with $r = 0.5$ (stable equilibrium value), $r = 1.5$ (stable equilibrium with perturbations alternating in sign), $r = 2.2$ (two-point cycle), $r = 2.6$ (four-point cycle), and $r = 4.0$ (chaos). For $r = 4$, simulations starting from population sizes of 750 and 750.1 are shown to demonstrate the butterfly effect; the two simulations are coincident for 15–20 generations, but then diverge. Despite the fact that the trajectory in the chaotic region is not *exactly* periodic it is often approximately periodic with some stochastic variation in the length of the periods. May (1981a) suggests that for large r the behavior of the model is almost periodic with a period of approximately $[\exp(r-1)]/r$. For $r = 4$ this gives an approximate period of 5.0 that agrees well with the realization.

The reason for the bifurcation from a single stable equilibrium to a stable two-point cycle at $r = 2$ is shown in Figure 2.6. The left half of this figure graphs $F(n)$ against n for $r = 1.5$, 2, and 2.2. The equilibrium at $n = K$ is the point at which the curve intersects the 45° line, and its stability is determined by the slope of the curve at this point, $F'(K)$. As r increases, $F(n)$ rises more steeply at the origin and so falls more steeply beyond the hump, so that $F'(K)$ passes from stability (> -1) to instability (< -1) as r increases beyond 2.

Now consider the behavior of $F_2(n) = F(F(n))$, such that $n_{t+2} = F_2(n_t)$. This function is called the second iterate, and it determines the behavior of the system in alternate generations. By the chain rule, the derivative of F_2 is

$$\begin{aligned}\frac{dF_2}{dn} &= \frac{dF(F(n))}{dn} = \frac{dF(F(n))}{dF(n)} \times \frac{dF(n)}{dn} \\ &= F'^2(n) \qquad \text{if } F(n) = n\end{aligned} \tag{27}$$

Thus $F_2'(K) = F'^2(K)$. The behavior of $F_2(n)$ shown in the right-hand plots of Figure 2.6 results from the facts that $0 < F_2'(K) < 1$ when $r < 2$ whereas $F_2'(K) > 1$ when $r > 2$. It can be seen that a stable two-point cycle is born at the precise parameter value ($r = 2$) for which the equilibrium at $n = K$ becomes unstable.

The stability of the two-point cycle alternating between $\hat{n}_1$ and $\hat{n}_2$, say, is determined by the derivatives of F_2 at these points. From the chain rule above it follows that

$$F_2'(\hat{n}_1) = F'(\hat{n}_1)F'(\hat{n}_2) = F_2'(\hat{n}_2) \tag{28}$$

so that these derivatives are the same. As r increases further, they will become more negative until they become less than –1. At this value of r the two-point cycle becomes unstable and a four-point cycle is born. This period-doubling cascade continues until the chaotic region is reached when $r =$ 2.692.

Chaos has attracted attention because of the remarkable fact that apparently random behavior can result from a simple deterministic model, but its importance in population biology is unclear; there are a few enthusiasts (for example, Schaffer and Kot, 1986), but as many skeptics. Hassell et al. (1976) examined 28 sets of insect population data, fitted a model similar to Equation 22 to them, and found that 26 of them fell within the region of stable equilibria, one (the Colorado potato beetle) was in the region of persistent periodic cycles, and only one (Nicholson's laboratory experiments on blowflies) fell in the chaotic region. These results suggest that chaotic behavior is unlikely to be widespread. On the other hand, the outbreak behavior of many forest insect pests resembles chaotic behavior quite closely. These species are often subject to delayed density-dependent regulation (Turchin, 1990), the effects of damage to the trees from an outbreak of the insects taking several years to express themselves, and it may be that this time lag predisposes the system to behave chaotically.

It should be noted that chaotic behavior does not occur in continuous-time models for one or two species, but can occur in such models with three or more species explicitly represented (Hastings and Powell, 1991). Similarly, it is much easier to generate chaos in discrete-time models for several interacting species with biologically plausible parameter values than in single-species models. It is traditional for simplicity to discuss chaos in single-species discrete-time models, in which chaotic behavior is seldom encountered with biologically plausible parameter values (unless there is an additional time lag due to delayed density dependence), but this may mask its real importance, since most biological systems involve interactions between several species.

There is also a major problem in distinguishing chaotic dynamics from a process generated by a strictly periodic underlying mechanism with random variability superimposed on it. Two sorts of random variability may occur: (1) errors of measurement that affect estimates of the population size but are not built in to the process (i.e., they do not affect the population size in subsequent years); (2) random variability in the parameters of the model (such as r and K) due to environmental fluctuations, whose effects are built into the process and affect the population size in subsequent generations. Such a process with stochastic variability built into it mimics a deterministic chaotic process, and it is likely to be very difficult, if not impossible, to distinguish between them from internal evidence from the time series itself, particularly in view of the short span of most ecological time series (Tong and Smith, 1992). Sugihara and May (1990) have developed a methodology for distinguishing between chaos and a periodic time series with errors of measurement that are not built into the process, but it does not address the biologically more important situation in which variability from environmental causes is built into the process.

TESTING FOR DENSITY DEPENDENCE

Density-dependent factors are thought by many ecologists to play a key role in regulating the size of animal and plant populations, and much effort has been put into identifying what these factors are, though some ecologists consider that they are of minor importance in determining fluctuations in population size compared with random environmental impacts (Strong, 1986). It is therefore important to be able to test for the existence of density dependence and to estimate its magnitude. I shall now consider some of the statistical problems that arise in assessing the importance of density-dependent regulation from a time series of population sizes in successive years. Delayed density dependence will not be considered. The material in this section will not be used subsequently and may be omitted on first reading.

If n_t is the population size at time t, we may write

$$n_{t+1} = C_t n_t \tag{29}$$

where C_t is the per capita growth rate at time t, which can be estimated by n_{t+1}/n_t. Taking the logarithm of the above equation, and writing $x_t = \log n_t$ and $c_t = \log C_t$, we obtain

$$x_{t+1} = c_t + x_t \tag{30}$$

where the logarithmic growth rate c_t can be estimated by $x_{t+1} - x_t$. It is usually more meaningful to consider the logarithmic growth rate, and it is also simpler, since differences are easier to consider statistically than ratios.

To represent density dependence suppose that c_t depends linearly on x_t:

$$c_t = \alpha - \beta x_t + e_t \tag{31}$$

where e_t is a random component representing environmental variability. Thus

$$x_{t+1} = \alpha + (1 - \beta)x_t + e_t = \alpha + \gamma x_t + e_t \tag{32}$$

Density independence is represented by $\beta = 0$ ($\gamma = 1$), and density dependence by $\beta > 0$ ($\gamma < 1$).

An obvious way to test for density dependence is to calculate the slope of the regression of x_{t+1} on x_t and to test whether it is less than 1. For example, Figure 2.7 plots x_{t+1} against x_t for the number of nesting pairs of Great Tits in Marley Wood near Oxford, England from 1948 through 1991. (I am grateful to C. M. Perrins for these data.) The ordinary least squares regression through these points is

$$x_{t+1} = 2.42 + 0.31x_t \tag{33}$$

indicating a slope substantially less than unity that suggests strong density-

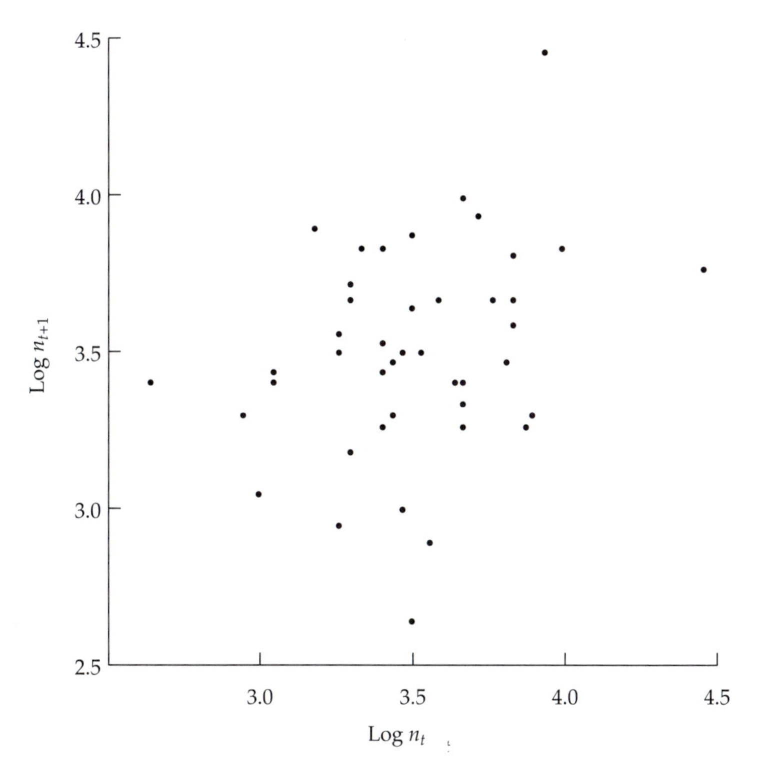

Figure 2.7. Log–log plot of the number of nesting pairs of Great Tits in successive years.

dependent regulation.

Unfortunately this procedure is statistically flawed because the assumptions underlying linear regression analysis are violated, and in fact the slope in the case of a density-independent random walk with no drift is usually less than 1. To get round this problem Pollard et al. (1987) proposed the following randomization test for density dependence given a time series of log population sizes $\{x_1, x_2, \dots, x_n\}$: (a) Find the slope of x_{t+1} on x_t. (b) Find the set of $(n-1)$ differences $\{d_1, d_2, \dots, d_{n-1}\}$ where $d_i = x_{i+1} - x_i$. (c) Form a random permutation of these differences, $\{d_1^*, d_2^*, \dots, d_{n-1}^*\}$. (d) Form a time series from these differences starting with x_1,

$$\{ x_1, x_2^* = x_1 + d_1^*, x_3^* = x_2^* + d_2^*, \dots, x_n^* = x_{n-1}^* + d_{n-1}^* \} \quad (34)$$

(e) Find the slope of x_{t+1}^* on x_t^* for this time series. (f) Repeat steps (c)–(e) 99 times, sort all the slopes (including the observed one) in rank order, and reject the hypothesis of density independence at the 5% level if the rank of the observed slope is 5 or less.

The rationale of this test is that, under density independence ($\gamma = 1$), $d_t = \alpha + e_t$, so that any ordering of these differences is equally likely since the errors e_t are independent. Thus under the null hypothesis the rank of the observed slope is equally likely to be 1, 2, ... , 100, so that a valid significance test at the 5% level is obtained by rejecting the null hypothesis whenever this rank is 5 or less. This type of significance test is known as a Monte Carlo test.

A problem with this type of test is that it is liable to some random variability, so that the same data might be judged significant with one set of simulations and not significant with another. This can be improved by doing more simulations (computer time permitting), for example by doing 499 simulations instead of 99 and rejecting the null hypothesis if the rank of the observed slope is 25 or less. Application of this test to the Great Tit data in Figure 2.7 shows highly significant evidence for density dependence.

When a significant effect of density dependence is found, it is of interest to estimate the magnitude of this effect. When $\gamma < 1$ the estimated slope (say, g) still has a downward bias, but it can be shown that an approximately unbiased estimate is given by $g + (1 + 3g)/n$, with standard error $[(1 - g^2)/n]^{1/2}$. For the Great Tit data the corrected slope is 0.35 with standard error 0.14. Thus there is substantial evidence for strong density dependence in this species.

One proviso should be noted. The above procedure assumes that the "errors" e_t reflect real changes in population size, due for example to climatic factors, which are incorporated into the process in subsequent years, rather than being errors in measuring the true population size; the existence of appreciable errors of measurement would lead to spurious significant results since they would induce a reduction in the slope of the regression. The Great Tit data have been carefully collected and there is no reason to suppose that they are contaminated with errors of measurement, but this possibility should always be considered.

FURTHER READING

May (1981a) provides a good introductory account of single species models. May and Oster (1976) is a readable discussion of chaotic behavior in simple ecological models; Hastings et al. (1993) is a recent review.

EXERCISES

1. (a) Differentiate Equation 5 with respect to t and hence verify that it is

the solution of Equation 4 with the initial condition that $n(0) = n_0$.
(b) Use a computer package, if available, to solve Equation 4 with initial condition $n(0) = a$.
(c) To derive this result yourself, write Equation 4 as

$$\frac{1}{n}\frac{dn}{dt} \equiv \frac{d \log n}{dt} = r$$

so that

$$\log n = \log n_0 + rt$$

2. The data shown in Figure 2.1 for the years 1937 through 1942 are as follows: {8, 30, 81, 282, 705, 1325}. Plot the natural logarithms of the numbers against time. Under exponential growth this should be a straight line with slope r. Does it look linear? Fit the best straight line and find its slope. Plot the log data and the fitted line on the same graph.

3. (a) Verify that Equation 8 satisfies Equation 7.
(b) Use a computer package, if available, to solve Equation 7 with initial condition $n(0) = a$.
(c) Plot Equation 8 against time for different initial values and parameter values and hence verify that $n = K$ is the globally stable equilibrium value.

4. Plot $f(n)$ against n for the model with the per capita growth rate given by Equation 9 for suitable parameter values and compare it with the plot for the corresponding logistic model.

5. Gause's data on *Paramecium caudatum* shown in Figure 2.2 for day 0 and for days 2 through 25 are as follows:
{2, 10, 10, 11, 21, 56, 104, 137, 165, 194, 217, 199, 201, 182, 192, 179, 190, 206, 209, 196, 195, 234, 210, 210, 180}.
(a) Plot the data.
(b) Estimate r and K in the logistic equation and plot the theoretical curve against the data. (As a quick and easy method, first estimate K as the average population size after day 9, since it seems roughly constant after then. To estimate r, show that $(K - n)/n$ is proportional to $\exp(-rt)$. Hence, when plotted against t, $\log[(K - n)/n]$ should give a straight line with negative slope r. This relationship is unreliable for observations near K, so data up to day 9 may be used to estimate r.)

6. Comparable data for day 0 and days 2 through 25 on another species, *Paramecium aurelia*, are as follows:
{2, 14, 34, 56, 94, 189, 266, 330, 416, 507, 580, 610, 513, 593, 557, 560, 522,

565, 517, 500, 585, 500, 495, 525, 510}
Fit a logistic model to these data.

7. The equilibria of the logistic model for which $f(n) \equiv r(1 - n/K)n = 0$ are at $n = 0$ and $n = K$. Evaluate $f'(n)$ at these points, and hence find their local stability.

8. Determine the local stability of the equilibria for the model in which the per capita growth rate is defined by Equation 9.

9. For the discrete-time logistic model the recurrence function is $F(n) = \exp[r(1 - n/K)]n$. Verify that the equilibria satisfying $F(n) = n$ are 0 and K. Find the derivative $F'(n)$ at these points and hence determine their local stability.

10. Investigate the behavior of the discrete-time logistic model by simulation for different parameter values.

11. I show below the number of nesting pairs of Great Tits and Blue Tits in Marley Wood near Oxford, England from 1948 through 1991. (I am grateful to C. M. Perrins for these data.) For each of these data sets, plot $\log n_{t+1}$ against $\log n_t$, estimate the slope of this regression by ordinary least squares, and find the approximately unbiased estimate with its standard error. If you have the *Mathematica* model solutions, use the package given there to do the significance test of Pollard et al. (1987) for density dependence.

    Great Tits: {21, 30, 31, 32, 20, 21, 31, 27, 24, 49, 27, 41, 51, 86, 43, 39, 54, 46, 45, 32, 33, 38, 30, 46, 36, 39, 26, 33, 14, 30, 26, 19, 27, 39, 28, 46, 39, 30, 34, 33, 48, 26, 35, 18}

    Blue Tits: {19, 26, 34, 34, 17, 14, 18, 13, 15, 32, 17, 20, 25, 44, 21, 41, 46, 47, 52, 33, 33, 32, 28, 35, 29, 47, 58, 64, 39, 85, 46, 33, 46, 66, 61, 52, 49, 52, 39, 45, 64, 57, 71, 49}

CHAPTER 3

Population Dynamics

Two-Species Models

Different species of plants and animals interact with each other as well as with the physical environment. The most important types of interspecific interaction are competition and predation. Competition occurs between similar species that compete for a common resource, such as food or territory, and each species has a negative impact on the other one. In a prey–predator interaction, on the other hand, predators have a negative impact on prey, but prey have a positive impact on predators. This qualitative difference between the two types of interaction gives rise to different dynamic behavior, that will be the focus of attention in this chapter, starting from the classical models of Lotka (1925) and Volterra (1931).

COMPETITION BETWEEN SPECIES

Figure 2.2 shows the logistic growth of *Paramecium caudatum*; Gause obtained similar results with the closely related species *Paramecium aurelia* in a single culture (see Exercise 2.6). Figure 3.1 shows what happened when Gause grew these two species together in a mixed culture, starting with two individuals of each species. Both species grow exponentially at first, but eventually *P. caudatum* declines in density and seems doomed to extinction under competition with *P. aurelia*.

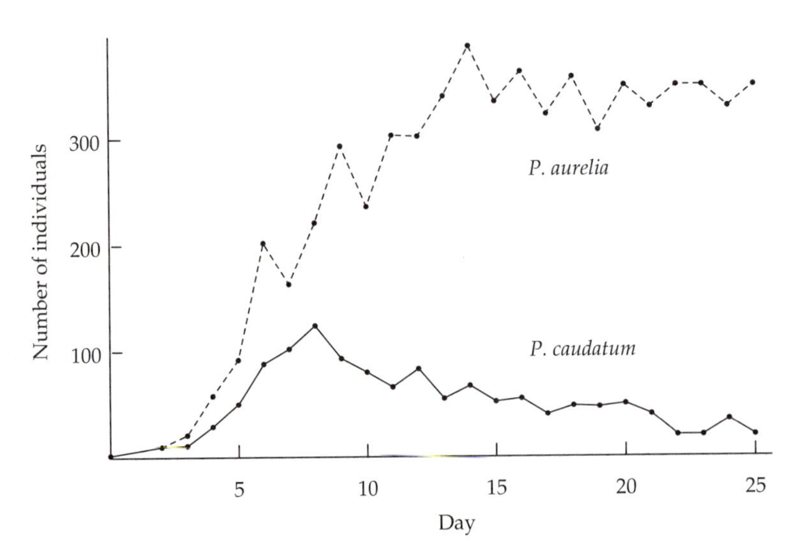

Figure 3.1. Growth of a mixed population of *Paramecium caudatum* and *P. aurelia*. Source: Gause, 1934.

The *Lotka–Volterra model* for this type of interaction generalizes the logistic model to two competing species with population sizes n_1 and n_2 as follows:

$$dn_1/dt = f_1(n_1, n_2) = r_1[1 - (n_1 + \gamma_{12}n_2)/K_1]n_1 \quad (1a)$$
$$dn_2/dt = f_2(n_1, n_2) = r_2[1 - (n_2 + \gamma_{21}n_1)/K_2]n_2 \quad (1b)$$

Here r_i is the intrinsic growth rate of species i in optimal conditions, K_i is the carrying capacity for this species in the absence of the other one, and γ_{ij} is a competitive coefficient representing the extent to which members of species j inhibit the growth of species i (for example, by using a common resource) relative to the extent to which members of species i inhibit their own growth. For example, if $\gamma_{12} = 0.1$, then a unit increase in the size of the second species will have one-tenth as large an inhibitory effect on the growth rate of the first species as the same increase in the first species.

Two ways of representing the behavior of this system graphically for a particular set of parameters are shown in Figure 3.2. Figure 3.2a is called a *vector field diagram*. The arrows are velocity vectors representing the speed and direction in which the population sizes are moving: the dots mark points in the $\{n_1, n_2\}$ plane and the attached arrow is the velocity vector $\{f_1, f_2\}$ at that point. The equilibrium values are the four points at which the

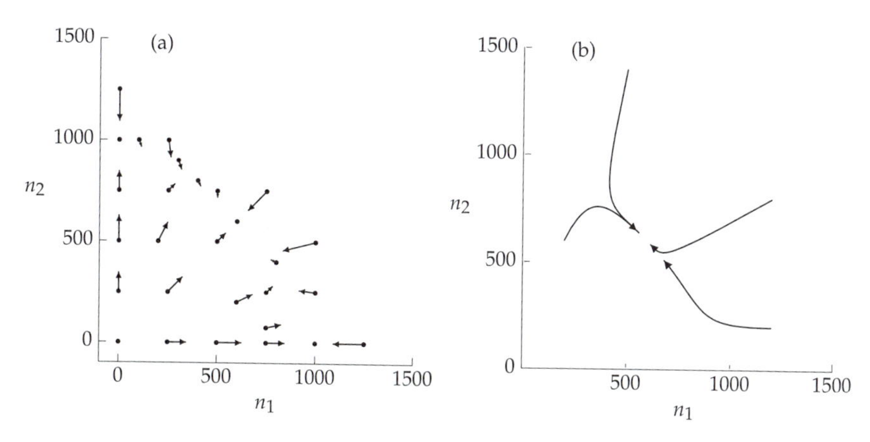

Figure 3.2. The Lotka–Volterra model of competition with $K_1 = K_2 = 1000$, $r_1 = r_2 = 0.5$, $\gamma_{12} = \gamma_{21} = 0.67$. (a) Vector field diagram; the arrows are velocity vectors. (b) Phase plane diagram showing four trajectories for this model over 20 generations.

arrow vanishes ({0, 0}, {0, 1000}, {1000, 0}, and {600, 600}). As we shall see in more detail shortly, the arrows converge and come to rest at the the unique stable equilibrium value {600, 600}, provided they are not on one of the boundaries.

Figure 3.2b is called the *phase plane diagram* of the system. It shows four trajectories obtained by solving Equation 1 numerically for 20 generations from a given starting point and plotting $n_1(t)$ against $n_2(t)$ in the $\{n_1, n_2\}$ plane; the arrow shows the direction of movement in time. Every trajectory (except those starting on the boundary) would ultimately tend towards the limiting value {600, 600}. The relation between the two diagrams is that if they were superimposed, the velocity vector would be tangent to the trajectory at any point on it.

To take the analysis further we first find a general expression for the equilibrium values by equating f_1 and f_2 to zero simultaneously. There are four equilibria, since in Equation 1a either n_1 or the term in square brackets may be zero, and likewise in Equation 1b. These equilibria are: (a) $\hat{n}_1 = \hat{n}_2 = 0$, with both species absent; (b) $\hat{n}_1 = K_1, \hat{n}_2 = 0$, with species 2 absent; (c) $\hat{n}_1 = 0, \hat{n}_2 = K_2$, with species 1 absent; and (d)

$$\begin{aligned} \hat{n}_1 &= (K_1 - \gamma_{12}K_2)/(1-\gamma_{12}\gamma_{21}) \\ \hat{n}_2 &= (K_2 - \gamma_{21}K_1)/(1-\gamma_{12}\gamma_{21}) \end{aligned} \tag{2}$$

with both species present. Note that (d) is only relevant when both equilibria are positive.

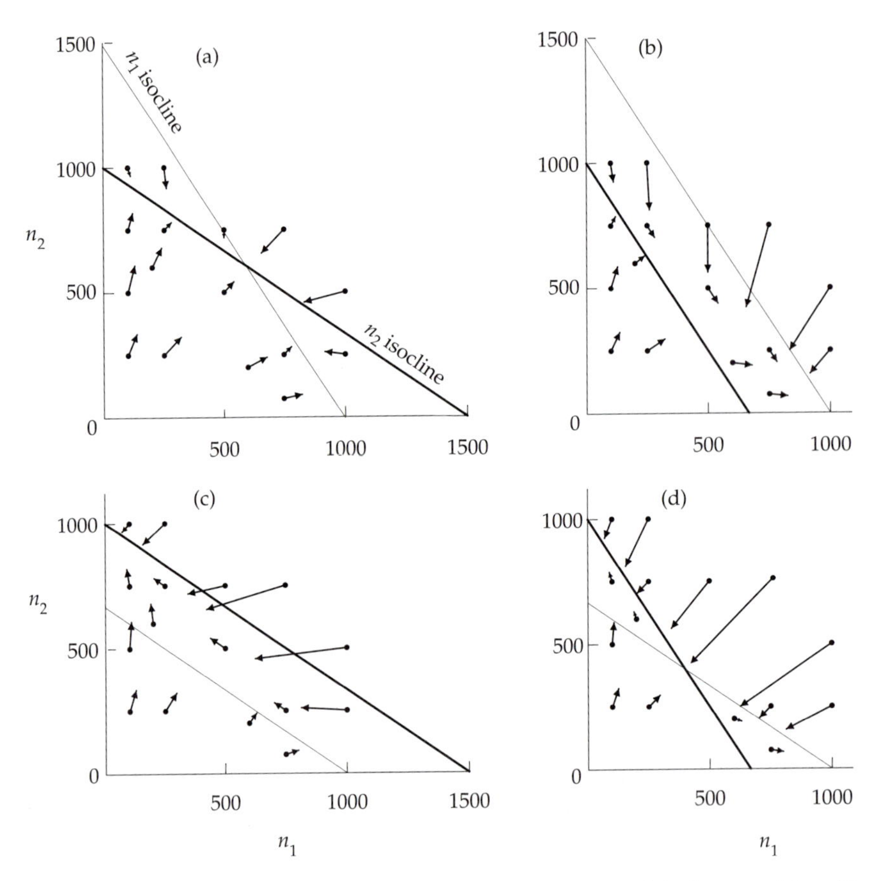

Figure 3.3. The four alternatives for the Lotka–Volterra model of competition. The thin line shows the n_1 isocline, the thick line the n_2 isocline; $K_1 = K_2 = 1000$ in all cases. (a) $\gamma_{12} = \gamma_{21} = 0.67$; (b) $\gamma_{12} = 0.67$, $\gamma_{21} = 1.5$; (c) $\gamma_{12} = 1.5$, $\gamma_{21} = 0.67$; (d) $\gamma_{12} = \gamma_{21} = 1.5$. The arrows are the velocity vectors when $r_1 = r_2 = 0.5$.

A graphical analysis of the stability of these equilibria is shown in Figure 3.3. Four qualitatively different situations arise according to whether K_1 is less than or greater than K_2/γ_{21} and whether K_2 is less than or greater than K_1/γ_{12}. The thin line in each graph is the n_1 isocline on which the term in square brackets in Equation 1a is zero; n_1 is stationary for values of $\{n_1, n_2\}$ on it. The thick line is the corresponding n_2 isocline from Equation 1b. The arrows are velocity vectors as in Figure 3.2a. The stability of the equilibria and the ultimate outcome can be inferred from the directions of these arrows. In Figure 3.3a, for example, if the system is outside the region between the two isoclines it tends to move inside this region, and once inside it tends to move toward the internal equilibrium where the two isoclines meet. Hence this is a stable equilibrium; the other three equilibria are unstable equilibria from which the system tends to diverge.

Though Figure 3.3 only shows the velocity vectors for four specific sets of parameter values, it covers the four qualitatively different types of behavior that can arise. The general directions of the arrows follow from the following facts: there is a horizontal component of motion to the left or the right (determined by the sign of f_1) depending on whether the current position is above or below the thin line (the n_1 isocline), and a vertical component down or up (determined by the sign of f_2) depending on whether it is above or below the thick line (the n_2 isocline). An arrow is the resultant of these two components.

The conclusions from the analysis in Figure 3.3 can be summarized as follows. (a) If $K_1 < K_2/\gamma_{21}$ and $K_2 < K_1/\gamma_{12}$ (that together imply that $\gamma_{12}\gamma_{21} < 1$) the internal equilibrium is stable, and both of the equilibria with only one species present are unstable; one expects coexistence of the two species. The quantity $\gamma_{12}\gamma_{21}$ is a measure of the relative importance of interspecific competition compared with intraspecific competition, so that coexistence occurs when the former is less strong than the latter. (b) If $K_1 > K_2/\gamma_{21}$ and $K_2 < K_1/\gamma_{12}$ there is no internal equilibrium, and species 1 always wins. (c) With the inequalities reversed, species 2 always wins. (d) If $K_1 > K_2/\gamma_{21}$ and $K_2 > K_1/\gamma_{12}$ (that together imply that $\gamma_{12}\gamma_{21} > 1$) the internal equilibrium is unstable, and both of the equilibria with only one species present are stable. The two species cannot coexist since interspecific competition is stronger than intraspecific competition, but which of them wins depends on the initial conditions; the species with the larger initial density (in appropriate units) will ultimately win. As expected, the equilibrium with both species absent is always unstable.

It is clear from Figure 3.3a that when a stable internal equilibrium exists it lies above the line joining K_2 on the n_2 axis to K_1 on the n_1 axis; in other words,

$$\frac{\hat{n}_1}{K_1} + \frac{\hat{n}_2}{K_2} > 1 \qquad (3)$$

However, this conclusion depends on the assumption that the isoclines are linear. Ayala (1972) showed that when two species of *Drosophila* (*D. pseudoobscura* and *D. willistoni*) were grown together they coexisted at equilibrium with population numbers of 399 and 657, respectively. When they were grown with only one species present, the equilibrium single-species population numbers were 772 and 1421. This result contradicts the above criterion for a stable internal equilibrium since

$$\frac{\hat{n}_1}{K_1} + \frac{\hat{n}_2}{K_2} = \frac{399}{772} + \frac{657}{1421} = 0.98 \tag{4}$$

A greater discrepancy was observed in other experiments.

The explanation is that in *Drosophila* the per capita rate of increase shows a markedly nonlinear decline with density so that the isoclines of zero growth are strongly curvilinear (Gilpin and Justice, 1972; Gilpin and Ayala, 1973). Figure 3.4 shows the estimated isoclines calculated from observed birth and death rates at different population densities of the two

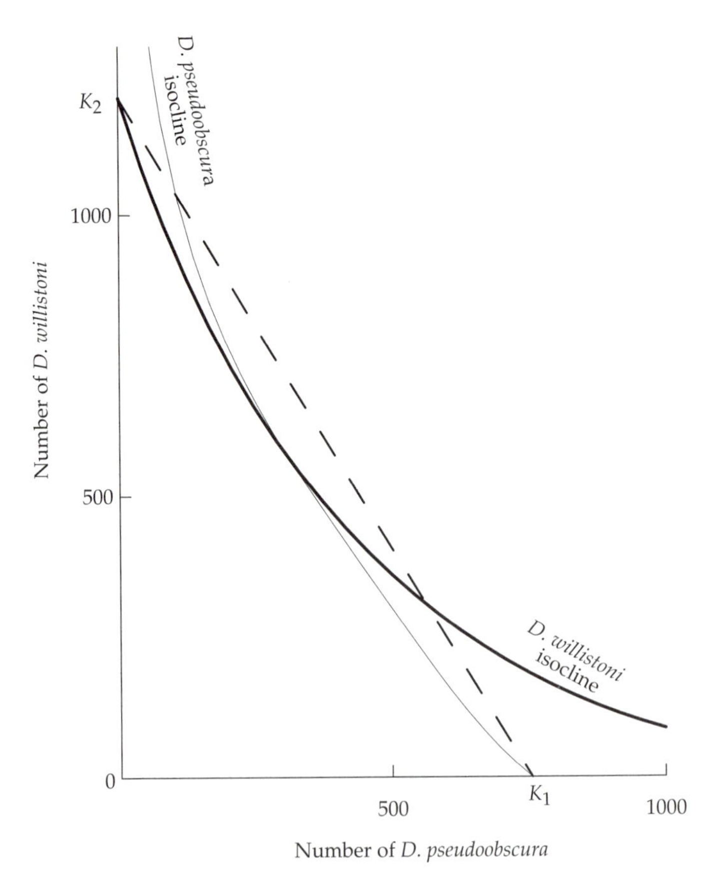

Figure 3.4. Estimated isoclines for *Drosophila pseudoobscura* and *D. willistoni*. After Gilpin and Justice, 1972.

species. The internal equilibrium where the isoclines intersect is below the dotted line joining K_1 and K_2. This equilibrium is nevertheless stable, as in Figure 3.3a; because the n_1 isocline crosses the n_2 isocline from above to below, the velocity vectors between the isoclines point toward the internal equilibrium where they intersect.

LOCAL STABILITY ANALYSIS FOR CONTINUOUS-TIME MODELS

This graphical analysis is illuminating but is not always so easy to apply. It is therefore useful in general to extend the algebraic local stability analysis developed in the last chapter to cover two or more variables. A detailed justification of the following account is given in Appendix H.

Consider the general continuous-time model with k variables

$$\mathrm{d}n_j/\mathrm{d}t = f_j(n_1, n_2, \ldots, n_k) \tag{5}$$

for $j = 1, \ldots, k$, with an equilibrium $\{\hat{n}_1, \hat{n}_2, \ldots, \hat{n}_k\}$ satisfying

$$f_j(\hat{n}_1, \hat{n}_2, \ldots, \hat{n}_k) = 0 \tag{6}$$

for $j = 1, \ldots, k$. Write $p_1, p_2, \ldots, p_k$ for small perturbations from the equilibrium position so that

$$n_j = \hat{n}_j + p_j \tag{7}$$

The model equations near the equilibrium can be approximated by the linear equations

$$\frac{\mathrm{d}p_j}{\mathrm{d}t} = \frac{\partial f_j}{\partial n_1} p_1 + \frac{\partial f_j}{\partial n_2} p_2 + \cdots + \frac{\partial f_j}{\partial n_k} p_k \tag{8}$$

where the partial derivatives are evaluated at the equilibrium. This set of k linear differential equations can be written in matrix notation as

$$\frac{\mathrm{d}}{\mathrm{d}t}\mathbf{p} = \mathbf{J}\mathbf{p} \tag{9}$$

where $\mathbf{p} = \{p_1, p_2, \ldots, p_k\}$ is the vector of perturbations and $\mathbf{J}$ is the Jacobian matrix of partial derivatives,

$$\mathbf{J} = \begin{bmatrix} \partial f_1/\partial n_1 & \partial f_1/\partial n_2 & \cdots & \partial f_1/\partial n_k \\ \partial f_2/\partial n_1 & \partial f_2/\partial n_2 & \cdots & \partial f_2/\partial n_k \\ \cdot & \cdot & & \cdot \\ \cdot & \cdot & \cdots & \cdot \\ \cdot & \cdot & & \cdot \\ \partial f_k/\partial n_1 & \partial f_k/\partial n_2 & \cdots & \partial f_k/\partial n_k \end{bmatrix} \tag{10}$$

The eigenvalues of **J** determine the local stability of the equilibrium.

The properties of eigenvalues and eigenvectors are discussed in Appendix F. A vector **e** is an eigenvector of a *k*-dimensional square matrix **A** and λ is the corresponding eigenvalue if

$$\mathbf{Ae} = \lambda \mathbf{e} \tag{11}$$

A $k \times k$ matrix generally has *k* distinct eigenvalues to each of which there is a distinct eigenvector. Note however that an eigenvector is only defined up to a multiplicative constant; if **e** is an eigenvector of **A** then so is 2**e** or **e**/3 or any other multiple of **e**. The eigenvalues are the roots of a polynomial equation of degree *k* and may therefore be real or complex.

It is shown in Appendix H that the solution of the set of linear differential equations (9) is

$$\mathbf{p}(t) = c_1 \exp(\lambda_1 t)\mathbf{e}_1 + c_2 \exp(\lambda_2 t)\mathbf{e}_2 + \cdots + c_k \exp(\lambda_k t)\mathbf{e}_k \tag{12}$$

where λ_j is the *j*th eigenvalue of **J**, $\mathbf{e}_j$ is the corresponding eigenvector, and the c_j's are constants chosen to satisfy the initial conditions. When $t = 0$ this equation can be written

$$\mathbf{p}(0) \equiv \mathbf{p}_0 = \mathbf{cR} \tag{13}$$

where **c** is the vector of constants and **R** is the matrix whose *j*th row is $\mathbf{e}_j$. Hence

$$\mathbf{c} = \mathbf{p}_0 \mathbf{R}^{-1} \tag{14}$$

(The constants can also be expressed in terms of the left eigenvalues as $\mathbf{c} = \mathbf{Lp}_0$; see Appendix H.)

If λ_j is real, the *j*th term in Equation 12 will decrease exponentially to zero or increase exponentially without limit according to whether λ_j is negative or positive. (See the worked example at the end of this section.)

If λ_j is complex, say $a + ib$, note that

$$\exp(a + ib)t = \exp at \exp ibt = (\exp at)(\cos bt + i \sin bt) \tag{15}$$

(See Appendix B.) Complex eigenvalues occur in conjugate pairs, $a \pm ib$, and choice of the constants in Equation 12 to satisfy the initial conditions will ensure that imaginary parts in the corresponding pair of terms drop out, leaving an expression that oscillates with period $2\pi/b$ because of the trigonometric terms in Equation 15 and that is damped or explosive according to whether a is negative or positive because of the exponential term. (See the worked example in Equations 30–32.)

Thus the criterion for the local stability of an equilibrium is that all the eigenvalues of the Jacobian matrix have negative real parts. The eventual behavior of the system near the equilibrium is determined by the dominant eigenvalue (with the largest real part). If this eigenvalue is real, the system will eventually converge toward or diverge from the equilibrium exponentially according to whether the eigenvalue is negative or positive, in the direction of the corresponding eigenvector. If the dominant eigenvalue is complex, the system will eventually oscillate about the equilibrium value with an amplitude that decreases or increases exponentially (damped or explosive oscillations) according to whether the real part of the eigenvalue is negative or positive. It must be remembered that this analysis only determines the behavior of the system near the equilibrium, that is to say, its local stability.

To recapitulate this important general rule: *an equilibrium of a model in continuous time is locally stable if and only if all the eigenvalues of the Jacobian matrix are negative, if real, or have negative real part, if complex.*

In several dimensions a computer program will usually be used to do the calculations, but for two variables there is a simple formula for finding the eigenvalues from the elements of the Jacobian matrix, that leads to a simple criterion for local stability. Write the Jacobian as

$$\mathbf{J} = \begin{bmatrix} a & b \\ c & d \end{bmatrix} \tag{16}$$

The determinant of this matrix is

$$D = \det \mathbf{J} = ad - bc \tag{17}$$

and the trace is defined as the sum of the diagonal elements

$$T = \text{trace}\, \mathbf{J} = a + d \tag{18}$$

The eigenvalues are

$$\frac{1}{2}\left(T \pm \sqrt{T^2 - 4D}\right) \tag{19}$$

The criterion for stability is that both eigenvalues have negative real parts, for which a necessary and sufficient condition is that the determinant be positive and the trace negative:

$$D > 0 \text{ and } T < 0 \tag{20}$$

(A proof of this result is given in Appendix H.)

To illustrate these results consider the internal equilibrium for case (a) of the Lotka–Volterra model of competition shown in Figure 3.3. The Jacobian matrix is

$$\begin{bmatrix} -0.3 & -0.2 \\ -0.2 & -0.3 \end{bmatrix} \tag{21}$$

The determinant and trace of this matrix are D = 0.05 and T = −0.6, so that the equilibrium is stable from Equation 20. Further information about the behavior of the system near the equilibrium can be obtained by calculating the eigensystem explicitly. The eigenvalues are −0.1 and −0.5 with eigenvectors {1, −1} and {1, 1}, respectively. (The latter are easy to calculate, up to a multiplicative constant, from their definition in Equation 11.) Hence,

$$\mathbf{p}(t) = c_1 \exp(-0.1t)\{1,-1\} + c_2 \exp(-0.5t)\{1,1\} \tag{22a}$$

or

$$\begin{aligned} p_1(t) &= c_1 \exp(-0.1t) + c_2 \exp(-0.5t) \\ p_2(t) &= -c_1 \exp(-0.1t) + c_2 \exp(-0.5t) \end{aligned} \tag{22b}$$

The constants are

$$\mathbf{c} = \mathbf{p}_0 \begin{bmatrix} 1 & -1 \\ 1 & 1 \end{bmatrix}^{-1} = \mathbf{p}_0 \begin{bmatrix} 0.5 & 0.5 \\ -0.5 & 0.5 \end{bmatrix} \tag{23a}$$

or

$$\begin{aligned} c_1 &= 0.5p_{10} - 0.5p_{20} \\ c_2 &= 0.5p_{10} + 0.5p_{20} \end{aligned} \tag{23b}$$

Figure 3.5a shows some of the trajectories of Equation 22 from different starting values. A perturbation starting on either of the eigenvectors stays on it (since this ensures that the other constant is zero), moving toward the equilibrium, but a perturbation anywhere else (ensuring that neither con-

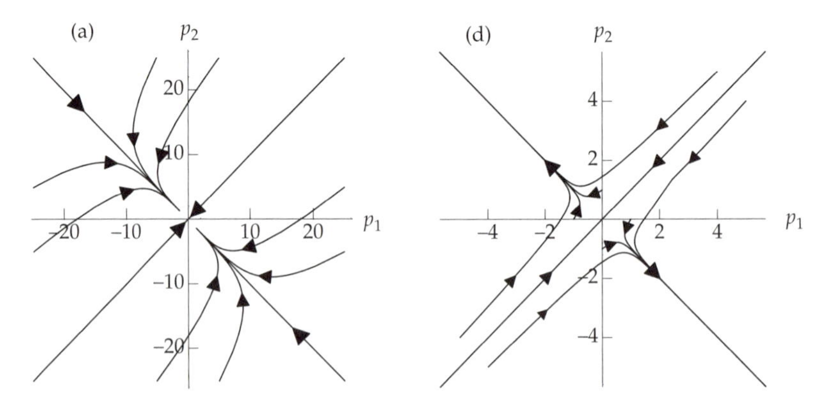

Figure 3.5. Phase plane diagram for small perturbations from the internal equilibrium under models (a) and (d) in Figure 3.3.

stant is zero) is drawn to the first eigenvector with the larger eigenvalue. This kind of stable equilibrium that arises in a two-dimensional system with a pair of negative, real eigenvalues, is called a stable node.

For case (d) of Figure 3.3 the Jacobian of the internal equilibrium is

$$\begin{bmatrix} -0.2 & -0.3 \\ -0.3 & -0.2 \end{bmatrix} \tag{24}$$

The determinant and trace of this matrix are $D = -0.05$ and $T = -0.4$, so that the equilibrium is unstable from Equation 20. The eigenvalues are +0.1 and –0.5, with eigenvectors {1, –1} and {1, 1} as before. Figure 3.5b shows some of the trajectories of the solutions analogous to Equation 22, but with +0.1 replacing the eigenvalue –0.1. A perturbation on the second eigenvector moves toward the equilibrium, but a perturbation anywhere else is drawn to the first eigenvector moving away from the equilibrium. This kind of unstable equilibrium, that arises in a two-dimensional system with a pair of real eigenvalues of opposite sign, is called a saddle.

If both eigenvalues are real and positive, any perturbation would move away from the equilibrium. This would be an unstable node. An example is the equilibrium at the origin.

Competition models usually give rise to real eigenvalues, but as we shall see in the next section prey–predator models often have complex eigenvalues, indicating oscillations.

PREY–PREDATOR MODELS

In a competitive interaction each species has a negative effect on the growth rate of the other. In a prey–predator interaction, predators have a negative effect on the growth rate of their prey, but prey have a positive effect on the growth rate of their predators. This asymmetry tends to give rise to oscillations in population numbers, that may either be damped, leading to a stable equilibrium value, or undamped, usually leading to self-perpetuating oscillations of characteristic period and amplitude called a *limit cycle*. The best known example of a limit cycle due ultimately to a prey–predator interaction is the ten-year cycle of many mammals in Canada, on which extensive historical records exist from the annual fur sales of the Hudson's Bay Company. Figure 3.6 shows data on the ten-year cycle in the numbers of lynx trapped in the Mackenzie River region between 1821 and 1934 (Elton and Nicholson, 1942).

The simplest model for the interaction between prey (species 1) and predator (species 2) is the classical Lotka–Volterra model:

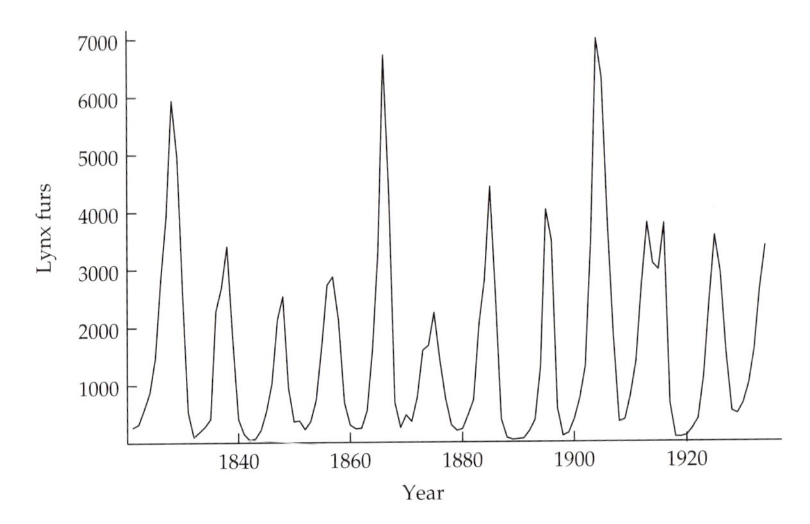

Figure 3.6. Lynx fur returns for the Mackenzie River region of Canada, 1821–1934. Source: Elton and Nicholson, 1942.

$$\begin{aligned} dn_1/dt &= r_1 n_1 - \alpha_1 n_1 n_2 \\ dn_2/dt &= -r_2 n_2 + \alpha_2 n_1 n_2 \end{aligned} \tag{25}$$

The prey species increases exponentially at rate r_1 in the absence of the predator, while the predator decreases exponentially at rate r_2 in the absence of prey. Predators encounter and consume prey at random, so that the total number of encounters is proportional to the product of their population densities, $n_1 n_2$, causing a corresponding depletion in the prey and an increase in the predator population.[1] The ratio α_2/α_1 is a conversion factor for converting prey units into predator units. The equilibrium with both species present is

$$\begin{aligned} \hat{n}_1 &= r_2/\alpha_2 \\ \hat{n}_2 &= r_1/\alpha_1 \end{aligned} \tag{26}$$

This simple model is of historic interest but is unrealistic in two respects. First, it allows the prey to grow exponentially without limit in

---

[1] This model of random encounter is analogous to the mass action law of chemistry in which the velocity of a chemical reaction is proportional to the product of the concentrations of the reactants.

the absence of the predator. This defect can be remedied by introducing a logistic density-dependent term in the prey equation:

$$\begin{aligned} dn_1/dt &= r_1(1-n_1/K_1)n_1 - \alpha_1 n_1 n_2 \\ dn_2/dt &= -r_2 n_2 + \alpha_2 n_1 n_2 \end{aligned} \tag{27}$$

An equilibrium with both species present exists, so that coexistence is possible, only when $K_1 > r_2/\alpha_2$; otherwise the maximum prey density is insufficient to support any predators.

The other defect in the Lotka–Volterra model is that the number of prey eaten by each predator is proportional to the prey abundance and increases without limit as the number of prey increase. It would be more realistic to allow for predator satiation by introducing a functional response curve $f(n_1)$ for the number of prey eaten per predator as a function of prey abundance. Holling (1965) suggested that a functional response of the type

$$f(n_1) = \alpha n_1/(1+\beta n_1) \tag{28}$$

is typical of many invertebrate predators; this function increases linearly at low prey densities but approaches an asymptotic value α/β at high densities. Incorporating both density dependence of the prey and predator satiation into the model, we obtain

$$\begin{aligned} dn_1/dt &= r_1(1-n_1/K_1)n_1 - \alpha_1 n_1 n_2/(1+\beta n_1) \\ dn_2/dt &= -r_2 n_2 + \alpha_2 n_1 n_2/(1+\beta n_1) \end{aligned} \tag{29}$$

I shall consider Equation 29 as a generic prey–predator model incorporating biologically plausible assumptions. (Arditi and Ginzburg [1989] and Akçakaya [1992] consider an alternative class of models in which the functional response is determined not by the number of prey, n_1, but by the prey/predator ratio, n_1/n_2.) The upper half of Figure 3.7 shows the behavior of this model with parameter values $r_1 = r_2 = 1$, $\alpha_1 = \alpha_2 = 0.01$, $\beta = 0.005$, and $K_1 = 500$. The equilibrium value with both species present is $\hat{n}_1 = 200$, $\hat{n}_2 = 120$. Figure 3.7a plots the trajectories of the prey and predator populations against time for 100 generations, starting from the initial values $n_1 = 300$, $n_2 = 50$ at time zero; the results were obtained by numerical solution of the equations using a computer program. There are damped oscillations about the equilibrium value, to which the system ultimately tends, with the predator population lagging behind that of the prey. In the phase plane diagram in Figure 3.7b the system spirals in toward the equilibrium value. This type of equilibrium is called a *stable spiral*. It is in fact globally stable and will attract the system from any starting value that includes both prey and predators.

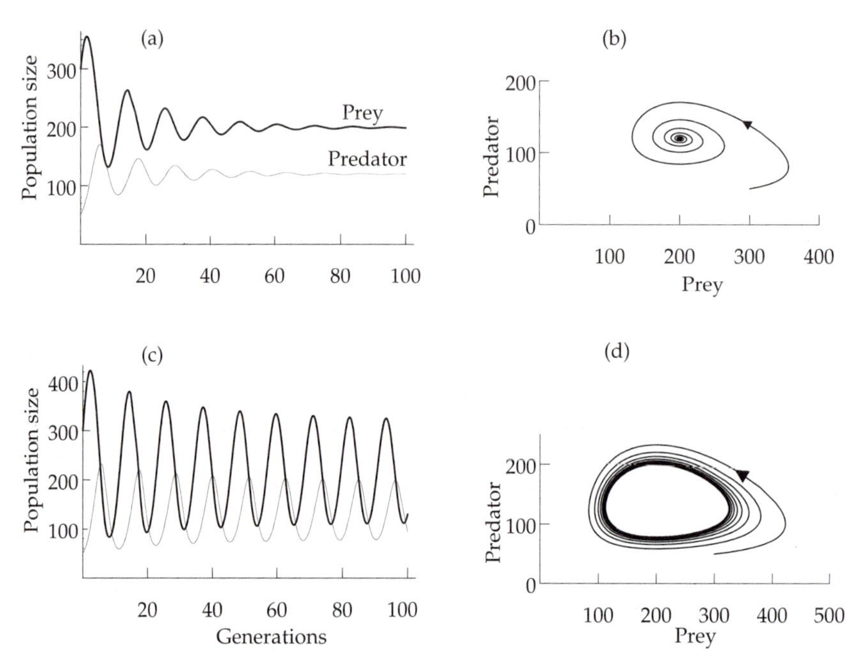

Figure 3.7. Behavior of the prey–predator model in Equation 29 with $r_1 = r_2 = 1$, $\alpha_1 = \alpha_2 = 0.01$, and $\beta = 0.005$, starting from $n_1 = 300$, $n_2 = 50$. (a) and (b) $K_1 = 500$; (c) and (d) $K_1 = 625$. Trajectories of prey and predator are plotted against time in (a) and (c); (b) and (d) are phase plane diagrams.

The lower half of Figure 3.7 shows the behavior of the model with the same parameter values except that K_1 is increased from 500 to 625, thus weakening the effect of density dependence in the prey. The oscillations are no longer damped but are sustained with a characteristic amplitude and period. The predator lags behind the prey as before, and in consequence the phase diagram spirals in toward a limiting curve, known as a *limit cycle*. The equilibrium value at $\hat{n}_1 = 200$, $\hat{n}_2 = 136$ is unstable, and if the system starts near this equilibrium, it will spiral out toward the limit cycle. The limit cycle is globally stable in the sense that the system will be attracted toward it from any starting value including both prey and predator.

Figure 3.7 shows the two main types of dynamic behavior found in prey–predator models, a stable spiral and a stable limit cycle The reason for oscillatory behavior in prey–predator models is most easily seen from the vector field diagram for the simple Lotka–Volterra model in Figure 3.8a. The prey isocline is the horizontal line $n_2 = r_1/\alpha_1$, and prey are increasing or decreasing (giving a velocity component to the right or left) according to whether the number of predators is below or above this line. The predator isocline is the vertical line $n_1 = r_2/\alpha_2$, and predators are increasing or decreasing (giving a velocity component up or down) according to whether the number of prey is to the right or left of this line. Thus the arrows chase each other round like a cat chasing its tail. The phase plane diagram in Figure 3.8b shows that trajectories for this model are closed orbits whose size (together with the amplitude of the corresponding oscillations) is determined by the initial conditions. However, this type of dynamic behavior is atypical.

The simplest way to explore the stability of prey–predator models is by local stability analysis. For example, the equilibrium in the upper half of Figure 3.7 has the Jacobian matrix

$$\begin{bmatrix} -0.1 & -1 \\ 0.3 & 0 \end{bmatrix} \tag{30}$$

The determinant and trace of this matrix are $D = 0.3$ and $T = -0.1$, so that the equilibrium is stable from Equation 20. Further information about the behavior of the system near the equilibrium can be obtained by calculating the eigensystem explicitly. The eigenvalues are $-0.05 \pm 0.545i$. The eigenvectors are $(-1.818 - 0.167i, i\}$ and $\{-1.818 + 0.167i, -i\}$, respectively. The constants are

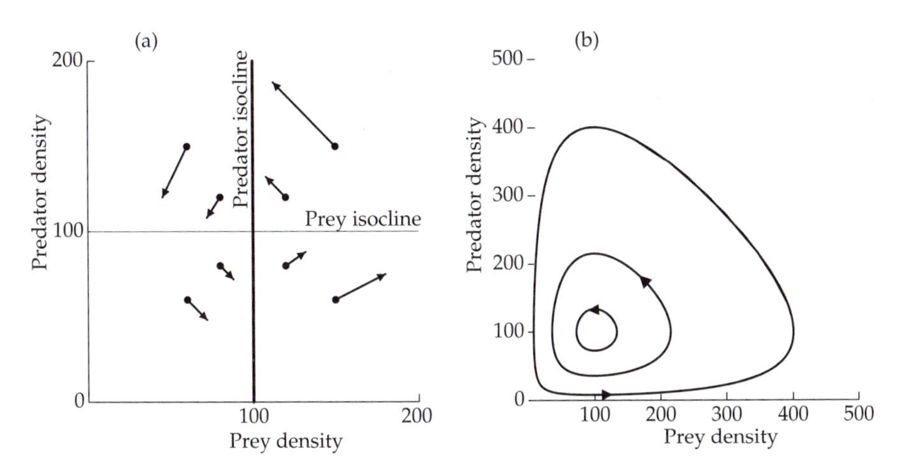

Figure 3.8. (a) Vector field diagram of Lotka–Volterra prey–predator model with $r_1 = r_2 = 1$ and $\alpha_1 = \alpha_2 = 0.01$; the arrows are velocity vectors. (b) Phase plane diagram showing three trajectories for this model.

$$\mathbf{c} = \mathbf{p}_0 \begin{bmatrix} (-1.818 - 0.167i) & i \\ (-1.818 + 0.167i) & -i \end{bmatrix}^{-1}$$
$$= \mathbf{p}_0 \begin{bmatrix} -0.275 & -0.275 \\ -0.046 - 0.5i & -0.046 + 0.5i \end{bmatrix} \tag{31a}$$

or

$$\begin{aligned} c_1 &= -0.275p_{10} - (0.046 + 0.5i)p_{20} \\ c_2 &= -0.275p_{10} - (0.046 - 0.5i)p_{20} \end{aligned} \tag{31b}$$

Substitution of Equation 15 into Equation 12 with these numerical values leads, after some algebraic reduction, to the final result:

$$\begin{aligned} p_1(t) &= \exp(-0.05t)[p_{10}\cos 0.545t - (0.09p_{10} + 1.83p_{20})\sin 0.545t] \\ p_2(t) &= \exp(-0.05t)[p_{20}\cos 0.545t + (0.55p_{10} + 0.09p_{20})\sin 0.545t] \end{aligned} \tag{32}$$

Thus there are damped oscillations near the equilibrium, with damping factor 0.05 and with period $2\pi/0.545$ = 11.5 years (or whatever units of time have been used). All these facts can be inferred from the eigenvalues. In addition, the full solution in Equation 32 contains information about the phase and relative amplitude of the oscillations.

In contrast, the equilibrium in the lower half of Figure 3.7 has eigenvalues $0.01 \pm 0.583i$, with positive real part. This shows that it is locally unstable and that there will be oscillations of increasing magnitude. Further investigation is needed to show that there is a stable limit cycle. It can in fact be shown in general that any locally unstable equilibrium of a continuous-time prey–predator model based on realistic biological assumptions gives rise to a stable limit cycle (Bulmer, 1976).

Investigation of the model in Equation 29 over a greater range of parameter values shows that density dependence of the prey (low values of K_1) promotes stability, whereas predator satiation (high values of β) promotes instability and leads to limit cycles.

The simplest explanation of the remarkable ten-year cycle shown in Figure 3.6 is that it is a limit cycle arising from a prey–predator interaction, with some environmental variability incorporated into the system. There seems no need to postulate chaotic behavior as suggested by Schaffer (1984). Lynx feed almost exclusively on snowshoe hares, that also show marked cyclical oscillations, and there is clear evidence that the lynx cycle is driven by the hare cycle; lynx starve and do not reproduce during hare lows. However, lynx predation is only a minor cause of hare mortality, and it seems likely that the hare cycle is not caused by the hare–lynx interaction but is due to periodic food shortage and is thus part of a plant–herbivore interaction between the hares and their food supply. The hare cycle then drives cyclical oscillations in a number of other species that are linked to them in the food chain (Bulmer, 1974, 1975).

MODELS OF INFECTIOUS DISEASE

Many infectious diseases are caused by very small organisms (viruses, bacteria, and protozoa, collectively called microparasites) that reproduce directly within the host. An infected host either recovers or dies within a short time and usually acquires immunity against reinfection after recovery, either permanently or for some time. The population dynamics of such infections can be modeled by classifying the host population into three types (susceptible, infected, and recovered) and constructing a set of differential equations for their frequencies.

Write x, y, and z for the numbers of susceptible, infected, and recovered hosts at time t. A simple model for their dynamics is

$$\mathrm{d}x/\mathrm{d}t = b(x + y + z) - \beta xy - dx \qquad (33a)$$

$$\mathrm{d}y/\mathrm{d}t = \beta xy - (D + \gamma)y \qquad (33b)$$

$$\mathrm{d}z/\mathrm{d}t = \gamma y - dz \qquad (33c)$$

This model is based on the following assumptions: (a) individuals are born at the per capita birth rate b into the susceptible class; (b) susceptible individuals are infected at a rate determined by the number of effective encounters between susceptible and infected individuals that follows a mass action law with rate constant β, the transmission coefficient; (c) susceptible and recovered individuals die at per capita rate d, while infected individuals die at rate $D > d$, the difference $D - d$ reflecting mortality due to the disease; (d) infected individuals recover at rate γ and are then immune for life (as is the case for measles). Age structure is ignored.

We also wish to incorporate some form of density dependence into the model so that the host population is kept constant at some total size N. One way of doing this is to assume that the birth rate b is instantaneously adjusted in response to population size by some unspecified mechanism so that the numbers of births and deaths are the same at this population size:

$$b(x + y + z) = d(x + z) + Dy \qquad (34)$$

Using this relationship in Equation 33 and eliminating z by substituting $z = N - x - y$, we find

$$dx/dt = d(N - x) + (D - d)y - \beta xy \tag{35a}$$

$$dy/dt = \beta xy - (D + \gamma)y \tag{35b}$$

where the parameters d, D, β, γ, and N are constant and independent of time.

I first consider a special case in which there are no births or deaths and infected individuals never recover, so that $z = 0$. Hence $x = N - y$, so that the equation for y becomes

$$dy/dt = \beta(N - y)y \tag{36}$$

This is the logistic equation with $r = \beta N$ and $K = N$. The number of infected individuals will increase in a logistic way until there are no susceptible individuals left. When the special assumptions of negligible birth, death, or recovery hold, the fit of the data to a logistic curve provides a direct test of the assumption that the net rate at which susceptible hosts become infected is proportional to the product of the numbers of susceptible and infected individuals. Anderson (1981) quotes data on infection of a coelenterate host by a protozoan pathogen that give an excellent fit to the logistic curve.

In the full model (Equation 35) the occurrence of births in the susceptible class prevents the elimination of that class. There are two possible equilibria, obtained by setting $y = 0$ or $y > 0$:

$$\hat{x} = N \qquad \hat{y} = 0 \tag{37a}$$

$$\hat{x} = (D + \gamma)/\beta \qquad \hat{y} > 0 \tag{37b}$$

The existence and stability of these equilibria depend on the magnitude of N, the total population size. If $N < (D + \gamma)/\beta$, the first equilibrium (with the disease eliminated) is the unique equilibrium and is stable; the second equilibrium does not exist since it would require that $\hat{x} > N$. (Note that in this case the right-hand side of Equation 35b is always negative for $x \leq N$ so that the number of infected individuals can never increase.) If $N > (D + \gamma)/\beta$, the first equilibrium is unstable while the second equilibrium (with the disease present) is stable. Thus the disease cannot persist in the host population unless the size of that population exceeds the threshold value $(D + \gamma)/\beta$.

The behavior of the model can be understood by considering the *reproductive rate* of the infection, $R(x)$, defined as the average number of secondary cases that one infected host gives rise to during its lifetime in a population containing x susceptible individuals. The average lifetime of an infected host is $1/(D + \gamma)$, the reciprocal of the rate at which individuals leave the infected class either through death or recovery; the rate of infection of susceptible hosts per unit time is βx; and the reproductive rate is their product:

$$R(x) = \beta x/(D + \gamma) \tag{38}$$

It is clear that an infection can only spread if its reproductive rate exceeds unity, so that it can only gain a foothold in an uninfected population if $R(N) > 1$.

These results can be summarized in the *threshold theorem: if the number of susceptible individuals is below the threshold value, $(D + \gamma)/\beta$, the infection cannot spread; if it is above the threshold value, the infection will spread until the number of susceptible individuals falls to the threshold value.*

The threshold theorem has an important implication for public-health immunization programs. The purpose of an immunization program against, for example, measles is to reduce the size of the unimmunized population below the threshold value; once this has been achieved the disease cannot spread even among the unimmunized, because there are too few of them. In a completely unimmunized population it has been estimated that the percentage of individuals susceptible to measles is a little under 10% of the population (Anderson and May, 1991). An immunization program must therefore aim to immunize well over 90% of the population to achieve its aim.

Some infectious diseases exhibit oscillations in their incidence with a period longer than that due to seasonal fluctuations. For example, Figure 3.9 shows the rather regular two-year cycle in the incidence of measles, with a peak every other year, in the period prior to mass vaccination. The model given by Equations 35 with parameter values suitable for measles shows oscillations about the equilibrium value with a period of 2.5 years, that is encouraging, but they are weakly damped. Anderson and May (1991) discuss a number of factors that may destabilize the equilibrium and convert the behavior of the system into a stable limit cycle leading to the observed behavior.

It has been assumed so far that some unspecified density-dependent factor operates to keep the total host population size constant. It is interesting from an ecological viewpoint to consider whether disease can itself act to regulate the host population size. Return to Equation 33 with $b > d$ so that in the absence of disease population size is increasing exponentially at rate $b - d$. There is a unique equilibrium value for this set of equations with $y > 0$, and it is a feasible equilibrium (with no negative components) provided that

$$b(d + \gamma) < d(D + \gamma) \tag{39}$$

When this inequality is satisfied the population size is stabilized at a constant value by the mortality directly due to the disease; when it is not sat-

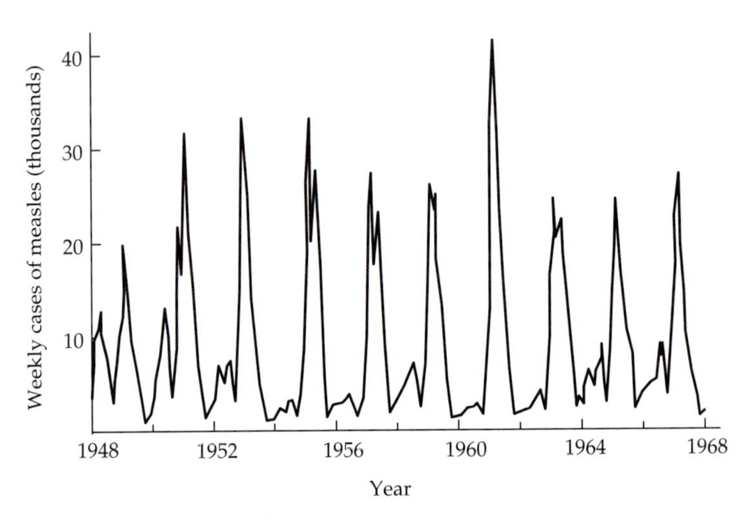

Figure 3.9. Weekly cases of measles in England and Wales, 1948–1968 prior to the introduction of mass vaccination. From Anderson and May, 1991.

isfied the population size grows exponentially in the presence of the disease, though not as fast as in its absence. Thus it is possible for disease to act as a density-dependent factor regulating the size of the host population.

LOCAL STABILITY ANALYSIS FOR DISCRETE-TIME MODELS

It may be more appropriate to represent populations with nonoverlapping generations by difference equations in discrete time rather than by differential equations in continuous time. A general model for k species in discrete time can be written

$$n_j(\mathrm{t}+1) = F_j[n_1(t), n_2(t), \ldots, n_k(t)] \tag{40}$$

for $j = 1, 2, \ldots, k$. The behavior of this model can be investigated by simulation, starting with a pair of initial values and iterating the process. Local stability analysis is the other useful technique for obtaining information about the stability of the equilibria. An equilibrium $\{\hat{n}_1, \hat{n}_2, \ldots, \hat{n}_k\}$ satisfies

$$F_j\{\hat{n}_1, \hat{n}_2, \ldots, \hat{n}_k\} = \hat{n}_j \tag{41}$$

for $j = 1, 2, \ldots, k$. Write $p_1, p_2, \ldots, p_k$ for small perturbations from an equilibrium position as before. The model equations near the equilibrium can be approximated by the linear equations

$$p_j(t+1) = \frac{\partial F_j}{\partial n_1} p_1(t) + \frac{\partial F_j}{\partial n_2} p_2(t) + \quad \cdots \quad + \frac{\partial F_j}{\partial n_k} p_k(t) \tag{42}$$

that can be written in matrix terminology as

$$\mathbf{p}(t+1) = \mathbf{J}\mathbf{p}(t) \tag{43}$$

where **J** is the Jacobian matrix of partial derivatives evaluated at the equilibrium. The solution of this equation is

$$\mathbf{p}(t) = \mathbf{J}^t \mathbf{p}(0) = c_1 \lambda_1^t \mathbf{e}_1 + c_2 \lambda_2^t \mathbf{e}_2 + \quad \cdots \quad + c_k \lambda_k^t \mathbf{e}_k \tag{44}$$

where the λ_j's are the eigenvalues of **J**, the $\mathbf{e}_j$'s are the eigenvectors (Appendix J) and the c_j's are chosen to satisfy the initial conditions when $t = 0$ (Equation 14). If λ_j is real, the jth term in Equation 44 will tend to zero if it is less than 1 in absolute value, and it will increase without limit otherwise. If λ_j is complex, say $a + ib$, write it in polar coordinates (absolute value and argument) as $\{r, \theta\}$ and note that

$$(a+ib)^t = r^t(\cos\theta t + i\sin\theta t) \tag{45}$$

(Appendix B). A pair of complex conjugate eigenvalues will give rise to a term that will oscillate with period $2\pi/\theta$, and whose amplitude will decrease (damped oscillations) or increase (increasing oscillations) according to whether $r < 1$ or $r > 1$.

Thus the criterion for the local stability of an equilibrium is that all the eigenvalues of the Jacobian matrix are less than 1 in absolute value. The behavior of the system near the equilibrium is determined by the dominant eigenvalue (with the largest absolute value). If this eigenvalue is real, the system will eventually converge toward or diverge from the equilibrium geometrically depending on whether its absolute value is less than or greater than 1, in the direction of the corresponding eigenvector; if it is negative the perturbations will alternate in sign each generation. If the dominant eigenvalue is complex, the system will eventually oscillate about the equilibrium value with an amplitude that decreases or increases exponentially (damped or explosive oscillations) depending on whether its absolute value is less than or greater than 1. It must be remembered that this analysis only determines the behavior of the system near the equilibrium, that is to say, its local stability.

To recapitulate this important general rule: *an equilibrium of a model in discrete time is locally stable if and only if all the eigenvalues of the Jacobian matrix are less than 1 in absolute value.*

In two dimensions a necessary and sufficient condition that both eigenvalues are less than 1 in absolute value (the analogue of Equation 20) is

$$2 > 1 + \det \mathbf{J} > |\text{trace } \mathbf{J}| \tag{46}$$

(see Appendix J).

HOST–PARASITOID MODELS

Nicholson (1933) and Nicholson and Bailey (1935) considered the following discrete-time model representing the life history of an insect parasitoid that lays eggs in, say, the larvae of a host species. The rules are that parasitoids search randomly for prey, that their fecundity is unlimited, and that each of them has searching efficiency a, called the area of discovery; this means that if the density of hosts per unit area in year t is $n_1(t)$, a parasitoid will on average find and lay an egg in $an_1(t)$ hosts. The number of eggs laid per unit area in year t is $an_1(t)n_2(t)$ where $n_2(t)$ is the density of parasitoids in that year, so that the average number of eggs laid per host is $an_2(t)$. Under the assumption of random search, the distribution of eggs in hosts follows the Poisson distribution with this mean, so that the chance of a host escaping parasitism altogether is $\exp[-an_2(t)]$, the zero term in this distribution. (The Poisson distribution is discussed in Appendix K.) We now suppose that each unparasitized host gives rise to R surviving offspring in the next year, while each parasitized host gives rise to 1 parasitoid next year, regardless of the number of eggs laid in it. Then,

$$\begin{aligned} n_1(t+1) &= Rn_1(t)\exp[-an_2(t)] \\ n_2(t+1) &= n_1(t)\left(1-\exp[-an_2(t)]\right) \end{aligned} \tag{47}$$

There is a unique equilibrium with both species present, but the eigenvalues are always complex with absolute value > 1, indicating instability with oscillations of increasing amplitude. Simulation of the full equations shows ever-increasing oscillations, not a limit cycle (Figure 3.10a). Both species could not coexist under these circumstances. Either the host would go extinct when it reached very low numbers, followed by the extinction of the parasitoid, or the parasitoid would go extinct when it reached very low numbers, followed by an unchecked increase of the host. Thus this is not a suitable model for a stable host–parasitoid interaction.

One possible factor stabilizing the interaction is that R, the rate of increase of the host, may be density-dependent. Another factor is that a, the area of discovery of the parasitoid, may decrease with increasing parasitoid density because of mutual interference between them. Hassell and Varley (1969) showed experimentally that in one system under laboratory conditions the area of discovery was inversely proportional to the square

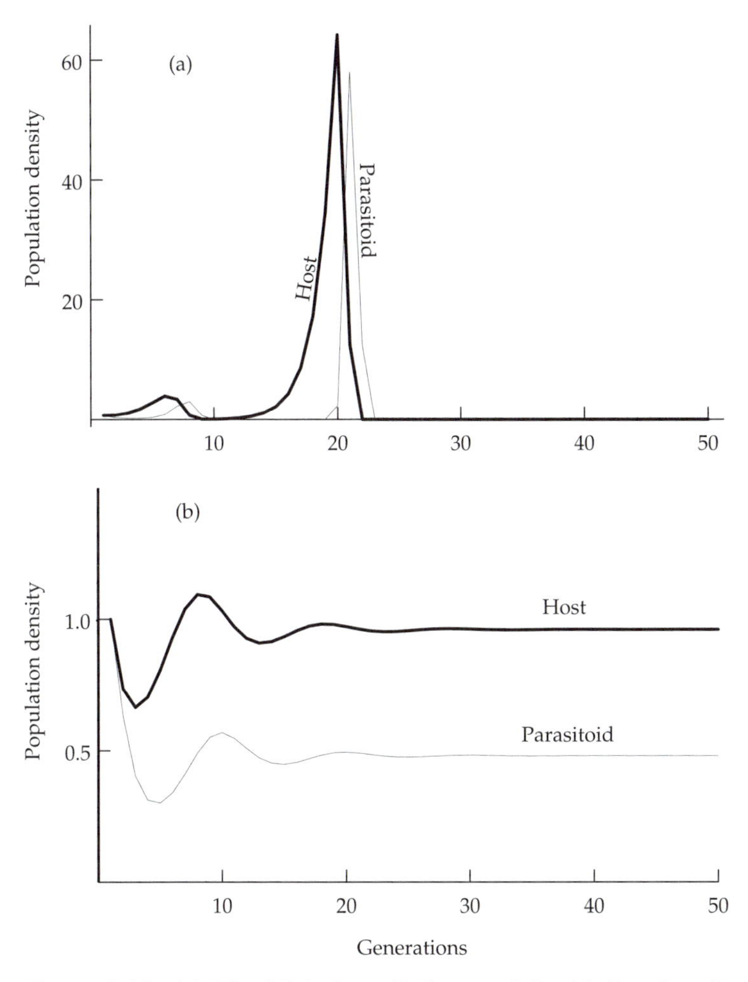

Figure 3.10. (a) The Nicholson–Bailey model with R = 2 and a = 1, (b) the Hassell–Varley model with R = 2 and Q = 1, both simulated for 50 generations starting with initial values of $n_1 = n_2 = 1$.

root of parasitoid numbers

$$a = Qn_2^{-0.5} \tag{48}$$

Substituting for a in the difference equations gives

$$\begin{aligned} n_1(t+1) &= Rn_1(t)\exp[-Qn_2^{0.5}(t)] \\ n_2(t+1) &= n_1(t)\left(1-\exp[-Qn_2^{0.5}(t)]\right) \end{aligned} \tag{49}$$

This gives a stable equilibrium when $R < 4.92$ (Figure 3.10b). Thus parasitoids can act as a density-dependent factor regulating the host population. The host–parasitoid interaction is a type of prey–predator interaction, so that the occurrence of oscillations is expected. The fact that these oscillations are ever-increasing rather than leading to a limit cycle when the equilibrium is locally unstable results from the time lag in discrete-time models.

FURTHER READING

Edelstein-Keshet (1988) provides a detailed account of the applications of differential and difference equations in biology; other useful mathematical books are Sandefur (1990), Abraham and Shaw (1992), and Strogatz (1994). Books emphasizing biological applications are Emlen (1973, 1984), Roughgarden (1979), Yodzis (1989), and Morin (1995); the last two discuss extensions to food webs and interactions between three or more species. May (1981b) is a good introductory account of models for two interacting species. Bailey (1975), Anderson (1981), and Anderson and May (1991) describe models of infectious disease. Hassell (1978, 1981) discusses arthropod prey–predator models, and Godfray (1994) reviews the biology of parasitoids.

EXERCISES

1. (a) Use pencil and paper to produce a vector field diagram like Figure 3.2a for the Lotka–Volterra model of competition with $K_1 = K_2 = 1000$, $r_1 = r_2 = 0.5$, and $\gamma_{12} = \gamma_{21} = 1.5$.
   (b) If you have a program for numerical solution of differential equations, produce a phase plane diagram like Figure 3.2b for the same model.

2. Verify the algebraic formulas for the four equilibria under the Lotka-Volterra competition model. Find algebraic formulas for the Jacobian matrices at these four equilibria. Verify that they have the numerical values given in Equations 21 and 24 for cases (a) and (d) of Figure 3.3, and find the eigenvalues and eigenvectors.

3. Use the criterion for local stability in Equation 20 together with the algebraic Jacobian matrices from Exercise 2 to find the conditions for local stability of the four equilibria.

4. Investigate algebraically the local stability of the equilibrium with both species present (Equation 26) of the Lotka–Volterra prey–predator model (Equation 25). What conclusion can be drawn about the behavior of the system?

5. Consider the generic prey–predator model of Equation 29 with $r_1 = r_2 = 1$, $\alpha_1 = \alpha_2 = 0.01$, and $\beta = 0.005$.
(a) Find the internal equilibrium and investigate its local stability numerically as a function of K_1.
(b) If you have a program for numerical solution of differential equations, investigate the behavior of the equations for values of K_1 predicted to give a stable and an unstable equilibrium.

6. Investigate the local stability of the equilibria in Equation 37 for the infectious-disease model in Equation 35 and verify the results stated in the text.

7. It is sometimes convenient to write Equation 35 in terms of relative rather than absolute numbers. Define $p = x/N$, $q = y/N$ so that

$$\mathrm{d}p/\mathrm{d}t = d(1-p) + (D-d)q - \beta Npq$$
$$\mathrm{d}q/\mathrm{d}t = \beta Npq - (D+\gamma)q$$

The following parameter values may represent a disease such as measles: $d = D = 1/70\ \mathrm{yr}^{-1}$ (life expectancy of 70 yr, no increase in mortality due to measles because of medical care); $\gamma = 50\ \mathrm{yr}^{-1}$ (infectious period $= 1/50$ yr $\cong$ 1 week); $\beta N = 500$ (estimated by equating observed relative frequency of susceptibles [about 0.1] with expected value $[(D+\gamma)/\beta N]$). (I have formulated the problem in this way because I do not know how to estimate β and N separately.)

Find equilibrium values of p and q under this model with $p < 1$. Investigate the local stability of this equilibrium by calculating the eigenvalues of the Jacobian matrix. Verify that they predict damped oscillations with a period of 2.5 years. (Note: You may use the results of the previous exercise with $N = 1$ since N is just a scale factor.)

8. (a) Find the equilibrium for the Nicholson-Bailey model (Equation 47). (This is most easily done with pencil and paper.)
(b) Without loss of generality take $a = 1$ since it acts as a scale parameter. (To see this rewrite the equations in terms of $x_1 = an_1$, $x_2 = an_2$.) Find the eigenvalues for a range of values of R between 1.01 and 5 and show that they are always complex with absolute value > 1.

(c) Simulate the model for different values of R (with $a = 1$) and show that there are oscillations of increasing amplitude rather than a limit cycle.

9. Repeat these calculations for the Hassell-Varley model (Equation 49) and show that there is stability for $R < 4.92$. (Without loss of generality it may be assumed that $Q = 1$. Why?)

CHAPTER 4

Age-Structured Populations

We return to consideration of a single population, but taking into account that birth and death rates may depend on age so that we should keep track separately of individual age classes. In this chapter I shall develop the basic theory of age-structured populations under the simplifying assumption that the age-specific fecundities and survival probabilities are constant. In the next chapter I shall apply this theory to discuss the evolution of life-history strategies, considering problems such as how often an animal or plant should breed, how many offspring it should have, and how long it should live.

THE LESLIE MATRIX

Consider a perennial plant or a bird species in a temperate climate that breeds once a year, and suppose that the population is censused each year just before the breeding season. A complete description of the population will include the numbers of individuals in each age group. For a dioecious species (i.e., one with separate sexes), such as birds, it should also count males and females separately, but it is usually sufficient to model only the female population on the assumption of *female demographic dominance*, which means that the number of births depends primarily on the number of breeding females in each age group.

We first define the following terminology:

$$\begin{aligned} n_x(t) &= \text{number of females of age } x \text{ in year } t \\ P_x &= \text{probability that a female aged } x \text{ in year } t \\ &\quad \text{survives until year } t+1 \\ &\quad \text{(age-specific survival probability)} \\ m_x &= \text{average number of female offspring} \\ &\quad \text{produced by a female aged } x \\ &\quad \text{(age-specific fecundity)} \\ f_x &= \text{number of these offspring surviving to age 1} \\ &= P_0 m_x \qquad \text{(net fecundity)} \end{aligned} \tag{1}$$

The age x takes integer values from 0 up to ω, the maximum lifespan. (Newborns are assigned age 0, and P_0 is their probability of survival to age 1.) It is assumed for simplicity that the survival and fecundity rates are independent of t. For a clonal, monoecious, or hermaphrodite species (such as many plants), the problem of distinguishing between the two sexes does not arise, and the above definitions would refer to all individuals aged x rather than to females only.

We shall now write down a set of recurrence relations for $n_x(t)$ in successive years for $x \geq 1$. For $x > 1$,

$$n_x(t+1) = P_{x-1} n_{x-1}(t) \tag{2}$$

This expresses the fact that females aged x next year are the survivors of females aged $x - 1$ this year. For $x = 1$, note that individuals aged 1 next year are the survivors of offspring born this year, so that

$$n_1(t+1) = f_1 n_1(t) + f_2 n_2(t) + \cdots + f_\omega n_\omega(t) \tag{3}$$

Write $\mathbf{n}(t)$ for the vector $\{n_1(t), n_2(t), \ldots, n_\omega(t)\}$. Equations 2 and 3 can be written in matrix form as

$$\mathbf{n}(t+1) = \mathbf{L}\mathbf{n}(t) \tag{4}$$

where

$$\mathbf{L} = \begin{bmatrix} f_1 & f_2 & f_3 & \cdots & f_{\omega-1} & f_\omega \\ P_1 & 0 & 0 & \cdots & 0 & 0 \\ 0 & P_2 & 0 & \cdots & 0 & 0 \\ 0 & 0 & P_3 & \cdots & 0 & 0 \\ \vdots & \vdots & \vdots & \cdots & \vdots & \vdots \\ 0 & 0 & 0 & \cdots & P_{\omega-1} & 0 \end{bmatrix} \tag{5}$$

The matrix **L** is called a *Leslie matrix* after P. H. Leslie who promoted its use in ecology and demography (Leslie, 1945, 1948). It has net fecundities in the first row, survival probabilities below the leading diagonal, and zero everywhere else.

For example, consider the data on female grey squirrels in Table 4.1. Annual survival was estimated by following marked individuals from the nestling stage, though data for later ages are based on very small sample sizes; fecundity was determined by measurements of litter size. The Leslie matrix is

$$\mathbf{L} = \begin{bmatrix} 0.32 & 0.57 & 0.57 & 0.57 & 0.57 & 0.57 & 0.57 \\ 0.46 & 0 & 0 & 0 & 0 & 0 & 0 \\ 0 & 0.77 & 0 & 0 & 0 & 0 & 0 \\ 0 & 0 & 0.65 & 0 & 0 & 0 & 0 \\ 0 & 0 & 0 & 0.67 & 0 & 0 & 0 \\ 0 & 0 & 0 & 0 & 0.64 & 0 & 0 \\ 0 & 0 & 0 & 0 & 0 & 0.88 & 0 \end{bmatrix} \qquad (6)$$

The Leslie matrix model is an age-classified model which assumes that age is the main variable affecting survival and fecundity rates. It may be that other variables such as size are better predictors of these rates than is age. This leads to the construction of a stage-classified matrix model in which the matrix has a more complex form than the Leslie matrix but can be analyzed in a similar way. For further information see Caswell (1989) or van Groenendael et al. (1988).

Table 4.1. Age-specific survival and fecundity of female grey squirrels in North Carolina.

Age x in years	P_x	m_x	f_x
0	0.25		
1	0.46	1.28	0.32
2	0.77	2.28	0.57
3	0.65	2.28	0.57
4	0.67	2.28	0.57
5	0.64	2.28	0.57
6	0.88	2.28	0.57
7		2.28	0.57

After Charlesworth, 1994.

ASYMPTOTIC BEHAVIOR

Equation 4 can be used to project the population into the future starting from an initial value $\mathbf{n}(0)$:

$$\begin{aligned}\mathbf{n}(1) &= \mathbf{L}\mathbf{n}(0)\\ \mathbf{n}(2) &= \mathbf{L}\mathbf{n}(1) = \mathbf{L}^2\mathbf{n}(0)\\ \mathbf{n}(3) &= \mathbf{L}\mathbf{n}(2) = \mathbf{L}^3\mathbf{n}(0)\end{aligned} \tag{7}$$

and so on. It is assumed in this chapter that the elements of the Leslie matrix remain constant over time. Since this may not be true over a long time span, the results must be treated with caution as forecasts of the future. In the context of human demography, for example, population projections into the distant future should be regarded as thought experiments of what *would* happen *if* the age-specific vital rates remain as they are today (a projection of current conditions into the future), rather than as forecasts of what *will* happen. In applying the Leslie matrix to the evolution of life-history strategies in the next chapter we shall incorporate density dependence into the model by supposing that some or all of the age-specific survival rates and fecundities are decreasing functions of some or all of the $n_x(t)$'s; the elements of the Leslie matrix will change until a stable equilibrium is attained with constant numbers in all age classes and thereafter remain constant.

Asymptotic growth rate and stable age distribution

Projections into the distant future involve calculating large powers of the Leslie matrix; they are most easily done by using its spectral decomposition (Appendix G). The Leslie matrix usually has a unique, real dominant eigenvalue λ_1, with corresponding eigenvector $\mathbf{e}_1$. (Exceptions to this rule are discussed in the subsection on periodic matrices.) In this case the asymptotic value of $\mathbf{n}(t)$ for large t is

$$\mathbf{n}(t) \cong c_1\lambda_1^t\mathbf{e}_1 \tag{8}$$

This means that the stable age distribution is proportional to $\mathbf{e}_1$ and that asymptotically all the age classes, and the total population size, increase at rate λ_1. The constant is

$$c_1 = \boldsymbol{\varepsilon}_1\mathbf{n}(0) \tag{9}$$

where $\boldsymbol{\varepsilon}_1$ is the dominant left eigenvector (standardized so that $\boldsymbol{\varepsilon}_1\mathbf{e}_1 = 1$). This means that the elements of the left eigenvector represent the relative importance of individuals of different ages in the initial population in determining the total population size in the distant future; they are called the *reproductive values* of these age groups. (Recall that λ is an eigenvalue of the matrix **L** with associated eigenvector [determined up to a multiplicative constant] if $\mathbf{Le} = \lambda\mathbf{e}$. A left eigenvector satisfies $\boldsymbol{\varepsilon}\mathbf{L} = \lambda\boldsymbol{\varepsilon}$; right and left eigenvalues are the same, while a left eigenvector is the right eigenvector of the transpose of **L**.)

For the squirrel data $\lambda_1 = 1.04$, so that when the stable age distribution is ultimately reached the population will increase by about 4% each year. One would usually expect a natural population to be nearly stationary in size if it is regulated by density-dependent factors; the observed eigenvalue is not significantly different from unity given the inaccuracies of the estimated survival and fecundity rates. The dominant right eigenvector (scaled to sum to 100) is

$$\mathbf{e}_1 = \{44, 20, 14, 9, 6, 4, 3\} \tag{10}$$

representing the stable age distribution, the percentages of the total population in different age classes at equilibrium. The dominant left eigenvector (scaled so that $\boldsymbol{\varepsilon}_1\mathbf{e}_1 = 1$) is

$$\boldsymbol{\varepsilon}_1 = \{1, 1.6, 1.4, 1.3, 1.2, 1.0, 0.5\}/119 \tag{11}$$

These figures represent the reproductive values of females of different ages in the initial population, so that a 2-year-old squirrel will have 1.6 times as many descendants in the distant future as a 1-year-old squirrel, and so on.

The special form of the Leslie matrix ensures that some simple formulas can be obtained for the dominant eigenvalue and right eigenvector. Suppose that asymptotically there is a stable age distribution so that all the age classes are increasing at the same rate λ. Write

$$l_x = P_0 P_1 \cdots P_{x-1} \tag{12}$$

for the probability that a newborn will survive to age x. The number of newborn female offspring born in year t is

$$n_0(t) = \sum_{x=1}^{\omega} n_x(t) m_x \tag{13}$$

Females aged x in year t are the survivors of individuals born in year $t - x$:

$$n_x(t) = n_0(t - x) l_x \tag{14}$$

If each of the age classes is increasing at geometric rate λ, then

$$n_0(t) = \lambda^x n_0(t-x) \tag{15a}$$

$$n_0(t-x) = \lambda^{-x} n_0(t) \tag{15b}$$

Substituting this into Equation 14 gives

$$n_x(t) = n_0(t)\lambda^{-x} l_x \tag{16}$$

Substitution of this expression into Equation 13 and division by $n_0(t)$ yields

$$\sum_{x=1}^{\omega} \lambda^{-x} l_x m_x = 1 \tag{17}$$

This important result provides an implicit equation for calculating λ.

It is often more convenient to calculate the dominant eigenvalue from Equation 17 than from the Leslie matrix, and in some cases it can be reduced to a simpler equation. In many birds the mortality and fecundity of adults are nearly constant and independent of age. Write a for the age of first reproduction, l_a for the probability that a newborn survives to this age, P for the annual survival rate in adults and m for the fecundity of adults. Since there is no upper age limit, Equation 17 is

$$\sum_{x=a}^{\infty} \lambda^{-x} l_a P^{x-a} m = \lambda^{-a} l_a m \sum_{x=0}^{\infty} (P/\lambda)^x = \frac{\lambda^{-a} l_a m}{1 - P/\lambda} = 1 \tag{18}$$

(See Appendix A for the formula for the sum of a geometric series.)

As an example, consider the investigation by Lande (1988) on the Northern Spotted Owl, whose demographic status is important because of the controversy between conservationists and logging interests in the Pacific Northwest of the United States. Females breed for the first time at age 3 and have a brood of average size 0.48 yearly (taking into account that they may not breed at all); with a 1:1 sex ratio, this gives $m = 0.24$. The probability that an owlet will survive to age 3 (l_a) is 0.0722, so that $l_a m$ = 0.01733, and the average adult survival probability is 0.942 at all ages greater than 3. Equation 18 is

$$\frac{0.01733\lambda^{-3}}{1 - 0.942\lambda^{-1}} = 1 \tag{19}$$

This gives a cubic equation in λ whose unique real root is $\lambda = 0.961$. This is not significantly different from unity, so that there is no evidence of a population decline in the owl's present habitat. (McKelvey et al. [1993] report a similar analysis based on different estimates of the parameter values.)

If a stable age distribution exists, Equation 16 shows that the proportion of females of age x ($x \geq 1$) is

$$c_x = \frac{n_x(t)}{\sum_{x=1}^{\omega} n_x(t)} = \frac{\lambda^{-x} l_x}{\sum_{x=1}^{\omega} \lambda^{-x} l_x} \propto \lambda^{-x} l_x \tag{20}$$

It is easier, and more informative, to use this explicit formula instead of the right eigenvector. When $\lambda = 1$ (stationary population size), c_x is proportional to l_x; the proportion of individuals of age x is proportional to the probability of survival to that age. When $\lambda > 1$ (increasing population size), c_x is weighted toward younger ages since there were fewer females alive several years ago to give birth to those who are old today; contrariwise, when $\lambda < 1$ (decreasing population size), c_x is weighted toward older ages.

The formula for the stable age distribution can sometimes be used to infer information about mortality from the observed age distribution of a natural population when direct information about age-specific female mortality is not available. Assume that the population is stationary ($\lambda = 1$) because of the action of density-dependent factors and that it has attained the stable age distribution. Then one can infer l_x from c_x instead of predicting c_x from l_x. Table 4.2 shows the age distribution of 172 breeding female Great Tits (*Parus major*) that had been ringed as nestlings near Oxford, England so that their exact ages were known. It can be assumed that females are reproductively mature at age 1 so that this represents the population age distribution. If this represents a stable age distribution with $\lambda = 1$, the ratio of successive age classes n_{x+1}/n_x is an estimate of P_x, the probability of survival from age x to age $x + 1$. Visual inspection suggests that this ratio is approximately constant, so that mortality is independent of age. This impression is confirmed by statistical analysis, the annual survival probability being estimated as 0.48.

Table 4.2. Age distribution of 172 female Great Tits.

Age (years)	1	2	3	4	5	6	7+	Total
Frequency	92	37	22	12	6	3	0	172

From Bulmer and Perrins, 1973.

Periodic matrices

It has been assumed so far that there is a unique dominant eigenvalue for the Leslie matrix so that the system tends asymptotically to a stable age distribution with each age class increasing at the same rate determined by this eigenvalue. This is true if the matrix is aperiodic, but not if it is periodic. I shall illustrate what this means by a simple example; a more complete treatment is given in the appendix to this chapter.

Bernardelli (1941) proposed a model for a hypothetical beetle that reproduces at age 3 and then dies; on average six female offspring in each brood survive to age 1. The annual survival probabilities at ages 1 and 2 are $\frac{1}{2}$ and $\frac{1}{3}$, respectively, leading to the Leslie matrix

$$\begin{bmatrix} 0 & 0 & 6 \\ \frac{1}{2} & 0 & 0 \\ 0 & \frac{1}{3} & 0 \end{bmatrix} \tag{21}$$

The *period* of a Leslie matrix is defined as the greatest common denominator of the ages at which reproduction is possible (the values of x with $f_x > 0$, giving positive entries in the first row of the matrix). A matrix with period 1 is called *aperiodic*; if the period is greater than 1 the matrix is called *periodic*. Most Leslie matrices (for example, the matrix in Equation 6) are aperiodic, which ensures that there is a unique dominant eigenvalue and that the system tends asymptotically to a stable age distributrion. In contrast, the Bernardelli matrix in Equation 21 is periodic with period 3. In consequence, it has three distinct eigenvalues with the same absolute value, one real and the other two complex, and the population does not settle down to a stable age distribution but gives cycles of period 3. The intuitive reason is that the population consists of three reproductively isolated year classes that behave independently of each other.

Most biological examples of periodicity are of this type, with individuals breeding once and then dying after a fixed number of years. A famous case is that of the periodical cicada, that breeds once after a period of development of exactly 17 years in northern states of the United States (13 years in southern states). A striking feature is that breeding is synchronized, only one of the 17 possible year classes existing in any area, though different year classes exist in different areas. I have suggested that this is caused by competitive exclusion between year classes due to predator satiation (Bulmer, 1977). The Leslie matrix model, with no explicit interactions between individuals of the same or different ages, is inadequate as an ecological model of this type of situation.

Continuous-time models

These ideas can be extended to organisms reproducing continuously by making x and t continuous variables. Write l_x for the probability that a female survives from birth to age x as before and let m_x be the rate of production of newborn females by a female of age x. By analogy with the discrete-time model we may anticipate a solution in which the population, after a long time, is increasing exponentially at rate r in all age groups, where r satisfies

$$\int_0^\infty e^{-rx} l_x m_x \, \mathrm{d}x = 1 \tag{22}$$

the analogue of Equation 17. By analogy with the discrete-time model, the unique real value of r satisfying Equation 22, which can be determined by solving this equation numerically, given the survival and fecundity functions, will describe the behavior of the population when it has settled down to a stable age distribution given by

$$c_x \propto e^{-rx} l_x \tag{23}$$

the analogue of Equation 20.

Lifetime reproductive success and generation time

It may be noted that the average number of female offspring that a newborn female will have during her life (her average lifetime reproductive success) is

$$R = \sum_{x=1}^{\omega} l_x m_x \tag{24a}$$

or

$$R = \int_0^\infty l_x m_x \, \mathrm{d}x \tag{24b}$$

in the discrete-time and continuous-time models, respectively. (Lifetime reproductive success in this sense is often called the net reproductive rate; the two terms are treated here as synonymous.) When $R = 1$, so that there is exact replacement, Equation 17 is satisfied with $\lambda = 1$ in the discrete-time model or Equation 22 with $r = 0$ in the continuous-time model, so that at equilibrium the population size is stationary. It is to be expected in a natural population that density-dependent factors operating on l_x and m_x will ensure that this is approximately the case. Lifetime reproductive success is used by some authors as a measure of fitness instead of the population growth rate λ or r in life-history studies since it is easier to measure; we shall discuss in the next chapter under what circumstances this is justified.

The mean age of mothers (the mean age at reproduction) is a natural measure of generation time. If we take a newborn offspring at random, the probability that its mother will be of age x is proportional to $c_x m_x$, where c_x is the proportion of females of age x and m_x is their fecundity. In a stationary population $c_x m_x$ is proportional to $l_x m_x$. Thus the mean age of mothers is

$$T = \sum_x x l_x m_x \Big/ \sum_x l_x m_x = \sum_x x l_x m_x \tag{25}$$

In a nonstationary population it is

$$T = \sum_x x \lambda^{-x} l_x m_x \tag{26}$$

by a similar argument.

Sensitivity analysis

An important use of the Leslie matrix is to calculate the population growth rate λ, and it is useful to know how sensitive λ is to changes in the parameters of the model. The change in λ due to unit change in one of these parameters,

$$s_p = \partial\lambda/\partial p \tag{27}$$

is called the *sensitivity* of λ to a change in p. Calculation of these sensitivities is useful in several contexts: (1) It increases understanding of the natural history of an organism, and can be used to evaluate management strategies for endangered species. (2) Estimated vital rates are subject to sampling errors leading to sampling error in the estimated value of λ, whose evaluation requires knowledge of the sensitivities. (3) The sensitivities measure the strength of selection on the vital parameters and are thus important in life-history theory, as will be seen in Chapter 5.

The simplest way to calculate a sensitivity is to change the parameter by, say, 10% and to find the change in λ. Sensitivity can be calculated analytically by implicit differentiation of Equation 17. Write Equation 17 as $g(\lambda) = 1$; the function g also depends of course on the life-history parameters. If we change one of these parameters from p to $p + dp$, then λ will change to $\lambda + d\lambda$ in such a way that $g(\lambda)$ remains constant at unity. Thus,

$$dg = \frac{\partial g}{\partial p} dp + \frac{\partial g}{\partial \lambda} d\lambda = 0 \tag{28}$$

so that

$$\frac{d\lambda}{dp} = -\frac{\partial g / \partial p}{\partial g / \partial \lambda} \tag{29}$$

Consider for example the effect on the geometric growth rate of the Northern Spotted Owl of a change in the age at first breeding from 3 to 4 years. The value of λ with first breeding at 3 years is 0.9608 from Equation 19. The comparable equation if owls delay breeding to age 4, assuming that the survival rate from age 3 to age 4 is 0.942, is

$$\frac{0.01733 \times 0.942\lambda^{-4}}{1 - 0.942\lambda^{-1}} = 1 \tag{30}$$

whose solution is $\lambda = 0.9604$. Thus age at first breeding makes a negligible difference to the population growth rate. (This is a fortuitous consequence of the demographic parameters of this example and is not true in general.) The other parameters in the model are continuous and their sensitivities can be found from Equation 29 using the function

$$g(\lambda) = \frac{\lambda^{-3} l_3 m}{1 - P/\lambda} \tag{31}$$

Table 4.3 shows the sensitivity of λ to changes in brood size (m), survival to age 3 (l_3), and adult survival (P) calculated in this way. The population growth rate is most sensitive to a change in adult survival, but this is already quite high and may have little possibility of further improvement.

Table 4.3. Sensitivity of λ to changes in life-history parameters in the Northern Spotted Owl.

Parameter	Sensitivity of λ
Brood size (m)	0.075
Survival to age 3 (l_3)	0.251
Adult survival (P)	0.962

FURTHER READING

Caswell (1989) is a thorough treatment of matrix population models. Charlesworth (1994) is an authoritative text on evolution in age-structured populations.

EXERCISES

1. (a) Use the recurrence relationship $\mathbf{n}(t+1) = \mathbf{L}\mathbf{n}(t)$ to find $\mathbf{n}(t)$ directly for 50 generations for the Leslie matrix in Equation 6, starting from an arbitrary value for $\mathbf{n}_0$. Find the growth rate in total population size and the age distribution in generations 1–50, and hence estimate the asymptotic growth rate and the stable age distribution.
(b) Find the eigenvalues of the Leslie matrix and the dominant eigenvector by using a computer program for finding eigenvalues and vectors. Compare the results for the dominant eigenvalue and eigenvector with your results from the previous part, and with the values given in the text.
(c) Find the dominant left eigenvector and compare it with the result given in the text.

2. (a) Find the dominant eigenvalue for the above example by solving Equation 17. (Hint: Plot

$$\sum_{x=1}^{\omega} \lambda^{-x} l_x m_x$$

against λ to identify the root approximately and then improve this estimate by trial and error; or use a program for finding roots of non-linear equations.)
(b) Find the stable age distribution from Equation 20 and compare the answer with that found in Exercise 1b.

3. Consider the age distribution of female Great Tits in Table 4.2. To test the hypothesis that the survival rate is constant, fit a geometric distribution, $P(x) = qp^{x-1}$, to the data by estimating $q = 1/\bar{x}$. Test the goodness of fit by a chi-square test, combining ages 6+, so that there are four degrees of freedom.
[Note: $P(x)$ is the probability of a female of age x, q is annual mortality, $p = 1 - q$, and $\bar{x}$ = mean age; $1/\bar{x}$ is an appropriate estimate of q since the mean of the theoretical geometric distribution is $1/q$.]

4. (a) Use the recurrence relationship $\mathbf{n}(t+1) = \mathbf{Ln}(t)$ to find $\mathbf{n}(t)$ directly for 50 generations for the matrix for Bernardelli's beetles in Equation 21, starting from an arbitrary value for $\mathbf{n}_0$.
(b) Find the eigenvalues of this matrix using a computer program for finding eigenvalues.

5. Estimate the generation time for the squirrel data in Table 4.1.

6. (a) Verify that the solutions of Equations 19 and 30 are $\lambda = 0.9608$ and 0.9604, respectively.
(b) Verify the results in Table 4.3.

APPENDIX 4.1. EIGENVALUES OF THE LESLIE MATRIX

I first show that the eigenvalues of the Leslie matrix are the roots of the equation

$$\sum_{x=1}^{\omega} \lambda^{-x} l_x m_x = 1 \tag{32}$$

Define

$$\mathbf{A} = \mathbf{L} - \lambda\mathbf{I} = \begin{bmatrix} f_1-\lambda & f_2 & f_3 & \cdots & f_{\omega-1} & f_\omega \\ P_1 & -\lambda & 0 & \cdots & 0 & 0 \\ 0 & P_2 & -\lambda & \cdots & 0 & 0 \\ 0 & 0 & P_3 & \cdots & 0 & 0 \\ \vdots & \vdots & \vdots & \cdots & \vdots & \vdots \\ 0 & 0 & 0 & \cdots & P_{\omega-1} & -\lambda \end{bmatrix} \tag{33}$$

The determinant of **A** can be expanded in terms of the elements of its first row as

$$|\mathbf{A}| = a_{11}|\mathbf{A}_{11}| - a_{12}|\mathbf{A}_{12}| + \cdots \pm a_{1\omega}|\mathbf{A}_{1\omega}| \tag{34}$$

where a_{ij} is the ijth element of **A** and $\mathbf{A}_{1j}$ is the matrix **A** with its first row and jth column deleted. This is a general formula for the expansion of the determinant of any matrix; the submatrices $\mathbf{A}_{1j}$ are called minors. A little experimentation will show that this expansion contains all the terms involved in the definition of the determinant of **A**, and with the correct signs. Now consider a typical minor, say $\mathbf{A}_{13}$:

$$\mathbf{A}_{13} = \begin{bmatrix} P_1 & -\lambda & 0 & 0 & \cdots & 0 & 0 \\ 0 & P_2 & 0 & 0 & \cdots & 0 & 0 \\ 0 & 0 & -\lambda & 0 & \cdots & 0 & 0 \\ 0 & 0 & P_4 & -\lambda & \cdots & 0 & 0 \\ \vdots & \vdots & \vdots & \vdots & \cdots & \vdots & \vdots \\ 0 & 0 & 0 & 0 & \cdots & P_{\omega-1} & \lambda \end{bmatrix} \tag{35}$$

There are nonzero terms immediately above the diagonal in the first row and immediately below the diagonal in rows 4 on, all other nondiagonal terms being 0. The matrix can be reduced to a diagonal matrix by elementary row operations without changing the diagonal elements. (Add λ/P_2 times row 2 to row 1, then P_4/λ times row 3 to row 4, then P_5/λ times row 4 to row 5, and so on.) These operations do not change the value of the determinant, which is therefore the product of the terms on the diagonal. Hence,

$$\begin{aligned} |\mathbf{A}| &= (f_1 - \lambda)(-\lambda)^{\omega-1} - P_1 f_2 (-\lambda)^{\omega-2} + \cdots \pm P_1 \cdots P_{\omega-1} f_\omega \\ &= (-1)^{\omega}(\lambda^{\omega} - l_1 m_1 \lambda^{\omega-1} - l_2 m_2 \lambda^{\omega-2} - \cdots - l_\omega m_\omega) \end{aligned} \tag{36}$$

Equating this expression to zero and division by λ^{ω} leads to Equation 32.

I now consider the roots of Equation 32 by an argument borrowed from Charlesworth (1994). First consider positive real roots. Write

$$g(\lambda) = \sum_{x=1}^{\omega} \lambda^{-x} l_x m_x \tag{37}$$

For $\lambda > 0$, $g(\lambda)$ is a decreasing function of λ, falling from very large values when λ is small to nearly zero when λ is large. Hence Equation 32 has a unique positive real root, where this function intersects the horizontal line $\lambda = 1$. Call this eigenvalue λ_1. All the other eigenvalues are either negative or complex.

Define the period of a Leslie matrix as the greatest common denominator of the ages at which reproduction is possible (the values of x with $f_x > 0$, giving positive entries in the first row of the matrix). A matrix with period 1 is called *aperiodic*; if the period is greater than 1 the matrix is called *periodic*.

I shall now prove the following theorem. If the matrix is aperiodic (the usual case), then all the other eigenvalues are less in absolute value than

λ_1, so that this is the unique dominant eigenvalue; this guarantees that the system asymptotically tends to a stable age distribution determined by the corresponding eigenvalue. However, if the matrix is periodic with period $b > 1$, there will be $b - 1$ additional (negative or complex) eigenvalues with absolute value λ_1 that ensure that the system asymptotically oscillates. To prove this theorem I shall consider negative and complex roots in turn.

Negative real roots will satisfy

$$g(-\lambda) = \sum_{x \text{ even}} \lambda^{-x} l_x m_x - \sum_{x \text{ odd}} \lambda^{-x} l_x m_x = 1 \tag{38}$$

with $\lambda > 0$. If $l_x m_x$ is zero for all odd values of x (that happens if the matrix is periodic with an even period), then $g(-\lambda) = g(\lambda)$, so that $-\lambda_1$ is an eigenvalue. Otherwise $g(-\lambda) < g(\lambda)$ for $\lambda > 0$, so that $g(-\lambda) < 1$ for $\lambda > \lambda_1$; in this case any negative root must be strictly less than λ_1 in absolute value.

Finally consider complex roots with absolute value r and argument θ. Writing $\lambda = r \exp(i\theta)$ we find that

$$g(\lambda) = \sum_{x=1}^{\omega} r^{-x} (\cos \theta x - i \sin \theta x) l_x m_x \tag{39}$$

Equating this to unity requires that

$$\sum_{x=1}^{\omega} r^{-x} \cos(\theta x) l_x m_x = 1 \tag{40a}$$

$$\sum_{x=1}^{\omega} r^{-x} \sin(\theta x) l_x m_x = 0 \tag{40b}$$

If the matrix is periodic with period b, these equations are satisfied with $r = \lambda_1$ and $\theta = 2\pi/b$, since $\cos(2\pi) = 1$, $\sin(2\pi) = 0$, so that there is a complex eigenvalue with absolute value λ_1. (In fact there are b roots with $r = \lambda_1$, with $\theta = 2\pi j/b$ [$j = 1, \ldots, b$]; the last is the positive real root λ_1, and if b is even, another is $-\lambda_1$.) If the matrix is aperiodic then

$$\sum_{x=1}^{\omega} r^{-x} \cos(\theta x) l_x m_x < \sum_{x=1}^{\omega} r^{-x} l_x m_x < 1 \qquad \text{if } r > \lambda_1 \tag{41}$$

since $\cos(\theta x) < 1$ unless θx is a multiple of 2π. In this case a complex root must have absolute value less than λ_1.

CHAPTER 5

Life-History Evolution

In an ideal world an organism would live forever ($l_x = 1$ for all x) and would produce an infinite number of offspring each generation (m_x = infinity for all x). In the real world this is not possible, and an organism will evolve to choose the best compromise between different longevity and fertility patterns open to it given the trade-offs imposed by external constraints. The study of life-history strategies concerns an *empirical* question, "What are the trade-offs and the constraints under which evolution operates?" and a *theoretical* question, "Given these trade-offs and constraints, what is the optimal strategy to adopt?" Finally, we want to understand whether the solutions adopted by plants and animals are optimal under what is known about the trade-offs and constraints involved.

In this chapter I begin by considering in detail the evolution of semelparity versus iteroparity (breeding once only versus breeding several times with fewer offspring per brood); the trade-off in this situation is sufficiently simple that it forms a useful paradigm for the development of life-history theory. I then show that life-history evolution in a constant environment under density-dependent regulation is based on the principle of maximizing lifetime reproductive success, and I apply this principle to the evolution of the age of reproductive maturity and the related problem of the allocation of resources between reproduction and growth, and to the evolution of senescence. Finally, I consider how the conclusions from these models are affected by environmental variability that entails that viability and fecundity vary unpredictably from one generation to the next.

THE EVOLUTION OF SEMELPARITY VERSUS ITEROPARITY

Some organisms breed only once in their life and then die, for example, annual plants and many insects; they are called *semelparous*. Others breed many times in successive breeding seasons, for example perennial plants and most birds and mammals, and are called *iteroparous*. Semelparous species can achieve a high reproductive output by allocating all their resources to reproduction; they sacrifice longevity for fecundity. Iteroparous species have lower reproductive output per year because they must allocate some of their resources to survival through the coming year.

Cole (1954) introduced the idea of studying the effects of selection on life-history phenomena by studying the effect of changes in the l_x and m_x functions on λ, the asymptotic geometric rate of increase defined by

$$\sum_{x=1}^{\omega} \lambda^{-x} l_x m_x = 1 \tag{1}$$

and assuming that selection acts to maximize λ. He obtained the following striking result in considering the advantages of semelparity versus iteroparity. Contrast a semelparous annual plant that produces b offspring with an immortal perennial plant that produces b^* offspring each year. The rate of increase of the annual plant is clearly b. For the perennial plant, the equation

$$\sum_{x=1}^{\infty} \lambda^{-x} b^* = \frac{\lambda^{-1} b^*}{1-\lambda^{-1}} = \frac{b^*}{\lambda - 1} = 1 \tag{2}$$

leads to the result that $\lambda = b^* + 1$. Thus the perennial is only at an advantage if

$$b^* + 1 > b \tag{3}$$

This leads to Cole's paradox: "For an annual species, the absolute gain in intrinsic population growth that could be achieved by changing to the perennial reproductive habit would be exactly equivalent to adding one individual to the average litter size." Thus, for our two plant species, producing 101 seeds at age 1 and then dying would be equivalent to producing 100 seeds each year forever. This raises the problem of why there should be any perennial plants. The same argument holds for animals with b interpreted as the number of females in a litter.

Gadgil and Bossert (1970) suggested that the paradox was due to an unrealistic assumption in Cole's treatment, that there is no mortality during the first year of life ($P_0 = l_1 = 1$). Allowing P_0 to be less than 1 but assuming $P_x = 1$ for $x > 0$ for the perennial, it follows from a similar argument that $\lambda = P_0 b$ for the annual and $P_0 b^* + 1$ for the perennial. Thus the perennial will be at an advantage if

$$b^* > b - 1/P_0 \tag{4}$$

This makes it more plausible for iteroparity to evolve. However, Gadgil and Bossert also remark that, in an annual species at constant population size under density-dependent regulation, $P_0 b = 1$. In this case, perennials will be at an advantage if $P_0 b^* > 0$, that is to say, if they produce any offspring. This raises the question why annuals should exist, though Gadgil and Bossert did not make this point.

This paradox arises from the unrealistic assumption that there is no adult mortality in perennials. Charnov and Schaffer (1973) introduced the first biologically plausible model allowing for both juvenile and adult mortality. As before P_0 is the proportion of juveniles surviving the first year; P_1 is the annual adult survival rate in perennials, assumed to be independent of age. As before $\lambda = P_0 b$ for the annual. For perennials

$$\sum_{x=1}^{\infty} \lambda^{-x} l_x m_x = P_0 b^* \sum_{x=1}^{\infty} \lambda^{-x} P_1^{x-1} = P_0 P_1^{-1} b^* / (\lambda P_1^{-1} - 1) = 1 \tag{5}$$

giving $\lambda = P_0 b^* + P_1$. This leads to the conclusion that the annual will be at a selective advantage over the perennial form when

$$(b - b^*) > P_1/P_0 \tag{6}$$

Since adult survival is likely to be much higher than juvenile survival, this requires a substantial excess of b over b^*.

The above result can be obtained in a simpler way by writing down the difference equations for the numbers of plants. Write $n_1(t)$ and $n_2(t)$ for the numbers of annual and perennial plants in year t. Then it is clear that

$$\begin{aligned} n_1(t+1) &= P_0 b n_1(t) \\ n_2(t+1) &= (P_0 b^* + P_1) n_2(t) \end{aligned} \tag{7}$$

so that their geometric rates of increase are $P_0 b$ and $P_0 b^* + P_1$, respectively. However, this alternative type of derivation, though completely equivalent, will not usually work out so simply for other life-history problems.

Density-dependent selection

So far we have assumed that the fecundities and survival rates are constant, but it would be more realistic to suppose that the population is regulated by density-dependent factors acting on at least some of these parameters. To be specific, suppose that density dependence acts on juvenile survival so that it declines exponentially with the number of juveniles

$$P_0 = p_0 \exp\{-\alpha[bn_1(t) + b^* n_2(t)]\} \tag{8}$$

In this formulation p_0 is the maximum juvenile survival in the absence of competition from other juveniles, and it is assumed that P_1, b, and b^* are constant.

To see what will happen, we could substitute this expression into Equation 7 and then simulate it by iteration for particular parameter values. However, we can obtain more insight from an analytical result obtained by *invasibility analysis*. We ask in turn: Can perennials invade a population dominated by annuals? Can annuals invade a population dominated by perennials?

A population of annuals will obey the difference equation

$$n_1(\mathrm{t} + 1) = p_0 \exp[-\alpha b n_1(t)] b n_1(t) \tag{9}$$

with an equilibrium value satisfying

$$p_0 \exp\{-\alpha b \hat{n}_1) b = 1 \tag{10}$$

at which point the population is stationary. (Note that Equation 9 has the form of the discrete-time logistic model in Equation 22 in Chapter 2 with $r = \log p_0 b$; I assume that this quantity is less than 2 so that the equilibrium is stable.) Consider the fate of a very small number of perennials introduced into this equilibrium population of annuals. The juvenile survival rate among both annuals and perennials is

$$\begin{aligned} P_0 &= p_0 \exp\{-\alpha[bn_1(t) + b^* n_2(t)]\} \\ &\cong p_0 \exp\{-\alpha b \hat{n}_1) = 1/b \end{aligned} \tag{11}$$

if $n_1(t) \cong \hat{n}_1$ and $n_2(t) \cong 0$. Hence

$$n_2(t+1) \cong (b^*/b + P_1) n_2(t) \tag{12}$$

The perennials will decrease in frequency (that is to say, when they are rare they will be unable to invade a population of annuals) when

$$b^*/b + P_1 < 1 \tag{13}$$

or equivalently when

$$b > \frac{b^*}{1 - P_1} \tag{14}$$

If the opposite inequality holds, perennials when rare will be able to invade a population of annuals.

Consider now the conditions under which annuals can invade a population dominated by perennials. A population of perennials will obey the difference equation

$$n_2(t+1) = \{p_0 \exp[-\alpha b^* n_2(t)]\, b^* + P_1\} n_2(t) \tag{15}$$

with an equilibrium satisfying

$$p_0 \exp\{-\alpha b^* \hat{n}_2) b^* + P_1 = 1 \tag{16}$$

at which point the population is stationary. Consider the fate of a very small number of perennials introduced into this equilibrium population of annuals. The juvenile survival rate among both annuals and perennials is

$$\begin{aligned} P_0 &= p_0 \exp\{-\alpha[bn_1(t) + b^* n_2(t)]\} \\ &\cong p_0 \exp\{-\alpha b^* \hat{n}_2) = \frac{1-P_1}{b^*} \end{aligned} \tag{17}$$

if $n_1(t) \cong 0$ and $n_2(t) \cong \hat{n}_2$. Hence,

$$n_1(t+1) \cong \frac{(1-P_1)b}{b^*} n_1(t) \tag{18}$$

Thus the annuals will increase in frequency when rare if and only if

$$\frac{(1-P_1)b}{b^*} > 1 \tag{19}$$

which leads to the same inequality as before:

$$b > \frac{b^*}{1-P_1} \tag{20}$$

We can conclude that, when the inequality in Equations 14 and 20 is satisfied, perennials cannot invade a population consisting predominantly of annuals whereas a few annuals can invade a population of perennials. Thus the annual form is at an unconditional advantage, and we should expect to find only annual plants under these circumstances. When the opposite inequality holds, the reverse is true and we should expect to find only perennials.

This conclusion does not depend on the form of the density-dependent function acting on juvenile mortality as long as it leads to a stable equilibrium. The result in Equation 14 follows directly from substituting the equilibrium condition that $P_0 b = 1$ in a population of annuals into Equation 6; similarly the result in Equation 20 follows from substituting the equilibrium condition that $P_0 b^* + P_1 = 1$ at equilibrium in a population of perennials into Equation 6.

The same argument leading to Equations 14 and 20 goes through if density dependence acts on adult survival or fecundity, but there is a complication. When density dependence acts only on juvenile mortality, the quantities b, b^*, and P_1 in Equations 14 and 20 are constants having the same values in both equations. Suppose, however, that density dependence acts on adult mortality so that it declines exponentially with the number of adults, as well acting on juvenile mortality in the same way as before. Then

$$\begin{aligned} P_0 &= p_0 \exp\{-\alpha[bn_1(t) + b^* n_2(t)]\} \\ P_1 &= p_1 \exp\{-\beta[n_1(t) + n_2(t)]\} \end{aligned} \tag{21}$$

The equilibrium population size $\hat{n}_1$ in a population of annuals differs from the equilibrium population size $\hat{n}_2$ in a population of perennials, so that P_1 in Equation 14 differs slightly from P_1 in Equation 20. Suppose for example that $p_0 = 0.02$, $p_1 = 0.5$, $b = 100$, $\alpha = 0.01$, and $\beta = 1$. It turns out that perennials when rare can invade a population of annuals when $b^* > 75$ and that when common they can resist invasion by annuals when $b^* > 79$. Thus one expects to see only annuals when $b^* < 75$, and only perennials when $b^* > 79$, but when b^* is between these limits one would expect to find a mixed population of both forms. For example, if $b^* = 77$, perennials when rare can increase in a population of annuals, but they cannot go all the way to fixation; there is in consequence a stable-equilibrium intermediate frequency (that is in fact 53% annuals and 47% perennials) at which the two forms are equally fit so that they remain there indefinitely. I shall discuss this idea of the equilibration of fitness under frequency-dependent selection in Chapter 7. The situation is analogous to competition between two species with slightly different resource usage. However, in life-history evolution the component of frequency-dependent selection is likely to be small, so that one usually expects to find a unique optimal strategy that is superior to all its alternatives. In the evolution of semelparity versus iteroparity density-dependent selection acting on juveniles seems the most likely form of population regulation; it leads to a unique optimal strategy.

Giant lobelias

Young (1990) has estimated the demographic parameters of two species of giant lobelia growing on Mt. Kenya. Both species take 40–60 years to grow to reproductive maturity. *Lobelia telekii* flowers once only and then dies, whereas *L. keniensis* is iteroparous, flowering about every ten years after reaching maturity. The two species occupy different habitats, the semelparous species *L. telekii* living on less favorable dry rocky slopes, the iteroparous species *L. keniensis* in more favorable moist valley bottoms.

The reproductive output (inflorescence dry weight) near the species boundary is 4–5 times higher in *Lobelia telekii* than in *L. keniensis*. Data on three *L. keniensis* populations are shown in Table 5.1; site A is a wet site well away from the *L. telekii* boundary; site B is a drier site near the boundary; and site C is an outlying site within the *L. telekii* range. The model developed above can be applied to this situation if we replace P_1, the adult survival rate in the iteroparous species, by the probability of survival from one flowering episode to the next, P^T, where P is the annual adult survivorship and T is the number of years between flowering episodes. Thus the annual species will be at an advantage if

$$\frac{b}{b^*} > \frac{1}{1-P^T} \tag{22}$$

The ratio b/b^* is between 4 and 5 in all three sites (though it probably decreases a little from A to B to C), so that the perennial is favored in site A, the annual is favored in site C, and the two are joint favorites in site B on the species boundary; these predictions agree with the observed geographical distribution. Young (1990) concludes that the annual species is favored in the harsher habitat because of the greatly increased mortality of the perennial species between flowering episodes, due both to the increased annual mortality and to the longer interval between flowering.

MAXIMIZATION OF LIFETIME REPRODUCTIVE SUCCESS

Lifetime reproductive success is defined as

$$R = \sum_{x=1}^{\omega} l_x m_x \tag{23}$$

Suppose for the moment that we are dealing with a clonal or monoecious

Table 5.1. Life-history parameters of three populations of *Lobelia keniensis*.

	Site		
Parameter	A	B	C
P = yearly adult survivorship	0.988	0.984	0.972
T = years between flowering	8	14	16
$1/(1 - P^T)$	10.9	4.9	2.7

Source: Young, 1990

organism, so that complications due to the existence of two sexes do not arise; l_x is the probability that a newborn will survive to age x, and m_x is the average number of offspring produced by an individual aged x. R is the average number of offspring that a newborn individual is expected to produce during its lifetime.

In the model of the last section, the lifetime reproductive success of an annual, as a seed, is P_0b while the lifetime reproductive success of a perennial is

$$P_0b^*\sum_{x=0}^{\infty} P_1^x = \frac{P_0b^*}{1-P_1} \tag{24}$$

Thus Equations 14 and 20 can be interpreted as saying that in a stationary population there is a selective advantage to the form with the higher lifetime reproductive success. This is not an accident. We shall now prove in general that, in a stationary population, maximizing R is equivalent to maximizing λ.

Since $\Sigma\lambda^{-x}l_xm_x$ is a decreasing function of λ, $\lambda < 1$ when $R < 1$ (and vice versa) and $\lambda > 1$ when $R > 1$ (and vice versa). Suppose that the population is fixed for a genotype G for a particular life history, and that density-dependent factors operate on the survival rates and fecundities until the population size is stationary, so that $\lambda = R = 1$. Now consider a mutant genotype G' that in these circumstances has geometric rate of increase λ' and lifetime reproductive success R'. The mutant genotype will invade the population if $\lambda' > 1$, that is equivalent to $R' > 1$. Thus maximizing R is equivalent to maximizing λ when the population size is stationary, but this is not true otherwise. In a rapidly increasing population there is greater selective emphasis on changes in life-history parameters in early life than in a stationary population (young individuals have relatively higher reproductive values), while the contrary is true in a decreasing population. These differences are not reflected in lifetime reproductive success R, which is therefore an inadequate measure of fitness in a nonstationary population. But if it is assumed that population size is stationary under the operation of density-dependent factors, then λ and R are equivalent measures of fitness that can be used interchangeably. Many authors prefer to develop life-history theory in terms of R rather than λ since it is often easier to deal with in theory and easier to measure in natural populations.

We saw in the last chapter how to calculate the sensitivity of the asymptotic growth rate λ to a small change in one of the life-history parameters p; this is a measure of the intensity of selection on this parameter. It is therefore of interest that there is a simple relationship in a stationary population between this quantity and the corresponding rate of change in the lifetime reproductive success. It follows simply from Equation 29 in Chapter 4 that, in a stationary population with $\lambda = R = 1$,

$$\frac{d\lambda}{dp} = \frac{dR/dp}{T} \tag{25}$$

where T is the generation time defined in Chapter 4's Equation 25. This relationship is intuitively sensible, since R is measured in generations (it is *lifetime* reproductive success), while λ is measured in years.

It has been assumed in this analysis that the appropriate definition of fitness in an age-structured population is given by the asymptotic geometric growth rate λ satisfying Equation 1. This is clearly the appropriate definition of fitness when comparing clones or parthenogenetic strains with different life-history parameters since the clone with the higher growth rate will eventually dominate the population. In a sexual species the situation is more complex since genes are broken up each generation by segregation and assortment, but population genetic analysis shows that λ remains the appropriate definition of fitness in the sense that one expects it to be maximized by natural selection (Lande, 1982; Charlesworth, 1990a, 1994); provided that adequate genetic variability is available, its detailed mechanism is likely to be relatively unimportant. The detailed analysis of this question is complex; I shall here give an account of the underlying principle in a simple situation.

Imagine a diploid, outcrossing, monoecious (hermaphrodite) plant species. Consider a character determined by a single locus with two alleles, with a common genotype, *aa*, and a rare genotype, *Aa*; the homozygote *AA* need not be considered under random mating since it will be extremely rare. Write $N_x(t)$ and $n_x(t)$, respectively, for the numbers of *aa* and *Aa* individuals of age x at time t. Let a population of *aa* individuals settle down under density-dependent selection until it is stationary with age-specific survival probabilities and fecundities P_x and m_x such that $\Sigma l_x m_x = 1$; the numbers of individuals of age x will be

$$N_x = l_x N / \Sigma l_x \tag{26}$$

where N is the total population size. Now introduce a small number of *Aa* individuals into the population and suppose that under these circumstances their survival probabilities and fecundities are P'_x and m'_x. For $x > 1$,

$$n_x(t+1) = P'_{x-1} n_{x-1}(t) \tag{27}$$

the analogue of Equation 2 in Chapter 4. The number of *Aa* individuals of age 1 in year $t + 1$ that arose from seedlings produced by *Aa* individuals in year t is

$$\frac{1}{2} P_0' \sum_{x=1}^{\omega} m_x' n_x(t) \tag{28}$$

(Since nearly all *Aa* plants will have been fertilized by pollen from *aa* plants, half their offspring will be *Aa*.) In addition we must take into account fertilization of *aa* plants by pollen from *Aa* plants. Assume that pollen production is proportional to fecundity. The fraction of pollen in year t coming from *Aa* plants is

$$\frac{\sum m_x' n_x(t)}{\sum m_x N_x + \sum m_x' n_x(t)} \cong \frac{\sum m_x' n_x(t)}{\sum m_x N_x} \tag{29}$$

The number of seedlings produced by *aa* plants each year is

$$\sum m_x N_x \tag{30}$$

The number of seedlings from *aa* plants fertilized by pollen from *Aa* plants in year t is the product of Equations 29 and 30, so that the number of *Aa* individuals of age 1 in year $t + 1$ that arose from seedlings produced by *aa* individuals in year t is

$$\frac{1}{2} P_0' \sum_{x=1}^{\omega} m_x' n_x(t) \tag{31}$$

Summing the contributions from Equations 28 and 31, we find that

$$n_1(t+1) = P_0' \sum_{x=1}^{\omega} m_x' n_x(t) \tag{32}$$

the analogue of Equation 3 in Chapter 4. From Equations 27 and 32 we see that the behavior of the *Aa* population can be described by a Leslie matrix with elements determined by P_x' and m_x', so that it will be able to invade an *aa* population only if its dominant eigenvalue exceeds unity. (I assume that the *Aa* population is initially sufficiently small that it achieves a stable age distribution before its numbers have grown large enough to affect the parameters affected by density-dependent regulation.)

We conclude that life-history evolution in a stationary population, whether sexual or asexual, can be based on the principle of maximizing lifetime reproductive success.

AGE AT MATURITY

The age at which individuals become reproductively mature is an important component of their life history, that one would expect to evolve to an optimum value. Assume that population size is stationary under the influence of density-dependent factors, so that the appropriate definition of fitness is lifetime reproductive success. How will age at maturity evolve under selection to maximize lifetime reproductive success?

Consider first a semelparous organism that breeds once at age t. If it has probability l_t of surviving to this age and has m_t offspring, its lifetime reproductive success is

$$R = l_t m_t \tag{33}$$

If m_t does not depend on t, R is maximized (with respect to t) by making t as small as possible, since l_t is a decreasing function of t: the earlier you reproduce, the greater the chance you live to do so. In many organisms, however, the number of eggs that a female can produce depends on her size and hence on her age, so that m_t is an increasing function of t. If m_t increases with age faster than l_t decreases, it will be optimal to postpone the age of reproduction beyond its minimum possible value.

McLaren (1966), quoted by Roff (1992), considered the adaptive significance of long life in the marine arrowworm *Sagitta elegans* in the Arctic. This organism is semelparous, reproducing parthenogenetically and then dying. In warmer waters it has several generations a year, but its food is only available to it in the Arctic for a very restricted season so that the life cycle there must be annual or some multiple thereof. In fact it is biennial, with individuals breeding in their second year, and the question is why it should not be annual. McLaren first rejected the suggestion that slow growth in the Arctic might reflect "natural selection at the level of the ecosystem," whereby lower rates of increase are selected to prevent overutilization of resources, since no explanation had been offered for how this group-selection process could overcome ordinary individual selection pressures. He then showed that the biennial life cycle is optimal in the Arctic by the following calculations. The average egg production at age 2 is $m_2 = 543$. From the observed relationship between egg production and size, together with the size distribution of 1-year-olds, he calculated that the average 1-year-old if it matured at this age could only on average produce 138 eggs (m_1). The numbers of 1- and 2-year-olds in a large sample were 3170 and 1216, respectively, from which one can calculate (assuming stationary population size) that

$$P_1 = \frac{l_2}{l_1} = \frac{1216}{3170} \tag{34}$$

Hence

$$\frac{l_2 m_2}{l_1 m_1} = \frac{1216}{3170} \times \frac{543}{138} = 1.51 \tag{35}$$

Thus individuals breeding at 2 years old have a substantially higher lifetime reproductive success than they would have if they reproduced in their first year and then died. By a similar though less direct argument McLaren showed that postponing reproduction to age 3 would give a slightly lower lifetime reproductive success than breeding at age 2. Thus the observed biennial cycle in the Arctic seems to be the optimal strategy.

Consider now an iteroparous organism that breeds for the first time at age t and then breeds each year while it is alive. Lifetime reproductive success (Equation 23) is

$$R = \sum_{x=t}^{\infty} l_x m_x$$

If l_x and m_x are independent of t, the age of maturity, R is maximized by maturing as early as possible (i.e., by equating t to the first age at which $m_x > 0$). However, it may be advantageous to postpone maturity if there is a cost of reproduction that causes a reduction in either survivorship or fecundity in subsequent years.

Roff (1984, 1992) has constructed the following model to apply to fish. (i) Fish grow according to the von Bertalanffy equation until maturity and then cease growth: that is to say,

$$\begin{aligned} L(x) &= L_\infty(1 - e^{-kx}) \qquad x \le t \\ L(x) &= L_\infty(1 - e^{-kt}) \qquad x \ge t \end{aligned} \tag{36}$$

where $L(x)$ is the length of a fish of age x and L_∞ is its maximum length. (ii) Fecundity is proportional to the cube of length:

$$m_x = c_1 L^3(x) \tag{37}$$

(iii) There is a very high mortality during the egg and early larval stage, the proportion surviving this stage being p. This is followed by a constant mortality rate M in the juvenile and adult periods, so that

$$l_x = pe^{-Mx} \tag{38}$$

Substituting these expressions for l_x and m_x into R, we find that

$$R = c\sum_{x=t}^{\infty} e^{-Mx}(1 - e^{-kt})^3 = c(1 - e^{-kt})^3 e^{-Mt} \tag{39}$$

using the formula for the sum of a geometric series, where c is a constant. Equating dR/dt to zero we find that the optimal age at maturity is

$$t = \frac{1}{k}\log\left(\frac{3k}{M}+1\right) \tag{40}$$

Roff (1984) gives estimates of k, M, and t for 30 species of fish. Figure 5.1 shows that there is good agreement between the observed age of maturity and the value predicted from Equation 40. Unfortunately, there are two simplifying assumptions in the theory that are known not to be true. First, it is assumed that the growth of fish ceases when they reproduce for the first time. In fact growth continues after maturity, though more slowly than before. Second, it has been assumed that the survival rate is unaffected by reproduction, whereas in fact survivorship is rather lower in mature fish after maturity than in juvenile fish. Thus the model has exaggerated the effect of reproduction on subsequent fecundity and ignored its effect on subsequent survival. These two deficiencies of the model work in opposite directions, and it may be that they cancel out. The good fit

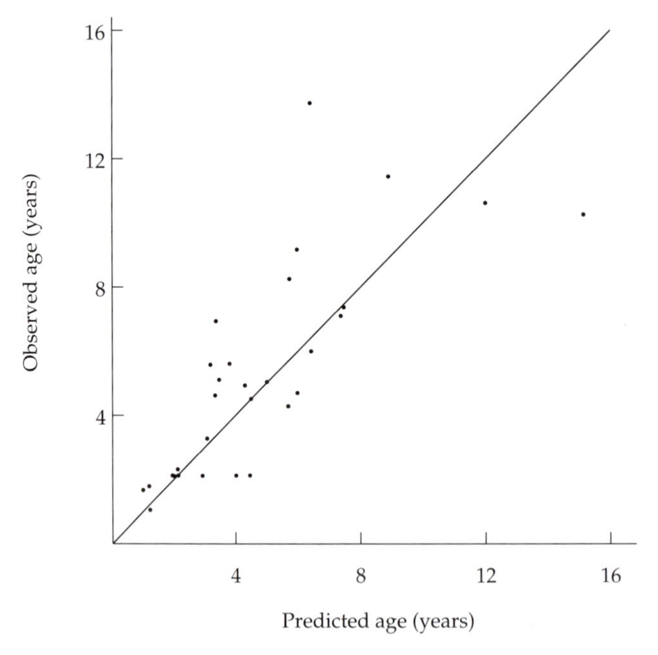

Figure 5.1. Observed and predicted age at maturity in 30 species of fish. After Roff, 1984.

achieved in Figure 5.1 suggests that this may be so. The situation illustrates the difficulty of testing life-history theory empirically.

RESOURCE ALLOCATION TO REPRODUCTION VERSUS GROWTH

A classical trade-off in life-history theory is the allocation of resources between reproduction and growth. The fitness of an annual plant is the number of offspring (fertilized seeds) it has produced by the end of the season. At any time during the season it can choose between allocating resources to somatic growth (that will give it a larger reproductive potential later) or to the development of reproductive structures. What is the optimal strategy of resource allocation through time that will maximize its fitness?

This problem is seen particularly clearly in colonies of social insects (Macevicz and Oster, 1976). Consider an annual eusocial insect colony (for example, a bumblebee or a wasp), founded by a single queen in the spring, increasing in size through the summer through the efforts of sterile workers, and releasing winged reproductives (males and queens) in the fall; fertilized queens overwinter and found colonies the following spring. What is the optimal allocation between producing more workers and producing reproductives through the season that will maximize the number of reproductives released at the end of the season? We shall for the moment ignore the fact that two sexes of reproductives are produced, and we shall refer to them both as queens; the optimal sex ratio among reproductives will be discussed in Chapter 10.

To formalize this problem in the simplest possible model write $W(t)$ and $Q(t)$ for the biomass of workers and queens at time t. At any time, the colony produces new individuals (that may be either workers or queens) at a rate proportional to the numbers of workers at that time; the mortalities of workers and queens are μ and ν respectively. These assumptions are embodied in the differential equations

$$\begin{aligned} dW/dt &= Ru(t)W(t) - \mu W(t) \\ dQ/dt &= R[1-u(t)]W(t) - \nu Q(t) \end{aligned} \tag{41}$$

where $u(t)$ is the proportion of the colony's resources allocated to producing workers rather than queens, $0 \le u(t) \le 1$. The constant R measures the return on unit effort, assumed the same whether allocated to producing workers or queens since they are measured in biomass. Counting the founding queen as a worker since she forages to feed the first brood and will not be one of the reproductives released at the end of the season, the initial conditions at the beginning of the season are $W(0) = 1$, $Q(0) = 0$. The problem is to choose the function $u(t)$ for times $0 \le t \le T$, where T is the time when the colony releases queens at the end of the season, so as to maximize the number of queens released at that time, $Q(T)$. In the terminology of control theory, W and Q are state variables satisfying the equations of motion in Equation 41, $u(t)$ is the control variable, and $Q(T)$ is the objective function. The problem is to choose the control variable so as to maximize the objective function. This problem can be solved by Pontryagin's maximum principle, discussed in Appendix 5.1. The solution is

$$\begin{aligned} u(t) &= 1 \qquad t < t^* \\ u(t) &= 0 \qquad t \ge t^* \\ t^* &= T - \log[R/(R - \mu + \nu)]/(\mu - \nu) \end{aligned} \tag{42}$$

That is to say, make only workers until the critical switching time t^* and thereafter make only queens.

This is known as a bang-bang solution, that Oster and Wilson (1978) characterize in an economic analogy as follows. Given the choice between reinvestment in the means of production (i.e., workers) or in using the means of production to manufacture a terminal product (i.e., queens), the optimal allocation of effort so as to produce the most product in a fixed time is to reinvest until the last possible moment and then switch entirely to product manufacture. No intermediate policy yields higher returns.

If we assume a bang-bang solution with a switching time from worker production to queen production at time $t = t^*$, it is straightforward to find the optimal switching time in Equation 42 by elementary calculus. The number of workers increases exponentially at rate $R - \mu$ to reach its maximum value of $\exp[(R - \mu)t^*]$ at time t^* and then decreases exponentially at rate μ. The differential equation for queen production is therefore

$$\mathrm{d}Q/\mathrm{d}t = Re^{Rt^*}e^{-\mu t} - \nu Q \qquad t \ge t^* \tag{43}$$

with initial condition $Q(t^*) = 0$. The solution is

$$Q(t) = Re^{Rt^*}\left[e^{-\nu(t-t^*)-\mu t^*} - e^{-\mu t}\right] \tag{44}$$

The optimal switching time is found by solving

$$\mathrm{d}Q(T)/\mathrm{d}t^* = 0 \tag{45}$$

as an equation in t^*.

Consider as an example the following data on the hornet *Vespa oriental-*

is quoted by Alexander (1982). Between May 15 and September 15 workers increased roughly exponentially from 10 to 250, so that $R - \mu = 0.027$ per day. Of young marked hornets, half had disappeared (died) after 30 days, so that $\mu = 0.023$ and $R = 0.050$. If queen mortality is ignored, the optimal strategy from Equation 42 is to switch from making workers to making queens 27 days before the end of the season. In fact, reproductives appeared over about 40 days between September 15 and October 30. The discrepancy may be due to uncertainty over the end of the season, that will be earlier in some years than others. A model that incorporates this uncertainty will predict a strategy of producing only workers until some fixed switching time t_1^*, then producing a mixture of workers and queens with a decreasing proportion of workers until a second switching time t_2^*, and finally producing only queens until the season ends. This is an example of bet-hedging, that will be discussed later in this chapter.

The maximum principle has the advantage of giving an explicit analytic solution but is difficult to implement except in the simplest circumstances. The alternative technique of dynamic programming is easier to understand and to implement, and it can provide numerical solutions for complicated models. It uses backwards induction on a discrete-time version of the model. The discrete-time difference equations for hornets, with time measured in weeks, are

$$\begin{aligned} W_t &= 0.35u_{t-1}W_{t-1} + 0.84W_{t-1} \\ Q_t &= 0.35(1 - u_{t-1})W_{t-1} + Q_{t-1} \end{aligned} \tag{46}$$

Starting at the end of the season (assumed constant at T weeks in all years),

$$Q_T = 0.35(1 - u_{T-1})W_{T-1} + Q_{T-1} \tag{47}$$

This is maximized by taking $u_{T-1} = 0$. We can now express Q_T in terms of values in week $T - 2$:

$$\begin{aligned} Q_T &= 0.35W_{T-1} + Q_{T-1} \\ &= 0.35(1.84 - 0.65u_{T-2})W_{T-2} + Q_{T-2} \end{aligned} \tag{48}$$

This is maximized by taking $u_{T-2} = 0$. We can continue backwards in the same way until eventually the coefficient of u becomes positive, when Q_T is maximized by taking $u = 1$, and it turns out that the coefficient of u is positive for all preceeding days. Thus the optimal strategy is to make only workers until a fixed time before the end of the season, and thereafter to make only queens. The effect of a variable season length can be investigated in a similar way, though the problem must be solved iteratively because of its greater complexity. (See Exercises 4 and 5. The closely related technique of stochastic dynamic programming is discussed in Chapter 6.)

The same argument can be used to model the allocation of resources between vegetative and reproductive growth in an annual plant (Cohen, 1971; Mirmirani and Oster, 1978; King and Roughgarden, 1982a,b). During the course of the season the plant can reinvest its resources into vegetative biomass that will enable it to produce more seeds later or it can produce seeds now; its objective is to produce as many seeds as possible by the end of the season. If the season ends at a fixed, predictable time this can be modeled by Equation 41, with the following reinterpretation of the symbols: W is vegetative biomass, Q is seed biomass, R is resource conversion efficiency, μ is loss of plant material (e.g., by grazing) and v is seed loss (e.g., by predation). Thus we expect a sudden switch from vegetative growth to reproduction, as observed in many annual plants. However, some desert annuals exhibit graded strategies with an intermediate period of both vegetative and reproductive growth followed by pure reproduction (Cohen, 1971). This can be explained as a bet-hedging response to uncertainty about the length of the season, that is determined by rainfall in deserts and varies greatly from year to year (King and Roughgarden, 1982b). Bet-hedging will be discussed more fully at the end of this chapter.

SENESCENCE

> Gather ye rosebuds while ye may,
> Old Time is still a-flying:
> And this same flower that smiles today,
> Tomorrow will be dying.
>
> —Robert Herrick (1591–1674), *To the Virgins, To Make Much of Time*

Aging is the deterioration of bodily function with advancing age and is evidenced by an increase in age-specific mortality or a decrease in age-specific fecundity with age. Most studies concern proximate, physiological causes of aging, but we are here concerned with the ultimate question of whether aging is just an inevitable consequence of wear-and-tear or whether it has evolved through trade-offs between age-specific selection pressures.

Evolutionary theories of aging are based on the fact that the selective effect of a mutant allele decreases with age: a mutant allele having an effect on the phenotype at an advanced age will have little effect on fitness because the animal will usually have died of something else by then. This can be quantified by considering the sensitivity of R to small changes in age-specific survival rates and fecundities. The sensitivity of R to a change in P_a, the probability of survival from age a to age $a + 1$, is

$$\frac{dR}{dP_a} = \frac{1}{P_a} \sum_{x=a+1}^{\omega} l_x m_x \tag{49}$$

As Charlesworth (1994) remarks, it is more natural to work with the logarithm of the survival probability, since independent factors affecting survival are expected to be additive on a logarithmic scale. On this scale,

$$\frac{dR}{d \log P_a} = \sum_{x=a+1}^{\omega} l_x m_x \tag{50}$$

The quantity on the right-hand side is the relative contribution to reproductive output made by individuals aged over a years, which is clearly a decreasing function of a. The sensitivity of R to a change in m_a, the fecundity at age a, is

$$\frac{dR}{dm_a} = l_a \tag{51}$$

that is simply the probability of survival to this age.

One mechanism through which senescence might evolve is that deleterious mutations manifesting themselves late in life can accumulate because selection against them is so weak—the *mutation-accumulation theory* proposed by Medawar (1952). Consider a rare deleterious allele with frequency p_t in generation t, with selection coefficient s against it, but with mutation rate to it at rate u from the wild-type allele. In the next generation the allele frequency will be decreased to $(1 - s)p_t$ through selection against it but increased by u through mutation toward it (since nearly all alleles are wild-type):

$$p_{t+1} = (1 - s)p_t + u \tag{52}$$

At equilibrium $p = u/s$, that will increase with age if s decreases with age. (The mutation rate u refers to germ-line mutations, so there is no reason why it should change with the age at which its effects are expressed.) This theory was tested in *Drosophila* by Rose and Charlesworth (1980) who argued that under this theory the genetic variance in the number of eggs laid should increase with the age of the fly. In fact, they found the genetic variance in fecundity to be constant, giving no support to the mutation-accumulation theory, but there is some ambiguity because of the need to transform the data before analysis to allow for the decrease in the mean number of eggs laid with age. On the other hand, Hughes and

Charlesworth (1994) have shown that the genetic variance in the rate of mortality increases dramatically with age in male *Drosophila*, in support of the mutation-accumulation theory.

The second evolutionary theory of senescence is the *antagonistic-pleiotropy theory* proposed by Williams (1957)—that there is a trade-off between early fecundity and later mortality. A "gather ye rosebuds" gene that has the dual effect of increasing fecundity early in life and decreasing survival later in life will be favored because of the much stronger selection pressure on its beneficial effect in early life than on its deleterious effects later. To test this theory, Rose and Charlesworth (1980) selected a line of flies for late reproductive output, using eggs laid by adults surviving at least 21 days from pupal emergence (that presumably also selects for longevity). Their results (see Table 5.2) show that there was a response to selection, accompanied by an increase in longevity and *a decrease in early laying*. This gives substantial support to the antagonistic-pleiotropy theory. There is of course no inconsistency in supposing that mutation accumulation determines the increase in male mortality, while antagonistic pleiotropy determines the decrease in female fecundity with age.

A submodel of the antagonistic-pleiotropy theory is the *disposable-soma theory* (Kirkwood and Rose, 1991). The idea is that the critical trade-off concerns the proportion of its resources that an organism allocates to the maintenance of the integrity of DNA in somatic cells. The maintenance of of error-free DNA is done by a mechanism called "proofreading" that is energetically expensive, and it may be that an organism will do better by allocating more energy to other activities at the expense of allowing some errors to accumulate in somatic DNA, while maintaining the highest accu-

Table 5.2. Results of selection for late reproductive output in *Drosophila*.

	Eggs layed (5-day total)	
Days	Selected line	Control line
1–5	422	551
11–15	392	323
16–20	287	239
21–25	183	137
Longevity (days):	30.2	26.8

After Rose and Charlesworth, 1980.

racy in germ-line DNA. A formal model can be constructed by expressing l_x and m_x as functions of the relative investment in somatic maintenance, s ($0 < s < 1$). There is a value of s that is large enough to prevent senescence, but it turns out that the optimal investment, maximizing R, is less than this critical value, so that senescence will evolve.

EVOLUTION IN A VARIABLE ENVIRONMENT: BET-HEDGING STRATEGIES

Natural populations are subject to selection pressures that are not constant, as assumed up to now, but that fluctuate over both space and time. In environments subject to temporal fluctuations between generations, there is general agreement that the relevant measure of fitness is its geometric rather than its arithmetic mean. Before considering the reason for using the geometric mean fitness, I discuss a simple example (Seger and Brockmann, 1987). Consider a haploid annual population that encounters wet and dry years with equal frequency, and let there be four different genotypes with the fitnesses in the two kinds of years shown in Table 5.3. A_1 and A_2 are specialists for wet and dry years, respectively, A_3 is a generalist that does equally well in both years, and A_4 is phenotypically polymorphic, developing the phenotype of A_1 (the wet-year specialist) 44 percent of the time and that of A_2 (the dry-year specialist) 56 percent of the time. (The reason for this choice for A_4 will appear later; note that the fitness of A_4 in a specified environment is calculated as the appropriate weighted arithmetic average of the fitnesses of A_1 and A_2 in that environment.) It is assumed that the phenotype is determined before any cues about the forthcoming climate are available. The arithmetic mean fitness of A_1 over years is $0.5 \times (1 + 0.58)$, while its geometric mean fitness is $(1 \times 0.58)^{0.5}$, the factor of 0.5 being determined by the fact that the two kinds of year occur with equal frequencies.

Suppose that the population is initially fixed for A_1 (the wet-year specialist) and that A_2 (the dry-year specialist) appears as a rare mutant. It will increase to fixation because it has higher geometric mean fitness than A_1. The population geometric mean fitness will be raised in consequence, but it will not be maximized by the fixation of A_2; the population geometric mean fitness reaches a maximum value of 0.795 when A_2 reaches a frequency of 56% and declines thereafter.

Now suppose that A_2 is fixed and that the generalist A_3 is introduced as a rare mutant. It has lower arithmetic mean but higher geometric mean fitness than A_2 and will therefore increase to fixation. A_3 has achieved a higher geometric mean fitness by reducing its variability in fitness at the

Table 5.3. Hypothetical fitnesses of four genotypes in two environments.

	Genotype			
Environment	A_1	A_2	A_3	A_4
Wet	1.00	0.6	0.785	0.776
Dry	0.58	1.0	0.785	0.815
Arithmetic mean:	0.790	0.800	0.785	0.796
Geometric mean:	0.762	0.775	0.785	0.795

After Seger and Brockmann, 1987.

expense of reducing its arithmetic average fitness at the same time. This is the characteristic feature of bet-hedging. I show in Appendix 5.2 why decreasing the variance of fitness increases its geometric mean even though the arithmetic mean may be reduced.

Finally A_4 exhibits a more sophisticated bet-hedging strategy that has been called "adaptive coin-flipping" (Cooper and Kaplan, 1982). This genotype spins a roulette wheel to decide whether to be a wet-year specialist (with probability p) or a dry-year specialist (with probability $1 - p$); p is chosen to be 0.44 because this value maximizes the geometric mean fitness. Note that this genotype is successful against any of the other three genotypes, whereas a genetically polymorphic mixture of wet-year and dry-year specialists with exactly the same phenotypic characteristics is not stable.

It is straightforward to verify these results by simulation. The reason for the success of the genotype with the highest geometric mean fitness is that fitnesses are multiplicative across generations. Consider the behavior of two alleles with gene frequencies p_t and $q_t = 1 - p_t$ in a haploid population with discrete, nonoverlapping generations; write W_{1t} and W_{2t}, respectively, for their absolute fitnesses in year t. Then

$$\begin{aligned} \frac{p_{t+1}}{q_{t+1}} &= \left(\frac{W_{1t}}{W_{2t}}\right)\frac{p_t}{q_t} \\ \frac{p_{t+2}}{q_{t+2}} &= \left(\frac{W_{1t}}{W_{2t}} \times \frac{W_{1,t+1}}{W_{2,t+1}}\right)\frac{p_t}{q_t} \end{aligned} \tag{53}$$

and so on, from which it follows that the genotype with the higher geometric mean fitness will ultimately win. The technical derivation of this result is rather subtle and is discussed in detail in Appendix 5.2.

Cohen (1966) has suggested that bet-hedging explains delayed germination of seeds of an annual plant in a variable environment. In the simplest form of Cohen's model there are two types of year, good and bad, that occur in a random sequence with probabilities p and $1 - p$, respectively. A fraction g of all seeds present germinates each year; of the seeds that do not germinate a fraction d decays each year, and a germinating seed produces s new seeds in a good year but none in a bad year. The problem is to find the optimal germination rate g when the decision to germinate must be taken before any information about the type of year is known (except the frequencies of good and bad years) so that no environmental tracking is possible.

The geometric mean fitness of a line with germination rate g is

$$G(g) = [(1-g)(1-d)]^{1-p}[(1-g)(1-d) + gs]^{p} \qquad (54)$$

Cohen supposed that selection would act on g to maximize G, with d, p, and s fixed; this leads to the optimal strategy

$$g = (sp + d - 1)/(s + d - 1) \qquad (55)$$

Cohen concludes that "we can expect to find the fraction that germinates every year to be approximately equal to the probability of producing a high yield" (i.e., $g \approx p$) since s is likely to be large. It should be remembered, however, that s is the average number of new seeds that a germinating seedling is likely to produce; if seedling mortality is high, there is no reason why s should be large, even though adult plants produce large numbers of seeds. The assumption that s is large amounts to assuming that the population is increasing rapidly in size and ignores the effect of density-dependent mortality in curbing population growth.

The simplest way of assessing the impact of density-dependent seedling mortality is to assume that g is given by the above equation but that s has been adjusted by density-dependent mortality of seedlings to ensure that $G = 1$. For given p and d Equation 55 can be solved simultaneously with the equation $G = 1$ to find s and g. The justification of this procedure is to suppose first that the population is fixed for a genotype with germination rate g and that s is adjusted by density dependence to ensure that $G = 1$. A rare mutant with germination rate g' will be able to invade the population if and only if $G(g') > 1$, with s still held constant at the same value as long as the mutant is rare. Hence the evolutionarily stable germination rate will be given by the original equation with s given by $G = 1$ (Bulmer, 1984). The optimal value of g turns out to be rather less than p, particularly when d is small.

This model can be extended in several ways to make it more realistic: incorporating density dependence by a more explicit model, allowing germination rates to vary in response to cues that are correlated with the quality of the forthcoming year, and adding spatial variation and migration. It can also be applied to other problems. Seger and Brockmann (1987) point out that insects living in unpredictable seasonal environments often face a diapause problem that is similar in many ways to the seed-dormancy problem. Animals enter diapause to avoid unfavorable conditions such as freezing or desiccation. For example, suppose that all adults will be killed by the first hard frost of the autumn, but that the day on which this will happen is unpredictable. Then a larva produced in the middle of the season may in some years have plenty of time to complete development, emerge, and reproduce without going into diapause, but in other years it would do better to stay in the cocoon and emerge the following spring. The optimal strategy is the bet-hedging strategy of emerging directly (i.e., not going into diapause) with probability $(sp - 2)/(s - 2)$, where p is the probability that a nondiapausing individual will successfully reproduce, giving rise to s surviving offspring the following year. Seger and Brockmann assume that the overwintering mortality d is the same for the larva making the decision of whether to emerge as it would be for the larva's offspring should the larva decide to emerge and reproduce; under this assumption there is no survival penalty for waiting and so d cancels out. The factor 2 rather than 1 appears because replacement in a sexual organism requires the production of two surviving offspring. If we consider only females and define s as the average number of surviving female offspring, assuming a sex ratio of unity, the probability becomes $(sp - 1)/(s - 1)$.

Several features of this diapause model should be noted. First, there should not be a sudden switch from all individuals emerging to all diapausing (the bang-bang strategy predicted by models that assume a completely predictable environment, in which the advantage suddenly flips from one strategy to the other), but a gradual increase in the proportion of insects diapausing. Second, the first few diapausing offspring should appear as soon as p drops below unity, that may be early in the season. Third, under this type of bet-hedging the strategy cannot be realized as a genetic polymorphism, but will only work if produced by a single genotype with variable diapause behavior.

In this account of selection in a variable environment I have confined myself to populations without age structure, avoiding the difficult extension of the theory to age-structured populations (Tuljapurkar, 1989, 1990). Nilsson et al. (1994) have suggested an alternative explanation of delayed germination, that it may benefit the mother plant by reducing sibling competition.

FURTHER READING

Roff (1992) and Stearns (1992) review the evolution of life histories. Charlesworth (1994) is a theoretical treatment of evolution in age-structured populations with a good discussion of life-history evolution. Harvey et al. (1991) is a symposium on the evolution of reproductive strategies. Rose (1991) and Partridge and Barton (1993) review the evolution of aging. Yoshimura and Clark (1993) discuss evolution in a varying environment.

EXERCISES

1. Consider the model for the evolution of semelparity versus iteroparity with density dependence acting on juvenile survival (see Equation 8) with $p_0 = 0.02$, $\alpha = 0.01$, $P_1 = 0.5$, and $b = 100$. Simulate the model using Equation 7 with $b^* = 45$ and 55, and verify that the invasibility criterion predicts the outcome of the simulation correctly.

2. Suppose that density dependence acts on juvenile survival as in Exercise 1 but that it also acts on adult survival according to Equation 21 with $p_1 = 0.5$ and $\beta = 1$. Simulate this model with $b^* = 70, 77$, and 85. Verify that the invasibility criterion predicts the outcome of the simulation correctly.

3. Consider a parthenogenetic semelparous organism with two possible strategies: (a) breed at age 1 and produce m_1 offspring, (b) breed at age 2 and produce m_2 offspring; the probability of survival to age 1 is P_0, and the probability of surviving from age 1 to age 2 in the second case is P_1. These parameters are constant, and there is no density-dependent regulation of population growth. Find the geometric growth rates of these two types, and the condition for the second to outgrow the first; evaluate this condition numerically if $m_1 = 2$ and $P_0 = 0.4$, 0.5, and 0.6. Find the lifetime reproductive success of each type and show, either generally or in this numerical case, that it gives the same criterion for the success of the second type in a stationary population, but not otherwise.

4. (a) Consider the dynamic programming problem outlined in Equations 46–48. Suppose that the end of the season is $T = 22$ weeks, and continue the induction back to time zero to find the optimal strategy throughout the season.

(b) Do these calculations again with time measured in days rather than weeks. Compare the answer with that obtained from Equation 42.

5. Return to the problem in Exercise 4 with time measured in weeks but suppose that the season does not always end in week 22 but has an equal chance of ending at any time between weeks 20 and 24. In week 24 the objective is to maximize q_{24}; in week 23 the season is equally likely to end this week or next week, and we assume from the bet-hedging principle that the objective is to maximize the geometric mean of the queens produced in weeks 23 and 24, that is equivalent to maximizing $q_{23}q_{24}$; in week 22 the season is equally likely to end in weeks 22, 23, or 24, and we assume that the objective is to maximize $q_{22}q_{23}q_{24}$; and so on. Unfortunately it is not possible to solve this problem directly by backwards induction because finding the value of u that maximizes the relevant expression for a particular week requires knowing the relative numbers of queens and workers in that week, which cannot be determined without knowing the values of u in previous weeks. This circularity can be broken by an iterative procedure. We first assume a trial solution for u in all weeks, calculate the numbers of queens and workers in different weeks, recalculate the optimal values of u on this basis, and then recalculate the numbers of queens and workers, continuing until the solution converges (hopefully) on a stable solution. Implement this procedure starting with the assumption that u is zero for weeks 18–24 and unity before then. (This exercise is quite difficult.)

6. Calculate the age-specific sensitivities of lifetime reproductive success to changes in the logarithm of the survival rate and in the fecundity at different ages from Equations 50 and 51 for the squirrel data in Table 4.1.

7. Consider the model of Seger and Brockmann (1987) described in the text and in Table 5.3. Verify the arithmetic and geometric mean fitnesses of the four genotypes. Use a random number generator to simulate the history of the population for 1000 years, starting with 25 individuals of each genotype and assuming that the total population size is regulated to contain 100 individuals each year.

8. (a) Consider Cohen's model for the delayed germination of seeds. Show that the germination rate g that maximizes the logarithm of the geometric mean fitness G in Equation 54 is given by Equation 55.
(b) Now assume that density-dependent factors act on s to make $G = 1$.

Find the optimal germination rate and the number of surviving seeds s for $d = 0.01$, 0.1, and 0.5 and for $p = 0.1$, 0.25, and 0.75. (For given d and p, the pair of equations $G = 1$ and

$$g = (sp + d - 1)/(s + d - 1)$$

can be solved to find s and g simultaneously using a suitable routine for the numerical solution of nonlinear equations.)

APPENDIX 5.1. PONTRYAGIN'S MAXIMUM PRINCIPLE

The control problem in continuous time can be formulated as follows. There is a vector of *state variables* $\mathbf{x}(t)$ and a vector of *control variables* $\mathbf{u}(t)$. The state vector obeys the given equations of motion

$$\mathrm{d}x_i/\mathrm{d}t = f_i(\mathbf{x}, \mathbf{u}) \tag{56}$$

with initial state $\mathbf{x}_0$ at time zero. The control vector can be chosen as any function of time provided that it always lies within the *control region* **U**. The objective is to choose the control vector so as to maximize the objective function J at the terminal time T, where J is defined as

$$J = \int_0^T I(\mathbf{x}, \mathbf{u})\,\mathrm{d}t + F(\mathbf{x}_T) \tag{57}$$

Thus J may depend both on the trajectory of the state variables and the control variables through the intermediate function I and on the final state at time T through the final function F.

The solution of this problem through Pontryagin's maximum principle goes as follows. For each of the state variables $x_i(t)$ define a corresponding *costate variable* $y_i(t)$ and then define the *Hamiltonian function*

$$H = I + \sum_i y_i f_i \tag{58}$$

The costate variables are to be defined to satisfy the differential equations

$$\mathrm{d}y_i/\mathrm{d}t = -\partial H/\partial x_i \tag{59}$$

with boundary conditions at terminal time

$$y_i(T) = \partial F/\partial x_i(T) \tag{60}$$

The solution of the control problem is the control vector **u** lying within the control region **U** that maximizes the Hamiltonian function H. For the derivation of this result see Intriligator (1971) or Clark (1990).

We now consider the problem discussed in the text of maximizing the number of reproductives produced by a colony of social insects at the end of the season. The state variables are the numbers of workers and reproductives, say $W(t)$ and $Q(t)$. The single control variable $u(t)$ is the proportion of effort that the colony allocates to producing new workers rather than reproductives, that must lie between 0 and 1, $0 \le u(t) \le 1$. The simplest model for the equations of motion, ignoring density dependence and temporal heterogeneity, is

$$\begin{aligned} dW/dt &= RuW - \mu W \\ dQ/dt &= R(1-u)W - \nu Q \end{aligned} \tag{61}$$

with initial conditions $W(0) = 1$ and $Q(0) = 0$, counting the founding queen as a worker; μ and ν (with $\mu > \nu$) are the mortalities of workers and reproductives and R is the rate of production of new workers (reproductives) resulting from allocation of unit effort per worker. The problem is to maximize the number of reproductives at the end of the season; the objective function is

$$J = Q(T) \tag{62}$$

so that there is no intermediate function, and the final function is

$$F = Q(T) \tag{63}$$

Following the recipe, the Hamiltonian is

$$\begin{aligned} H &= y_1 (Ru - \mu) + y_2[R(1-u)W - \nu Q] \\ &= (y_1 - y_2)RWu + \text{terms not involving } u \end{aligned} \tag{64}$$

Thus the Hamiltonian is *linear* in u. The choice of u between 0 and 1 that maximizes u is $u = 1$ if $y_1 > y_2$ and $u = 0$ if $y_1 < y_2$. The costate variables satisfy the differential equations

$$\begin{aligned} dy_1/dt &= -y_1(Ru-\mu) - y_2R(1-u) \\ dy_2/dt &= \nu y_2 \end{aligned} \tag{65}$$

with final conditions

$$\begin{aligned} y_1(T) &= 0 \\ y_2(T) &= 1 \end{aligned} \tag{66}$$

From Equation 66 it follows that $u(T) = 0$; the colony should make only queens at the end of the season. We now consider whether the bang-bang solution

$$u(t) = 0 \qquad t^* \leq t \leq T \tag{67a}$$

$$u(t) = 1 \qquad 0 \leq t \leq t^* \tag{67b}$$

provides a consistent answer to the problem.

Suppose that $u = 0$ (make only reproductives) when $t > t^*$. The solution of Equations 65 and 66 with $u = 0$ is

$$\begin{aligned} y_2(t) &= e^{\nu(t-T)} \\ y_1(t) &= R(e^{\nu(t-T)} - e^{\mu(t-T)})/(\mu - \nu) \end{aligned} \tag{68}$$

Thus

$$y_1/y_2 = R(1 - e^{(\mu-\nu)(t-T)})/(\mu - \nu) \tag{69}$$

At the switchpoint t^*, we must have $y_1 = y_2$, whence

$$t^* = T - \log[R/(R - \mu + \nu)]/(\mu - \nu) \tag{70}$$

Since y_1/y_2 is a decreasing function of t, $y_1 < y_2$ when $t > t^*$, so that $u = 0$ is the optimal strategy after this time.

Suppose now that $u = 1$ (make only workers) when $t < t^*$. The solution of Equation 65 with $u = 1$ satisfying

$$y_1(t^*) = y_2(t^*) = e^{\nu(t^*-T)} \tag{71}$$

is

$$\begin{aligned} y_2 &= e^{\nu(t - T)} \\ y_1 &= e^{\nu(t^* - T) - (R - \mu)(t - t^*)} \end{aligned} \tag{72}$$

so that

$$y_1/y_2 = e^{-(R-\mu+\nu)(t-t^*)} \tag{73}$$

Since y_1/y_2 is a decreasing function of t, $y_1 > y_2$ when $t < t^*$, so that $u = 1$ is the optimal strategy for all times less than t^*.

Thus the optimal strategy is to make only workers until the switching time t^* given by Equation 70 and only reproductives thereafter. The costate variables are given by Equation 72 before t^* and by Equation 68 afterwards.

We assumed above that the season ended at a fixed time T each year. In fact the season is likely to end later in some years than others, so that the end of the season can be considered as a random variable with probability density function $f(t)$ (with $0 \leq t \leq T$, where T is the latest possible date). In a stochastic environment the bet-hedging argument shows that the geometric mean fitness is maximized, that is equivalent to maximizing the expectation of the logarithm of the number of queens produced when the season ends. Thus the objective function to be maximized is

$$J = \int_0^T \log Q(t) f(t)\,\mathrm{d}t \tag{74}$$

so that the intermediate function is

$$I = \log Q(t) f(t) \tag{75}$$

and the final function is zero. The equations of motion are given by Equation 61. The Hamiltonian is

$$\begin{aligned} H &= \log Qf + y_1(Ru - \mu) + y_2[R(1-u)W - \nu Q] \\ &= (y_1 - y_2)RWu \ + \text{ terms not involving } u \end{aligned} \tag{76}$$

The costate variables satisfy

$$\begin{aligned} \mathrm{d}y_1/\mathrm{d}t &= -y_1(Ru - \mu) - y_2 R(1-u) \\ \mathrm{d}y_2/\mathrm{d}t &= -\frac{f}{Q} + \nu y_2 \end{aligned} \tag{77}$$

with final conditions

$$\begin{aligned} y_1(T) &= 0 \\ y_2(T) &= 0 \end{aligned} \tag{78}$$

Since the Hamiltonian is linear in the control variable u, we must have $u = 1$ when $y_1 > y_2$ and $u = 0$ when $y_1 < y_2$, but u may take intermediate values when $y_1 = y_2$. It turns out that the solution in a stochastic environment often turns out to be of the form: (i) make workers only ($u = 1$) until some switching time t_1^*, (ii) make a mixture of workers and queens ($0 < u < 1$) between t_1^* and a second switching time t_2^*, (iii) make queens only after t_2^*. This is the optimal solution to the problem if the costate variables satisfy

$$\begin{aligned} &y_1 > y_2 && 0 < t < t_1^* \\ &y_1 = y_2 && t_1^* < t < t_2^* \\ &y_1 < y_2 && t_2^* < t < T \end{aligned} \tag{79}$$

King and Roughgarden (1982b) discuss this problem in the context of the allocation of resources between vegetative and reproductive growth in annual plants in a variable environment.

APPENDIX 5.2. SELECTION IN A VARIABLE ENVIRONMENT

First consider a single genotype with absolute fitness W, that varies randomly from year to year. Thus W is a nonnegative random variable, and we define its mean and variance and those of its logarithm as

$$\begin{aligned} E(W) &= M \\ \mathrm{Var}(W) &= V \\ E(\log W) &= \log G \\ \mathrm{Var}(\log W) &= V^* \end{aligned} \tag{80}$$

M is the arithmetic mean of the distribution, and G is the geometric mean. For example, suppose that W takes the values 1.5, 1, and 0.6 with probabilities 0.25, 0.5, and 0.25, respectively. Then

$$\begin{aligned} M &= 1.025 \\ G &= 0.974 \\ V &= 0.1019 \\ V^* &= 0.1056 \end{aligned} \tag{81}$$

Note that $G < M$. (Since the logarithmic function is concave, Jensen's inequality ensures that $G \leq M$, with equality only if $V = 0$.) By the delta technique (taking expectations in a Taylor-series expansion of log W about $W = M$),

$$\begin{aligned} E(\log W) &\cong E\left(\log M + \frac{W-M}{M} - \frac{(W-M)^2}{2M^2} \right) \\ &= \log M - \frac{V}{2M^2} \end{aligned} \tag{82}$$

$$G = \exp[E(\log W)] \cong M \exp\left(-\frac{V}{2M^2} \right)$$

In the above example, this approximation gives 0.976 rather than 0.974 for the geometric mean. Thus the geometric mean is less than the arithmetic mean by an amount that is determined by the coefficient of variation, $\sqrt{V}/M$. This explains why it is possible to increase the geometric mean fitness by decreasing the variance of fitness even though the arithmetic mean fitness may be reduced.

Suppose now that $\{W_t\}$, $t = 1, 2, \ldots, T$, are identically and independently distributed random variables with the distribution of W, representing the fitnesses in years 1 through T. We consider the distribution of their product

$$Y = W_1 W_2 \cdots W_T \tag{83}$$

which determines the population size after T years:

$$n_T = Y n_0 \tag{84}$$

Its expected value is

$$E(Y) = E(W_1 W_2 \cdots W_T) = E(W_1)E(W_2) \cdots E(W_T) = M^T \tag{85}$$

All our other results are asymptotic and depend on considering the asymptotic properties of the sum

$$\log Y = \log W_1 + \log W_2 + \cdots + \log W_T \tag{86}$$

By the central limit theorem, log Y will become approximately normally distributed for large T with mean T log G and variance TV^*. Thus Y has approximately a lognormal distribution with the appropriate parameters. From the properties of this distribution (Aitchison and Brown, 1957), it follows that

$$\begin{aligned}
&\text{median}(Y) \cong G^T \\
&E(Y) = M^T \cong G^T e^{TV^*/2} \\
&\text{Var}(Y) \cong G^{2T} e^{2TV^*} \\
&\text{Coefficient of variation } \cong e^{TV^*/2} \\
&\text{Skewness} \cong e^{3TV^*/2}
\end{aligned} \tag{87}$$

Thus both the coefficient of variation and the skewness increase without bound as T becomes larger. This is strange behavior that requires delicate treatment.

We now calculate the probability that Y is less than some arbitrary positive constant c. Thus

$$\begin{aligned}
P_c \equiv \Pr[Y < c] &= \Pr\left[\frac{\log Y - T\log G}{\sqrt{TV^*}} < \frac{\log c - T\log G}{\sqrt{TV^*}}\right] \\
&\cong \Phi\left(\frac{-\sqrt{T}\log G}{\sqrt{V^*}}\right)
\end{aligned} \tag{88}$$

where $\Phi(\cdot)$ denotes the standard normal distribution function. Hence for any positive value of c, P_c tends to zero if $G > 1$ and to unity if $G < 1$. Thus if $G > 1$, P_c tends to zero for any c however large, whereas if $G < 1$ it tends to 1 for any positive c however small. We conclude that Y tends in probability to 0 or ∞ according to whether the geometric mean is less than or greater than 1. Note that if $G < 1 < M$, then Y tends in probability to zero but its expected value tends to infinity!

Finally, consider the behavior of two alleles, A_1 and A_2, with gene frequencies p_t and $q_t = 1 - p_t$ respectively, in a haploid population with discrete, nonoverlapping generations. Write W_{1t} and W_{2t} for their absolute fitnesses in year t, being realizations of the random variables W_1 and W_2. The recurrence relation for the ratio of the gene frequencies is

$$\frac{p_{t+1}}{q_{t+1}} = \left(\frac{W_{1t}}{W_{2t}}\right)\frac{p_t}{q_t} \tag{89}$$

so that

$$\frac{p_t}{q_t} = \left(\frac{W_{1,t-1}}{W_{2,t-1}}\right) \times \left(\frac{W_{1,t-2}}{W_{2,t-2}}\right) \times \cdots \times \left(\frac{W_{1,1}}{W_{2,1}}\right) \frac{p_1}{q_1} \tag{90}$$

A_1 or A_2 will go to fixation according to whether $E[\log (W_1/W_2)]$ is greater or less than unity. But

$$\begin{aligned} E[\log(W_1/W_2)] &= E[\log W_1 - \log W_2] \\ &= E(\log W_1) - E(\log W_2) \\ &= \log G_1 - \log G_2 \end{aligned} \tag{91}$$

Thus the allele with the higher geometric mean fitness will, almost certainly, go to fixation.

This account ignores the effect of population density on the fitnesses of the two alleles, but the argument in the last paragraph goes through unchanged in the presence of density dependence provided that it acts in the same way on the fitnesses of the two alleles so that W_1/W_2 is unaffected by population density.

CHAPTER 6

Foraging Theory and Resource Management

How doth the little busy bee
    Improve each shining hour,
And gather honey all the day
    From every opening flower!

—Isaac Watts (1674–1748), *Against Idleness and Mischief*

Getting enough to eat is very important to animals, so that searching for food (foraging) in an efficient way is an important component of fitness. Foraging theory addresses the question of how animals should forage so as to maximize their fitness and has had considerable success in explaining empirical observations of foraging behavior. Humans also have the problem of exploiting natural resources (fisheries, forests) in an optimal way, but our behavior is molded not by natural selection but by economic and political considerations. The theory underlying the optimal management of renewable resources by man will be considered in the second part of this chapter.

FORAGING THEORY

Foraging theory is based on the assumption that foraging behavior has been molded by natural selection to maximize fitness. To develop a theory we must specify two things: (i) How do different possible behaviors affect fitness? (ii) What is the decision variable that the predator can choose? It is assumed that selection acts on the decision variable to maximize fitness, subject to other constraints that must also be incorporated into the model.

There are two important types of foraging model, the patch model and the prey model. In the patch model the predator feeds by foraging in patches that become depleted the longer it stays, and it has to decide when to leave a patch; the decision variable is the length of time the predator stays in a patch. In the prey model the predator has to decide which prey items to include in its diet, so that the decision variable is whether or not to pursue different types of prey. It is often plausible to suppose that fitness is proportional to the net rate of energy gain, so that animals are selected to maximize the latter quantity. I shall here develop the patch model and then the prey model under this assumption about fitness; I shall then consider why the average net rate of energy gain may not be an appropriate measure of fitness in a variable environment; and I shall finally describe the technique of stochastic dynamic programming, that makes it possible to construct detailed models of decision-making by animals that lead directly to the maximization of fitness without using a surrogate measure of fitness like the rate of energy gain.

The patch model and the marginal-value theorem

Consider an animal that feeds by foraging in patches. The idea is that the longer the animal stays in the patch, the more the resource is depleted and the instantaneous rate of energy gain falls. The penalty for staying too long in a patch is resource depletion; the penalty for staying too short a time is the time and energy spent in traveling between patches. What is the optimal time to stay in order to maximize the rate of energy gain? In this model the behavior that an animal can choose is the length of time it stays in the patch, and the currency to be maximized (assumed to be proportional to fitness) is the net rate of energy gain.

Suppose first that all patches are of equal quality. Let the net gain of energy from staying time t in the patch be $g(t)$. This function allows for the energy spent in searching for food and also the energy spent in moving from one patch to the next, so that $g(0) < 0$; it is the net gain in energy from the time of arrival at the patch to the time of arrival at the next patch at time $t + T$, where T is the travel time to the next patch. It is assumed that $g(t)$ is concave because of resource depletion. The rate of energy gain by an individual spending time t in each patch is

$$R = \frac{g(t)}{T + t} \tag{1}$$

We now assume that an individual acts so as to maximize its rate of energy gain. This is a shorthand way of saying that the animal's fitness is proportional to its rate of energy gain and that selection has acted to maximize fitness. This is often (though not always) a reasonable assumption. Differentiating R with respect to t and equating this derivative to zero to obtain the maximum, we find that

$$g'(t) = \frac{g(t)}{T+t} \tag{2}$$

This means that an animal should leave when the marginal (i.e., instantaneous) rate of energy gain $g'(t)$ has fallen to the average rate of energy gain from moving to a new patch, R. Charnov (1976) called this the *marginal-value theorem*.

This theorem is illustrated graphically in Figure 6.1 with $g(t) = -1 + 2t - 2t^2$, $T = 2$; $g(0) = -1$ is the energy cost of travel to a new patch that takes 2 time units. If an animal habitually leaves a patch at time t_0, a line from –2 on the t axis intersecting the curve at $[t_0, g(t_0)]$ will have slope R, the average rate of gain by that animal. The line with maximal slope of any line touching or intersecting the curve will be tangent to the curve, like the upper dotted line in Figure 6.1. The slope of this line is equal to the derivative of the curve where they touch.

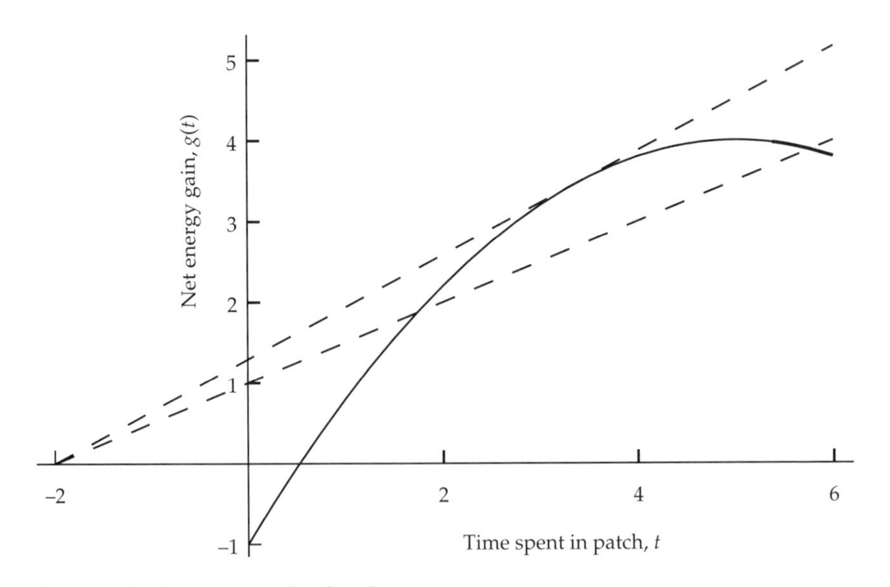

Figure 6.1. The marginal-value theorem.

This model can be extended to the situation with several patch types, a patch of type i occurring with probability P_i and having gain function $g_i(t)$, with $g_i(0)$ being the same for all patches. Assume that the predator can recognize the type of patch it is in. If it leaves a patch of type i after time t_i then

$$R = \frac{\sum_i P_i g_i(t_i)}{T + \sum_i P_i t_i} \tag{3}$$

To justify this formula consider an animal that has just visited n different patches, encountering a patch of type i n_i times. It will have gained $\Sigma n_i g_i(t_i)$ energy units during time $nT + \Sigma n_i t_i$ so that its rate of energy gain is

$$R = \frac{\sum_i p_i g_i(t_i)}{T + \sum_i p_i t_i} \tag{3a}$$

where $p_i = n_i/n$ is the proportion of type-i patches visited. For large n, p_i will tend to the corresponding probability P_i, giving Equation 3. Finding the time t_i that maximizes R by equating to zero the derivative of g_i with respect to t_i leads to the equation

$$\frac{dg_i(t_i)}{dt_i} = R \tag{4}$$

Thus the predator should leave when the marginal rate of energy gain within the patch equals the average overall rate of gain—the marginal-value theorem as before.

Best and Bierzychudek (1982) applied the logic of this argument to explain the foraging behavior of bumblebees on foxgloves. The inflorescence of the foxglove (Fig. 6.2) has new flowers at the top and old flowers at the bottom. It is protandrous (newly opened flowers functioning first as males, then as females), so that the inflorescence consists of (from top to bottom) closed buds, newly-opened buds, male flowers, female flowers, and maturing seed capsules. The caloric reward of a flower declines linearly with position from bottom to top. To the bees each inflorescence represents a patch. Some inflorescences will have been previously visited by bees on the same day and will be empty; others will be full. Best and Bierzychudek predicted that bees should begin with the lowest flower, that contains most nectar and forage upward, neither skipping nor revisiting any flower, until the energy return expected in the next flower

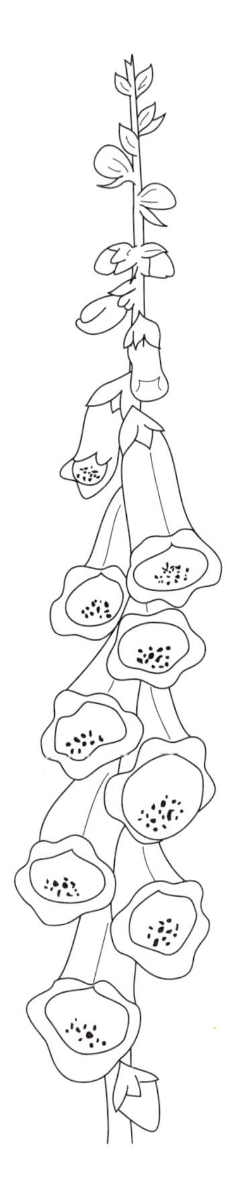

Figure 6.2. The vertical inflorescence of the foxglove. From Best and Bierzychudek, 1982.

is lower than the average return from a new inflorescence. All these predictions were confirmed except that bees skip some flowers because they go straight up the inflorescence rather than spiral around it; the authors suggest that spiraling would carry too much risk of revisiting a flower. When they found an empty inflorescence, the bees left immediately. It is interesting to observe that it is in the plant's interest to have bees visit female flowers first, pollinating them with pollen from the previous plant, and then male flowers, acquiring pollen that they will take to the next plant. Thus the pattern of nectar presentation by the plant has coevolved with the foraging behavior of its pollinators.

The prey model and diet choice

Another type of choice a predator makes is which prey items to include in the diet. The idea is that some prey items may not be worth pursuing because their energetic value to the predator is too small in comparison with the time taken to catch and consume them. To analyze this situation, suppose that there are n prey items and define the following quantities for the ith type:

$$
\begin{aligned}
h_i &= \text{handling time (time taken to catch and consume prey)} \\
E_i &= \text{net energetic gain from catching and consuming this prey} \\
\lambda_i &= \text{rate of encounter with this type of prey} \\
p_i &= \text{relative frequency that predator attacks this type of prey}
\end{aligned}
\tag{5}
$$

The first three quantities are fixed parameters of the model; the problem is to find the p_i's that maximize the rate of energy gain by the predator.

When a predator finds a prey of type *i*, it may choose either to catch and consume it or to continue searching, and the p_i's define the relative frequencies of the first choice for different types of prey; these frequencies define the range of behaviors open to the predator. The currency to be maximized is the rate of energy gain, as before.

Let *s* be the energy lost per second while searching. In *S* seconds of searching, the total time (including the time spent catching and consuming prey) for a given choice of the p_i's is

$$T = S + S\sum_i \lambda_i p_i h_i \tag{6}$$

The net energy gained is

$$G = S\sum_i \lambda_i p_i E_i - Ss \tag{7}$$

The net rate of energy gain is

$$R = \frac{G}{T} = \frac{\sum_i \lambda_i p_i E_i^*}{1 + \sum_i \lambda_i p_i h_i} - s \tag{8}$$

where $E_i^* = E_i + sh_i$. (E_i^* is the difference in energy gain between eating the prey [and gaining E_i] and ignoring the prey [and "gaining" $-sh_i$]; it measures the net energy gain from attacking rather than ignoring the prey. Compare the concept of "opportunity cost" introduced later in this chapter.) Differentiating *R* with respect to p_j to find the optimal probability of attacking prey of type *j* we find that

$$\begin{aligned}\frac{\partial R}{\partial p_j} &\propto \lambda_j E_j^*\left(1 + \sum_i \lambda_i p_i h_i\right) - \lambda_j h_j \sum_i \lambda_i p_i E_i^* \\ &= \lambda_j \left[E_j^*\left(1 + \sum_{i \neq j} \lambda_i p_i h_i\right) - h_j \sum_{i \neq j} \lambda_i p_i E_i^*\right]\end{aligned} \tag{9}$$

where the constant of proportionality is positive. The term in square brackets is independent of p_j so that *R* is maximized by taking $p_j = 1$ when this term is positive and $p_j = 0$ when it is negative.

Thus a particular prey type should either always be taken ($p_j = 1$) or never be taken ($p_j = 0$). The following rules for determining which prey items should be included in the diet are derived in the appendix to this chapter. Define the profitability of prey j as

$$\rho_j = E_j^*/h_j \tag{10}$$

and rank the prey in order of their profitability so that $\rho_1 > \rho_2 > \cdots > \rho_n$. No prey item with a negative profitability should be included in the diet, which is intuitively obvious, and we shall assume that at least the first prey item has positive profitability. If there are only two prey types, the rule is to take only the first type if

$$\rho_2 < \frac{\lambda_1 E_1^*}{1 + \lambda_1 h_1} \tag{11}$$

and to take both types otherwise. With n prey types, the types which should be included in the diet can be determined by the following algorithm suggested by the above result. Add the types to the diet in order of increasing rank until for the first time

$$\rho_{k+1} < \frac{\sum_{i=1}^{k} \lambda_i E_i^*}{1 + \sum_{i=1}^{k} \lambda_i h_i} \tag{12}$$

The first k items are included in the diet, and the rest are excluded.

These results make three predictions: (1) The zero–one rule that prey types are either always taken or never taken as long as conditions remain constant. (2) Prey are ranked in order of their profitability. (3) The inclusion of a prey in the diet depends on its profitability and on the characteristics, including the abundance, of more profitable prey but is independent of its own abundance. These rules can be justified intuitively as follows: (1) If it is advantageous to chase a rabbit today, then it should be advantageous to chase it tomorrow under identical circumstances. (2) This is obvious. (3) The reason that less profitable prey should be ignored is that by attacking them the predator would be losing the opportunity of finding a more profitable prey type. This opportunity obviously depends on the abundance of the more profitable types, but the decision of whether or not to attack a prey that has just been encountered should not be affected by the likelihood of encountering a similar prey in the future.

Hanson and Green (1989) tested the predictions of the prey-choice model in experiments with pigeons offered two prey types. They found that nearly all opportunities with the rich prey type were accepted, and that as the encounter rate with the rich prey type decreased, the pigeons accepted more opportunities with the poor prey type. Both of these results agree with the predictions of the prey-choice model. The acceptance of the poor prey type often, but not always, approximated the zero–one rule; possible reasons for departures from the zero–one rule are discussed in the next subsection. However, selection of the poor prey type changed when the encounter rate with only the poor prey type was changed, in violation of the model prediction. The reason for the failure of this prediction is unclear.

The prey model: Central-place foraging

Many animals search for prey and then bring it back either to the nest or to a perch from which foraging trips are made. This situation gives rise to models of *central-place foraging* that can be subdivided into single-prey loader and multiple-prey loader models according to whether the animal brings back a single prey or many prey on each trip. Central-place foraging is a special case of the prey model with the added constraint that the predator must bring the prey back to the same place each time.

As an example of the single-prey loader model, consider the study of Krebs and Avery (1985) on bee-eaters bringing prey back to their nests in the Camargue region of southern France. They observed foraging trips by bee-eaters traveling different distances from their nest and recorded the type of prey brought back. The prey types were subsequently categorized for simplicity into big (mainly dragonflies) and small (mainly bees in one year, small dragonflies in the other). The prediction to be tested is that, with increasing travel distance, small prey items should be excluded from the diet of the nestlings because it is not worthwhile returning from distant sites with a small prey.

The terms h_i, E_i, λ_i, and p_i are defined as above in Equation 5. It is assumed that the quantity to be maximized is the average rate of energy delivery to the young, so that energy lost by the parent while searching, s, is ignored, but the round-trip travel time from the nest to the foraging site, τ, is an important new element of the problem. In S seconds of searching at the site (excluding handling time and travel time), the total time (including these two factors) for a given choice of the p_i's is

$$T = S + S\sum_i \lambda_i p_i (h_i + \tau) \tag{13}$$

and the net energy gained is

$$G = S\sum_i \lambda_i p_i E_i \tag{14}$$

so that the rate of energy delivery to the young is

$$R = \frac{G}{T} = \frac{\sum_i \lambda_i p_i E_i}{1 + \sum_i \lambda_i p_i (h_i + \tau)} \tag{15}$$

The results of the prey model hold if we replace E_j^* by E_j and h_j by $h_j + \tau$.

For two prey types, large and small as considered above, on the assumption that large prey are more profitable, it follows from making these replacements in Equation 11 that the birds should bring back only large prey when

$$\tau > \tau_{crit} = \frac{E_2/\lambda_1 + E_2 h_1 - E_1 h_2}{E_1 - E_2} \tag{16}$$

and they should bring back both types when $\tau < \tau_{crit}$. Data for two years are given in Table 6.1. Observed behavior was in accord with prediction, except that there was a gradual drop in the proportion of small prey with travel time rather than an abrupt switch. The travel times at which the percentage of small prey fell below 50% were 33 seconds in 1981 and 60 seconds in 1982. Several factors may account for the gradual change (rather than abrupt switch) in behavior, that has been found in many studies. It may be, for example, that different birds have different thresholds because they have different handling times, or because they have imperfect knowledge of the abundance or profitability of the prey, or because they are processing this information differently (and sometimes imperfectly). Given these considerations, it would be rather remarkable if an abrupt threshold were observed since one cannot expect real birds to be identical, omniscient, or perfect. This question is discussed further by McNamara and Houston (1987).

Table 6.1. Foraging by bee-eaters on large and small prey.

Year	E_1 (mg dry wt)	E_2 (mg dry wt)	$\lambda_1 h_1$ (min^{-1})	h_1 (sec)	h_2 (sec)	τ_{crit} (sec)
1981	138	30	0.35	19	10	40
1982	218	41	0.19	20	12	63

After Krebs and Avery, 1985.

Risk-sensitive foraging

Most foraging models maximize the average net rate of energy gain. This assumes that a reward of 10 grams is twice as desirable as one of 5 grams, but this may not always be the case. It may be that there is a nonlinear relationship between the gain in fitness from a food item and its energetic equivalent. Denote by $w(x)$ the gain in fitness from an energetic gain of x units. In some cases $w(x)$ might be a concave function: a well-fed animal might not gain twice as much from 10 grams as from 5 grams because of the difficulty or cost of storing the extra energy. Such an animal would prefer getting 5 grams with certainty to having an even chance of getting nothing or 10 grams; this behavior is called *risk-averse*, because it results in a willingness to sacrifice some loss of mean energy gain if there is a corresponding *reduction* in the variance. On the other hand, a hungry animal that needs 8 grams to survive the night will find 5 grams worthless and would prefer an even chance of nothing or 10 grams over the certainty of 5 grams; this behavior is called *risk-prone* because it results in a willingness to sacrifice some loss of mean energy gain if there is a corresponding *increase* in the variance. More formally, suppose that x has mean value ξ and expand $w(x)$ in a Taylor series about ξ:

$$w(x) \cong w(\xi) + (x - \xi)w'(\xi) + \frac{(x - \xi)^2}{2} w''(\xi) \tag{17}$$

Taking expected values on both sides, we find that the average fitness is

$$E[w(x)] \cong w(\xi) + \frac{\mathrm{Var}(x)}{2} w''(\xi) \tag{18}$$

If $w(x)$ is concave [with $w''(x) < 0$] a high variance in x will decrease the average fitness leading to risk aversion; if $w(x)$ is convex [with $w''(x) > 0$] a high variance in x will increase the average fitness leading to risk-prone behavior.

An animal's apparent preference or utility function $u(x)$ can be estimated empirically. Caraco et al. (1980) report laboratory experiments on risk-sensitive foraging preferences in Yellow-eyed Juncos. A bird was given the choice, after an initial training period, of getting, say, two seeds with certainty at one feeding station or having an even chance of getting zero seeds or six seeds at another, and this experiment was repeated with different numbers of seeds. To find the utility function $u(x)$, where x = number of seeds, two points on the curve can be chosen arbitrarily since scale and location can be chosen arbitrarily: they chose $u(0) = 0$, $u(6) = 1$. They then found the rest of the curve as follows. Do experiments with zero or six seeds in the risky station, with expected utility 0.5, against $x = 0, 1, 2, 3,$

4, 5 in the certain station, and suppose the bird prefers the risky station when $x = 1$, is indifferent when $x = 2$, and prefers the certain station for $x \geq 3$; then $u(2) = 0.5$. One can then do another experiment with two or six seeds in the risky station (with expected utility 0.75) and find, say, that the bird is indifferent between this and the certainty of three seeds, so that $u(3) = 0.75$. In this way the whole utility function can be estimated, though some interpolation may be necessary because of the restriction to integral numbers of seeds.

Caraco et al. (1980) estimated the utility curves for five birds that had been starved for one hour before the experiment and found a concave curve (risk aversion) in every case, while in a group of four birds that had been starved for four hours the curve was convex (risk proneness). The curves for one typical bird in each group are shown in Figure 6.3. These results have been confirmed in a more rigorous experiment by Caraco et al. (1990).

The idea of risk-sensitive foraging has no connection with the idea of bet-hedging considered in Chapter 5. In bet-hedging, fitness varies between generations, and animals are selected to adopt the strategy with the highest geometric mean fitness. In risk-sensitive foraging, the variability is within a generation, and animals are selected to adopt the strategy with the highest arithmetic mean fitness (i.e., to maximize the expected fitness), but this is not the same as maximizing the expected rate of energy gain if there is a nonlinear relation between fitness and energy gain. In risk aversion an animal may sacrifice some loss of mean energy gain if there is a sufficient reduction in the variance of energy gain because this will increase its expected fitness [Equation 18 with $w''(x) < 0$]. In bet-hedging an animal may sacrifice some expected fitness if there is a sufficient reduction in the variance of fitness because this will increase its geometric mean fitness [Equation 82 in Appendix 5.2]. There is a superficial resemblance between these two situations, but the underlying rationale is quite different.

Stochastic dynamic programming

It was seen in the previous section that fitness may be related in a nonlinear way to energy gain and that the shape of the fitness curve may depend on the state of the animal. Thus maximizing expected fitness may not be the same as maximizing the expected rate of energy gain, particularly in a stochastic environment. The new technique of stochastic dynamic programming (whose application in evolutionary ecology is reviewed by Houston et al., 1988 and by Mangel and Clark, 1988) makes it possible to construct detailed models of decision-making by animals that lead directly

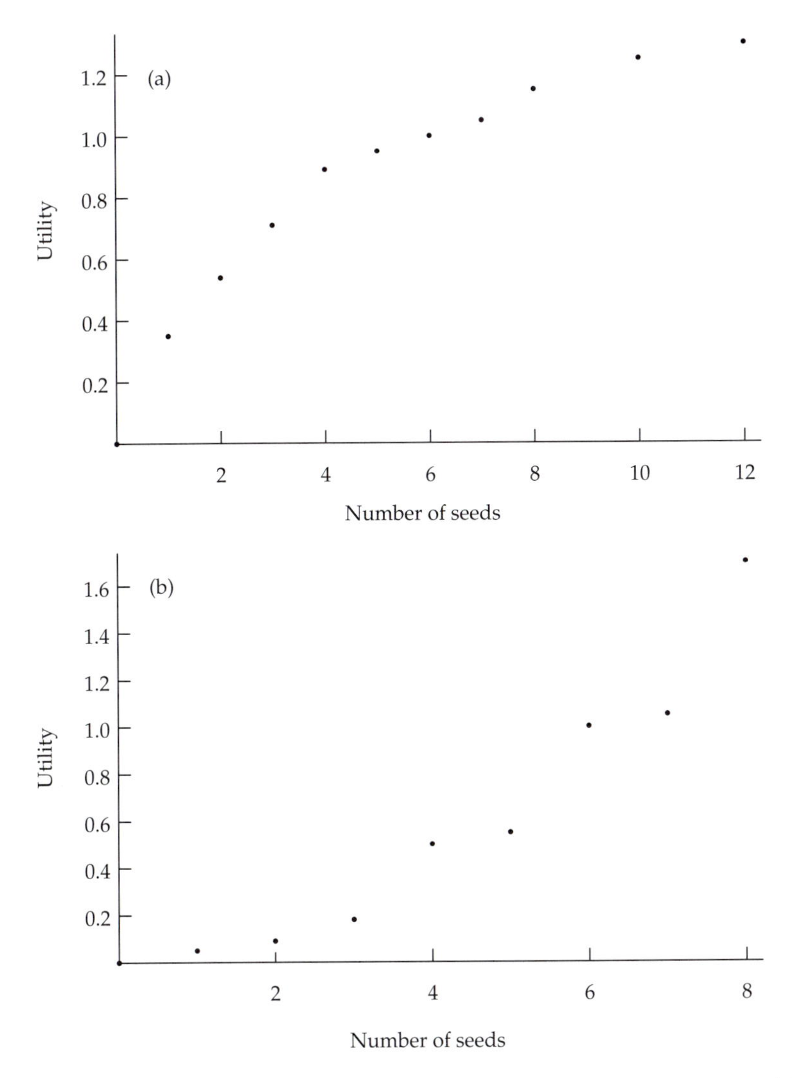

Figure 6.3. Utility curves for (a) a recently-fed bird, (b) a hungry bird. After Caraco et al., 1980.

to the maximization of fitness, rather than using a surrogate measure of fitness like the rate of energy gain. The price to be paid for greater realism in the model is more complexity in its analysis, so that results will be obtained by computer simulation rather than analytically. The basic idea is to use backwards induction.

Table 6.2. Parameter values for a patch-foraging model.

Habitat, i	β_i	λ_i
1	0.00	0.4
2	0.01	0.6
3	0.05	0.8

Consider the following simple patch-foraging model for an animal that can choose one of three habitats in which to forage on a particular day. Each habitat is characterized by two parameters: β_i = probability of death due to predation and λ_i = probability of finding a food item (assuming that not more than one can be found in a day) These parameters take the numerical values shown in Table 6.2. Thus habitat 1 has zero risk and low reward, habitat 3 has high risk and high reward, and habitat 2 is intermediate. The problem for the forager is how to balance the trade-off between these two factors when the riskier habitat is also the more rewarding in food.

To address this question we introduce the state variable X_t representing an animal's bodily energy reserves on day t and assume that X_t cannot fall below zero or rise above an upper limit of 5 units. Energy reserves are increased by daily food consumption (Z_t) and decreased by daily metabolic expenditure that will be taken as 1 unit:

$$X_{t+1} = X_t - 1 + Z_t \tag{19}$$

provided that X_{t+1} is replaced by zero if it becomes negative and by 5 if it exceeds this value. Z_t is a random variable whose distribution is determined by the animal's choice of habitat on day t; we shall suppose that a food item is worth two energy units, so that Z_t takes values 2 with probability λ_i and 0 with probability $1 - \lambda_i$ if the animal chooses habitat i. The animal dies if it is preyed upon or if $X_t = 0$ (starvation).

In a nonreproductive period (say, over winter), assume that maximizing fitness means maximizing survival probability from time $t = 1$ to time T (the time horizon, say, the beginning of the breeding season). The optimal strategy on a particular day will depend both on the animal's current energy reserves and on the time remaining till the beginning of the breeding season, and can be calculated by backwards induction from time T.

Suppose, for example, that $T = 100$. On day 99 the animal should choose habitat 1 if its energy reserves are 2 or more, since it will then survive with certainty for one day, but it should choose habitat 3 if its energy reserves are 1 since it must find a prey item in order to survive; its chance of survival in this case is 0.76.

We can now move back to day 98. Suppose that the energy reserves on that day are 2. If the animal chooses habitat 3, it will be preyed upon with probability 0.05, its energy reserves will fall to 1 on day 99 with probability $0.2 \times 0.95 = 0.19$ and will increase to 3 with probability $0.8 \times 0.95 = 0.76$; its chances of survival to day 100 would be $0.19 \times 0.76 + 0.76 = 0.9044$. One can calculate in a similar way that, if the animal chose habitat 1 or 2, its chances of survival to day 100 would be 0.856 and 0.89496, respectively, so that habitat 3 is optimal. One can repeat these calculations for other values of the energy reserve on day 98 and then move back to day 97, day 96, and so on.

These calculations are laborious but are easily computerized in a language such as *Mathematica* that supports recursive programming. Define the following quantities:

$$\begin{aligned} J(x,t) &= \text{probability of survival from day } t \text{ to day 100} \\ &\quad\ \text{given that } X_t = x \text{ and that the animal} \\ &\quad\ \text{uses the best strategy from then on} \\ H(x,t) &= \text{optimal habitat choice on day } t \text{ given that } X_t = x \\ P_i(x,t) &= \text{probability of survival from day } t \text{ to day 100} \\ &\quad\ \text{given that } X_t = x \text{ and that the animal} \\ &\quad\ \text{chooses habitat } i \text{ on day } t \\ &\quad\ \text{and uses the best strategy from then on} \end{aligned} \tag{20}$$

These quantities can now be calculated by backwards induction starting with day 100, using the following equations. The recurrence relation for $P_i(x, t)$ is

$$P_i(x,t) = (1-\beta_i)\left[\lambda_i J(x_i^*, t+1) + (1-\lambda_i)J(x-1, t+1)\right] \tag{21}$$

where x_i^* is the lesser of x_i and 5. The functions J and H are defined as

$$\begin{aligned} J(x,t) &= \max_i P_i(x,t) \\ H(x,t) &= i \text{ which maximizes } P_i(x,t) \text{ above} \end{aligned} \tag{22}$$

The final condition is

$$J(x,100) = 1 \qquad 1 \le x \le 5 \tag{23}$$

Hence the optimal strategy on day 1 (at the beginning of the winter) can be calculated. It is usually found that this strategy is insensitive to the length of the time horizon, provided that it is not very short, so that the rather arbitrary feature of fixing an exact time horizon is unimportant. In the above example, the optimal habitat choice on day 1 for energy reserves {1, 2, 3, 4, 5} is {3, 3, 2, 2, 1}; this remains the optimal set of choices until day 94, within a few days of the time horizon. The optimal choice on any day is strongly dependent on the animal's state; if its energy reserves are high it can afford to play it safe, but if they are low it pays to take risks.

Houston et al. (1988) and Mangel and Clark (1988) discuss several other applications of this technique, including the optimal size of hunting groups for lions, the reason for the dawn chorus of birds, the daily vertical migration of aquatic organisms, and the foraging behavior of orb-weaving spiders.

MANAGEMENT OF RENEWABLE RESOURCES

We humans also have the problem of exploiting natural resources (fisheries, whales, forests) in an optimal way. Much theoretical has been done on this subject, and there is only space here to give a very brief account of the basic ideas, based on the excellent book of Clark (1990). The main point to be made is that the criterion of optimality is more arbitrary, and more open to argument, in this context. In models of optimal foraging by animals we assume that they are selected to maximize their expected fitness, which is a well-defined concept, and we explore the theoretical consequences of this assumption. In the human context it is not obvious what an optimal level of exploitation should be, as is evidenced by conflicts between environmental and economic interests. I shall first consider what the level of harvesting should be under a simple model in order to maintain the maximum sustainable yield, and I shall then discuss whether this is an economically justifiable optimality criterion. This discussion throws some light on the underlying reasons for the conflict of interest between environmental interests and those whose livelihood is at stake in a commercial fishing or logging industry.

The Schaefer model

Consider a natural population whose size in the absence of exploitation (fishing, whaling, hunting) would follow the logistic law (Equation 7 in Chapter 2). Suppose now that the population is harvested at rate $h(n)$ so that

$$\frac{\mathrm{d}n}{\mathrm{d}t} = f(n) = r\left(1 - \frac{n}{K}\right)n - h(n) \tag{24}$$

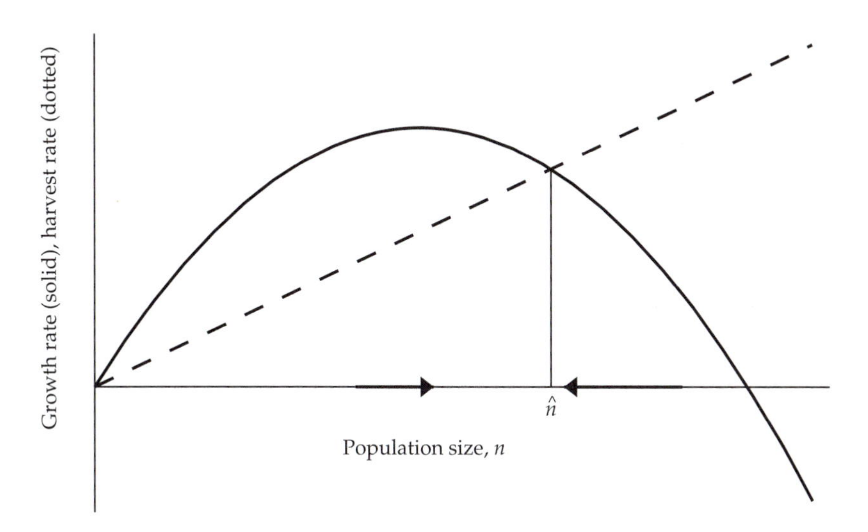

Figure 6.4. The Schaefer model, with harvesting rate proportional to population size. The solid curve shows the natural increase in the absence of fishing, the dashed line is the harvest rate; the actual increase is the difference between them.

Under the *catch per unit effort* model, assume that $h(n) = qEn$, where E is the fishing effort (e.g., number of vessel days per unit time) and q is a constant called the "catchability" coefficient. It is convenient to absorb q into E, so that $q = 1$. Then

$$\frac{\mathrm{d}n}{\mathrm{d}t} = f(n) = r\left(1 - \frac{n}{K}\right)n - En \tag{25}$$

Assume that E is constant. This is usually called the Schaefer model after M. B. Schaefer. There is a unique nonzero equilibrium at $\hat{n} = K(1 - E/r)$ (provided that $E < r$). Figure 6.4 shows that this equilibrium is stable since the growth rate is positive when $n < \hat{n}$ and negative when $n > \hat{n}$. The equilibrium harvest or sustainable yield at this point is

$$Y = E\hat{n} = KE\left(1 - \frac{E}{r}\right) \tag{26}$$

The sustainable yield is maximized when $E = r/2$, giving an equilibrium population size of $\hat{n} = K/2$ and a maximum sustainable yield (MSY) of $rK/4$.

Why harvest to maintain maximum sustainable yield?

It might seem that the problem of optimal exploitation of a renewable resource is solved. One should make an effort just sufficient to maintain the maximum sustainable yield; any other effort will in the long run yield less. Under the Schaefer model this effort is $r/2$, half the maximal *per capita* growth rate of the exploited species, and the population size under this level of exploitation will be half its maximal value in the absence of exploitation. This gives the maximum sustainable yield because the natural rate of increase, that provides the sustainable harvest, is maximal when $n = K/2$.

Why then are many species in danger of extinction through overexploitation? Clark (1990) argues forcibly that MSY is a defective concept because it addresses only the benefits of resource exploitation and ignores the costs. He writes: "This fundamental flaw means that MSY is virtually useless for descriptive theories of renewable resource exploitation." To explain this point I shall consider two economic models, the open-access fishery and the sole-owner fishery, in which costs as well as benefits are considered.

In the open-access model it is assumed that anyone is free to exploit the fishery. Consider the Schaefer model and assume a constant price p per unit of biomass. The revenue from effort E is

$$R = pY = pKE\left(1 - \frac{E}{r}\right) \tag{27}$$

The cost is

$$C = cE \tag{28}$$

The economist H. S. Gordon argued that at equilibrium the total cost must equal the total revenue, $C = R$, so that the equilibrium effort is

$$\hat{E} = r\left(1 - \frac{c}{pK}\right) \tag{29}$$

Cost here means "opportunity cost," including the cost of *not* undertaking the most profitable alternative activity. With this understanding, an effort $E > \hat{E}$ cannot be maintained because total costs exceed total revenues, and some fishermen would withdraw to move to more profitable jobs. On the other hand, if $E < \hat{E}$, then fishermen will be drawn in from other jobs. The equilibrium population size at $\hat{E}$ is $\hat{n} = c/p$.

The critical factor in the open-access fishery is the cost/price ratio, c/p. If this ratio exceeds K there will be no fishing; if it is between K and $K/2$, $\hat{E}$ will be less than the effort needed to give the maximum sustainable yield;

and if it is less than $K/2$, $\hat{E}$ will exceed the MSY effort, that may be considered to be biological overfishing. The problem is that as the price of fish rises and technological innovations reduce costs, there is a continual tendency for the falling cost/price ratio to lead to more severe overfishing.

In the sole-owner fishery it is assumed that the fishing effort is determined by a single individual or institution rather than by a large number of individuals competing with each other. The revenue and cost, R and C, are the same as before, but it may be assumed that the owner tries to maximize $R - C$, the net revenue. To do this solve

$$\begin{aligned}\frac{\mathrm{d}}{\mathrm{d}E}(R-C) &\equiv \frac{\mathrm{d}}{\mathrm{d}E}\left[pKE\left(1-\frac{E}{r}\right)-cE\right] \\ &= pK - 2pK\frac{E}{r} - c = 0\end{aligned} \tag{30}$$

so that

$$\hat{E} = \frac{r}{2}\left(1-\frac{c}{pK}\right) \tag{31}$$

This is half the equilibrium effort under the open-access model and never leads to biological overfishing. However, the model in this simple form has not incorporated any discount for future rewards at the expense of current rewards. If this were done it would lead to a higher fishing effort, and the open-access fishery can be shown to be equivalent to a sole-owner fishery with complete discounting for future rewards, that is to say, considering only current rewards and ignoring future rewards.

To illustrate the importance of discounting future rewards in a sole-access fishery, consider the example of harvesting the Antarctic blue whale by a single organization with exclusive rights to hunt this species. Whaling scientists have estimated that the carrying capacity of the environment for blue whales in the absence of hunting is $K = 150{,}000$ and that the maximum sustainable yield is MSY = 2000 whales per year (corresponding to $r \cong 0.05$); the standing population under the MSY policy is 75,000. If a blue whale is worth $10,000, the MSY policy will produce an annual income of $20 million. But if the organization kills all the 75,000 blue whales this year it will realize a lump sum of $750 million, producing an annual income if invested at 5% interest of $37.5 million, and an income if invested at only 3% of $22.5 million. There is a strong economic incentive on the organization to realize their capital (the whale population) and invest the proceeds in some more profitable endeavor. The problem is that whales, because of their low population growth rate, produce a low return on capital.

In conclusion there are two main reasons why economic considerations may lead to overexploitation of a natural resource. In a monopoly situation sole-owners of a resource stock tend to view it as a capital asset that is expected to earn dividends at the current rate; otherwise the owner would dispose of the stock. On the other hand, in an open-access situation in which the resource is the common property of a group of competing users, each of these users will consider only his own interests and will fail to take into account the costs that his actions may impose on other users, that may well lead to overexploitation.

FURTHER READING

Stephens and Krebs (1986) and Krebs and Kacelnik (1991) review foraging theory. Houston et al. (1988) and Mangel and Clark (1988) explore stochastic dynamic programming. Clark (1990) is an excellent account of resource management.

EXERCISES

1. In Best and Bierzychudek's study of bumblebees and foxgloves, the flowers on an inflorescence were given a score from 1 through 10, with 1 at the bottom and 10 at the top. A bumblebee's net energy intake from visiting n flowers was calculated as:

$$R(n) = \frac{P\sum_{i=1}^{n} E(i) - E_{\mathrm{b}} - PE_{\mathrm{w}}(n-1) - PnE_{\mathrm{f}} - E_{\mathrm{e}}(1-P)}{T_{\mathrm{b}} + PT_{\mathrm{w}}(n-1) + PnT_{\mathrm{f}} + T_{\mathrm{e}}(1-P)}$$

where

P = proportion of nonempty inflorescences = 0.375
$E(i)$ = caloric reward for ith flower position = 20.19 – 1.70 i
E_{b} and T_{b} are energy and time flying between plants:
E_{b} = 0.09 cal, T_{b} = 4.4 sec
E_{w} and T_{w} are energy and time flying between two flowers on same plant:
E_{w} = 0.07 cal, T_{w} = 3.3 sec
E_{f} and T_{f} are energy and time emptying a full flower:
E_{f} = 0.02 cal, T_{f} = 14.7 sec

E_e and T_e are energy and time checking an empty flower:
$E_e = 0.01$ cal, $T_e = 8.9$ sec

Justify the formula for $R(n)$. Calculate $R(n)$ for $n = 1, 2, \ldots, 10$ and hence find the value of n that maximizes $R(n)$. Compare $R(n)$ for different values of n with $[E(n) - E_w]/(T_f + T_w)$, the marginal value, and determine whether the criterion of equating R to the marginal value leads to the optimal strategy found above. The bees actually left after visiting on average 4.55 flowers; are they doing the right thing?

2. Derive Equation 8 from Equations 6 and 7, and hence obtain Equation 9. What is the constant of proportionality? When there are only two prey types, show that Equation 11 gives the condition under which only the first type should be taken. Hence obtain the condition given in Equation 16 for only large prey to be brought back to the nest under central-place foraging.

3. The following are the estimated utilities of 0, 1, 2, … , 8 seeds for the hungry bird shown in Figure 6.3b: {0, 0.05, 0.09, 0.18, 0.50, 0.55, 1.00, 1.05, 1.70}. The bird is offered the following choice: (a) throw a coin and get 2 or 7 seeds according to whether it is tails or heads; (b) throw the coin twice and get 1, 4, or 8 seeds according to whether there are 0, 1, or 2 heads. Which should the bird prefer?

4. Consider the patch-foraging model discussed in the text (Equations 20–23). Write a program to find the optimal habitat given energy state x and time t for $x = 1$ through 5 and $t = 99$ back to 80. At what time does the optimal strategy become insensitive to the time horizon?

5. Verify by a local stability analysis that the nonzero equilibrium under the Schaefer model is always stable. Derive the formulas for the optimal effort, the equilibrium population size at this point, and the maximum sustainable yield.

6. It has been estimated that the Antarctic fin whale has a carrying capacity (in the absence of exploitation) of 400,000 whales with a maximum annual growth rate of $r = 0.08$. Suppose that under open-access an equilibrium would be reached at a population size of 40,000 whales. What would be the equilibrium population size under a sole-owner fishery with no discount for future rewards? Find the sustainable yield under these two conditions, and compare it with the maximum sustainable yield.

APPENDIX 6.1. PREY CHOICE

The choice of items to include in the diet is determined by the quantity

$$E_j^*\left(1+\sum_{i\neq j}\lambda_i p_i h_i\right)-h_j\sum_{i\neq j}\lambda_i p_i E_i^* \tag{32}$$

The jth prey type should be included in the diet if this quantity is positive, and excluded if it is negative; call this the choice criterion. To investigate which prey types should be taken, we first consider the choice between two prey types. The criterion tells us that neither type should be taken when E_1^* and E_j^* are both negative. Assume that at least E_1^* is positive, define $\rho_j = E_j^*/h_j$ = profitability of prey j, and rank the prey in order of their profitability so that $\rho_1 > \rho_2$. The condition for prey type 1 to be included and for type 2 to be excluded from the diet is

$$\begin{aligned} &E_1^* > 0 \\ &\rho_2 < \frac{\lambda_1 E_1^*}{1+\lambda_1 h_1} \end{aligned} \tag{33}$$

The condition for prey type 2 to be included and for type 1 to be excluded from the diet is

$$\begin{aligned} &E_2^* > 0 \\ &\rho_1 < \frac{\lambda_2 E_2^*}{1+\lambda_2 h_2} \end{aligned} \tag{34}$$

The second part of this condition contradicts the assumption that $\rho_1 > \rho_2$. The condition for both types to be taken is

$$\begin{aligned} &\rho_1 > \frac{\lambda_2 E_2^*}{1+\lambda_2 h_2} \\ &\rho_2 > \frac{\lambda_1 E_1^*}{1+\lambda_1 h_1} \end{aligned} \tag{35}$$

If the prey are ordered by their profitability and at least the first of them has positive profitability, the rule is to take only the first type if

$$\rho_2 < \frac{\lambda_1 E_1^*}{1+\lambda_1 h_1} \tag{36}$$

and to take both types otherwise.

We now turn to the general case with n prey types. The types that should be included in the diet can be determined by the following algorithm suggested by the above result. Rank the prey in order of their profitability defined above so that $\rho_1 > \rho_2 > \cdots > \rho_n$. Add the types to the diet in order of increasing rank until for the first time

$$\rho_{k+1} < \frac{\sum_{i=1}^{k} \lambda_i E_i^*}{1 + \sum_{i=1}^{k} \lambda_i h_i} \tag{37}$$

The first k items are included in the diet, and the rest are excluded. This satisfies the choice criterion for the following reasons.

Define

$$\begin{aligned} A &= \sum_{i=1}^{k} \lambda_i E_i^* \\ B &= 1 + \sum_{i=1}^{k} \lambda_i h_i \end{aligned} \tag{38}$$

For an excluded item with $j \geq k + 1$, the criterion that it be excluded is

$$E_j^* B - h_j A < 0 \tag{39a}$$

or

$$\rho_j < A/B \tag{39b}$$

which is satisfied since $\rho_j \leq \rho_{k+1} < A/B$. For an included item with $j \leq k$, the criterion that it be included is

$$E_j^* (B - \lambda_j h_j) > h_j (A - \lambda_j E_j^*) \tag{40a}$$

or

$$\rho_j > A/B \tag{40b}$$

which is also satisfied.

CHAPTER 7

Frequency-Dependent Selection

In the last two chapters (on life-history evolution and optimal-foraging theory) we saw how the strategy with the highest fitness will evolve through natural selection. We supposed that fitness is determined by factors in the physical and biotic environment extrinsic to the population that is evolving. In discussing life-history evolution we considered the effect of density dependence whereby fitness will decrease as population density increases; but we ignored (except for a passing comment) the effect of frequency dependence whereby the fitness of a strategy may depend on its frequency *relative* to that of other strategies even though *total* population size remains fixed.

Frequency-dependent selection takes two main forms: negative and positive frequency dependence. In negative frequency dependence the fitness of a strategy declines as its relative frequency increases, while in positive frequency dependence fitness is an increasing function of frequency. Negative frequency dependence, which favors rare genotypes, tends to promote genetic variability and is of interest to population geneticists as a potential explanation for genetic polymorphism.

Hori (1993) has reported an interesting example of negative frequency dependence in scale-eating cichlid fish from Lake Tanganyika. The mouths of these fish are asymmetrical, opening to the right in some individuals and to the left in others because of an asymmetry in the joint of the jaw. They are predators on the scales of other fish, that they approach from behind. Right-mouthed individuals always attack the victim's left flank, and left-mouthed ones attack the right flank. Mouthedness is probably determined by a single genetic locus with two alleles, with right-mouthedness dominant over left-mouthedness. Prey are alert to attacks by predators, and when right-mouthed predators are common they are more alert to the possibility of attack from their left than from their right flank, and vice versa. Thus predators of the rarer type, whichever it is, will be more successful than those of the common type and will increase in frequency until the two types are equally successful. In fact the two types are equally frequent in Lake Tanganyika and have remained so over an 11-year period. This is a clear example of negative frequency-dependent selection leading to the maintenance of genetic polymorphism for mouthedness. Other examples are the maintenance of color polymorphism of prey by selective predation on common prey types leading to a selective advantage of rare types (Allen, 1988) and the maintenance of a large number of self-sterility alleles in plants with gametophytic self-incompatibility (in this system, pollen of genotype S_i cannot successfully fertilize any female plant that bears the S_i allele, so that pollen with a rare genotype will have the highest reproductive success).

In this chapter I shall first discuss genetic models of frequency dependence, and I shall then consider situations in which negative frequency dependence leads to an "ideal free distribution" in which the frequencies of individuals adopting different strategies stabilize when the fitness of each of the strategies adopted with nonzero frequency is the same. Situations that can be discussed more naturally in the context of evolutionary game theory will be considered in the next chapter, though there is a close connection between the two chapters, and some topics could equally well have been considered in either of them.

GENETIC ANALYSIS OF FREQUENCY-DEPENDENT MODELS

The simplest discrete-time model of frequency-dependent selection considers a locus with two alleles under random mating and supposes that the fitness of each of the three genotypes depends linearly on its frequency (Table 7.1). The three genotypes might, for example, produce red, yellow, and green snails subject to negative frequency-dependent predation. The frequency of the A_1 allele in the next generation is

$$p' = (w_{11}p^2 + w_{12}pq)/\overline{w} \qquad (1)$$

since A_1A_1 individuals transmit A_1 to all their offspring and A_1A_2 individuals

Table 7.1. A simple population genetic model of frequency-dependent selection.

	Genotype		
	A_1A_1	A_1A_2	A_2A_2
Frequency before selection	p^2	$2pq$	q^2
Fitness	$w_{11} = 1 - sp^2$	$w_{12} = 1 - 2spq$	$w_{22} = 1 - sq^2$
Frequency after selection	$w_{11}p^2/\overline{w}$	$2w_{12}pq/\overline{w}$	$w_{22}q^2/\overline{w}$

Notes: $\overline{w} = w_{11}p^2 + 2w_{12}pq + w_{22}q^2$; p and $q = 1 - p$ are the frequencies of alleles A_1 and A_2.

transmit A_1 to half of their offspring.[1] The change in the gene frequency is

$$\Delta p = p' - p = pq[p(w_{11} - w_{12}) + q(w_{12} - w_{22})]/\overline{w} \qquad (2)$$

Figure 7.1 plots Δp against p for (a) $s = 0.1$ and (b) $s = -0.1$. Under negative frequency-dependent selection ($s > 0$) there is a stable equilibrium at $p = 0.5$, so that genetic polymorphism is maintained; under positive frequency-dependent selection ($s < 0$) the equilibrium at $p = 0.5$ is unstable, and the stable equilibria are at fixation for one of the alleles, so that genetic variability is eliminated. (The graphical stability analysis, analogous to Figures 2.3 and 2.4, is strictly valid only for continuous-time models but can be used for discrete-time models under weak selection when the difference equation is nearly equivalent to a differential equation.)

Under this model, with $s < 0$, the frequencies of the three genotypes at equilibrium are 0.25, 0.5, and 0.25. The excess of heterozygotes is determined by the genetic constraint of the Hardy–Weinberg law; heterozygotes are less fit at equilibrium because they are more frequent than either of the homozygotes, but there is no way in which selection can reduce their frequency because segregation occurs each generation. If the color polymorphism were determined by a locus with three alleles, for example, with red dominant to yellow dominant to green, and with fitness declining linearly with frequency in each phenotypic class, the equilibrium would be with all three phenotypes at the same frequency of 0.33 because selection can act on the three allele frequencies in such a way that the three phenotypic classes are equally abundant and therefore equally fit.

The theory is greatly simplified by assuming asexual inheritance in that like begets like. This assumption also avoids artificial consequences of

[1] In a discrete-time model p' denotes the gene frequency in the next generation; in a continuous-time model it would denote the rate of change of gene frequency.

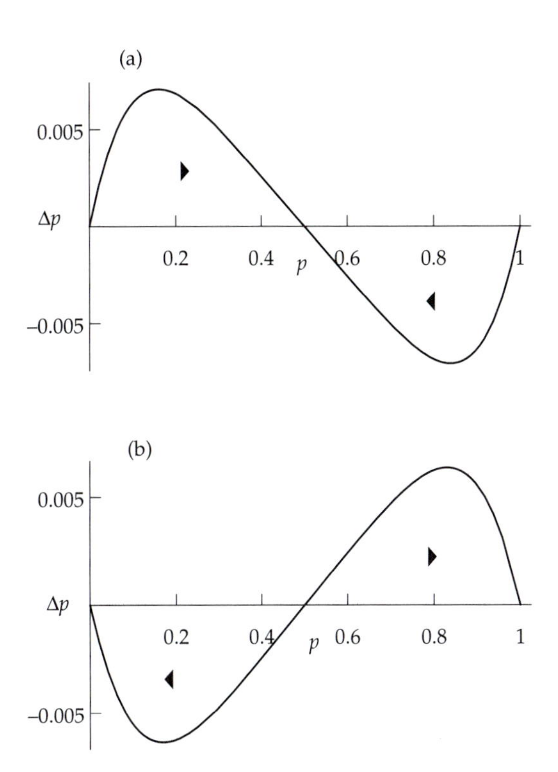

Figure 7.1. Stability analysis for a population genetic model of (a) negative and (b) positive frequency-dependent selection.

genetic constraints in sexual systems, but thought must be given to whether the result predicted by the asexual model is achievable in a sexual system. In general, suppose that individuals belong to one out of k different types (strategies). Write n_i for the number of i-strategists at a particular time and p_i for their relative frequency. The fitness of an i-strategist is w_i, a function of the p_i's, and the average fitness in the population is

$$\overline{w} = \sum_{i=1}^{k} w_i p_i \tag{3}$$

We now consider the evolution of the system in continuous time, following the treatment of Taylor and Jonker (1978). Under weak selection the continuous-time analogue of the discrete-time model

$$n_i(t+1) = w_i n_i(t) \tag{4a}$$

is

$$dn_i/dt = (w_i - 1)n_i \tag{4b}$$

Hence

$$\frac{dp_i}{dt} = \frac{d(n_i/\Sigma n_j)}{dt} = (w_i - \overline{w})p_i \tag{5}$$

At equilibrium (with $dp_i/dt = 0$), either

$$p_i = 0 \qquad \text{or} \qquad w_i = \overline{w} \tag{6}$$

Thus all strategies present at equilibrium with nonzero frequency must have the same fitness. The intuitive reason is that under frequency dependence the frequencies of the different types will be adjusted until they have the same fitness. This fact can be used to find all possible equilibria.

The local stability of these equilibria can be investigated in the usual way by finding the eigenvalues of the Jacobian matrix. One can allow for the fact that relative frequencies must sum to unity by defining p_k as a function of the other frequencies

$$p_k = 1 - p_1 - p_2 - \cdots - p_{k-1} \tag{7}$$

After substitution for p_k in the w_i's and in $\overline{w}$ from this relationship, the Jacobian matrix is the square matrix of order $k - 1$ whose ijth element (for $i, j = 1, \ldots, k - 1$) is

$$\frac{\partial}{\partial p_j}(w_i - \overline{w})p_i \tag{8a}$$

evaluated at the equilibrium. Alternatively, the ijth element of the Jacobian matrix can be calculated, without substitution for p_k, from the formula

$$\frac{\partial}{\partial p_j}(w_i - \overline{w})p_i - \frac{\partial}{\partial p_k}(w_i - \overline{w})p_i \tag{8b}$$

in which p_k is treated as an implicit function of the other frequencies.

For example, suppose that p_1, p_2, and p_3 are the frequencies of red, yellow, and green snails having fitnesses

$$w_i = 1 - sp_i \tag{9}$$

so that

$$\begin{aligned} \overline{w} &= 1 - s(p_1^2 + p_2^2 + p_3^2) \\ w_i - \overline{w} &= s(p_1^2 + p_2^2 + p_3^2 - p_i) \end{aligned} \tag{10}$$

There are seven possible equilibria satisfying Equation 6: an equilibrium with all three morphs present at the same frequency of 0.33, three equilibria with two morphs present at a frequency of 0.5 and the third absent, and three monomorphic equilibria. The eigenvalues of the Jacobian matrix calculated from Equation 8 are both $-\frac{1}{3}s$ evaluated at the equilibrium with all three morphs present, so that it is stable if $s > 0$ (negative frequency dependence); they are both s for any of the monomorphic equilbria, that are therefore stable if $s < 0$ (positive frequency dependence); and they are $\frac{1}{2}s$ and $-\frac{1}{2}s$ for the equilibria with two morphs present, that are therefore never stable.

A third type of frequency dependence is encountered in models for host–parasite interactions, that resemble prey–predator models and of which the following is a simple example (Seger, 1988). There is one host species and one parasite species, both haploid. Generations are discrete and nonoverlapping, and the two species reproduce synchronously. Host and parasite individuals encounter each other randomly at rates proportional to their relative frequencies. The host has a resistance gene with two alleles (r_1 and r_2) with frequencies p and $1 - p$, while the parasite has a virulence gene with two alleles (v_1 and v_2) with frequencies P and $1 - P$, such that r_1 hosts are more resistant to v_2 than to v_1 parasites, while r_2 hosts are more resistant to v_1 than to v_2 parasites. The fitnesses of r_1 and r_2 hosts are modeled as

$$\begin{aligned} w_1 &= 1 - sP \\ w_2 &= 1 - s(1 - P) \end{aligned} \tag{11a}$$

while the fitnesses of v_1 and v_2 parasites are modeled as

$$\begin{aligned} W_1 &= 1 - t(1 - p) \\ W_2 &= 1 - tp \end{aligned} \tag{11b}$$

where s and t are selection pressures. Thus selection depends on the gene frequency in the other species rather than in one's own species, the parasite gene being selected to match the host gene, and the host gene being selected to avoid matching that of the parasite. There is no mutation in the host, but v_1 mutates to v_2 and vice versa in the parasite at rate m. The recurrence relations for the two gene frequencies are

$$\begin{aligned} p' &= w_1 p / \overline{w} \\ P' &= W_1 P / \overline{W} + m(1 - 2P) \\ \overline{w} &= 1 - spP - s(1 - p)(1 - P) \\ \overline{W} &= 1 - tp(1 - P) - t(1 - p)P \end{aligned} \tag{12}$$

(The assumption that the effects of selection and mutation are additive is justified if they are both weak forces.)

Because of the symmetry of the model there is an internal equilibrium at $\hat{p} = \hat{P} = 0.5$. If the mutation rate m is above a critical value this equilibrium is stable, but if the mutation rate is below this value, as is more likely, the internal equilibrium becomes unstable, and the system ultimately converges to a stable limit cycle centered around it; in the extreme case with $m = 0$ the system spirals outward toward the boundaries. Thus this type of frequency-dependent selection, that is important in some current models of sexual selection and the maintenance of sex (see Chapters 11 and 12), maintains genetic variability in both species, but there is a tendency for cyclical dynamics because of the time delay generated by a prey–predator interaction.

IDEAL FREE DISTRIBUTIONS

Consider a group of animals foraging in a patchy environment. The patches may differ in quality, and the reward to an animal foraging in a particular patch may depend on the number of competitors also foraging there as well as on its intrinsic quality. The reward to an animal foraging in the ith patch can be represented as $w_i(n_i)$, depending both on the patch and on n_i, the number of animals in it. We shall suppose that w_i is a *decreasing* function of n_i since the animals are competing with each other within each patch; in other words we assume that there is negative frequency-dependent selection. How should the animals distribute themselves between the patches if they are free to move between them without penalty?

The answer is that at equilibrium they should be distributed so that the reward in every patch in which there are any animals is the same:

$$w_1(n_1) = w_2(n_2) = \cdots = C \quad \text{for all } i \text{ with } n_i > 0 \tag{13a}$$

Suppose, on the contrary, that $w_1(n_1) < w_2(n_2)$. Then it will pay animals to move from the first to the second patch, thus increasing the reward in the first patch and decreasing it in the second, until the rewards in the two patches are the same. This equilibrium will be stable if movement occurs in continuous time so that overshoot does not occur. The reward in an empty patch must be less than or equal to the constant reward in the non-empty patches; otherwise it would pay animals to move into it:

$$w_i(n_i) \le C \quad \text{if } n_i = 0 \tag{13b}$$

Fretwell and Lucas (1970) called this the *ideal free distribution*, since the assumptions are that the foragers act in an *ideal* manner to maximize their reward and are *free* to move between patches without hindrance; it would exclude, for example, the case in which a strong animal can physically exclude weaker competitors from the best patch. As Milinski and Parker (1991) observe, the same idea occurred at about the same time to several workers, and has since been applied to many other situations in which there is negative frequency dependence so that the fitness of an individual adopting the ith strategy (in the above case, foraging in the ith patch) is a decreasing function of the frequency of other individuals adopting the same strategy. In this general situation one would expect the frequencies of individuals adopting different strategies to equilibrate so that the fitness of each of the strategies adopted with nonzero frequency is the same. This may occur either by individuals learning the rewards from different strategies (as one might expect for foraging in a patchy environment) or by being genetically programmed to adopt a particular strategy as we assumed in the previous section; the model discussed there will also serve in the present context. In this section I shall first describe the original application of the ideal free distribution to foraging in a patchy environment, and I shall then discuss extensions of the model to competition between males for females.

Foraging in a patchy environment

To proceed further with the analysis of foraging in a patchy environment we must specify the reward functions w_i. The simplest model is the continuous-input model in which resource items are added continuously into each patch, at rate q_i into the ith patch, and are consumed immediately, being divided equally between the individuals in the patch. The reward function is

$$w_i(n_i) = q_i/n_i \tag{14}$$

This model is appropriate for many experimental tests of ideal-free-distribution theory, and it also arises in some field situations, such as fish competing for food drifting downstream or waders competing for worms rising to the surface of the mud (Sutherland and Parker, 1985). For the reward in different patches to be the same it is clear that n_i must be proportional to q_i:

$$n_i \propto q_i \tag{15}$$

This is called the input-matching rule: at equilibrium the number of competitors in each patch matches the input to that patch. In a well-known test of this rule, Milinski (1979) put six sticklebacks into a tank, into which waterfleas were introduced as prey at a rate of one waterflea every four seconds at one end of the tank and at twice that rate at the other end. After about four minutes, during which the fish sampled the two feeding stations, they equilibrated in the predicted ratio, with two fish at the less profitable end and four fish at the other end.

Sutherland (1983) has considered an alternative model in which competition occurs not by depletion of the prey population, but by interference between predators that reduces their searching efficiency. This model is appropriate for a situation like Oystercatchers searching for mussels on the seashore. Suppose that the density of prey in the ith patch is q_i and that a predator searches and finds all the prey in an area a_i in each second of searching time. Each prey item found requires h seconds (handling time) to consume (for example, opening and eating a mussel). Thus in S seconds of searching time a predator finds $a_i q_i S$ prey items in a total time $S + h a_i q_i S$ so that the gain rate in the ith patch is

$$w_i = a_i q_i / (1 + h a_i q_i) \tag{16}$$

To model the effect of interference, suppose that

$$a_i = Q n_i^{-m} \tag{17}$$

where m is a measure of the degree of interference. In Chapter 3 we saw that Hassell and Varley (1969) found that this was a good description of interference of insect parasitoids with $m = 0.5$ (see Equation 3.48); $m = 0$ means no interference, and $m = 1$ is very strong interference. Thus

$$w_i(n_i) = \frac{Q q_i n_i^{-m}}{1 + Q h q_i n_i^{-m}} \tag{18}$$

For this to be the same in all patches requires that $q_i n_i^{-m}$ is constant, so that

$$n_i \propto q_i^{1/m} \tag{19}$$

With very high interference ($m = 1$) this gives the input matching rule (Equation 15), but as the degree of interference decreases to more plausible levels the predators become more aggregated in the more productive patches.

It has been assumed that all individuals have the same competitive ability, but this is unlikely to be the case. Parasitized sticklebacks forage less actively than healthy ones, and individual oystercatchers differ in the degree of interference they experience. The effects of individual differences in competitive ability have been considered by Sutherland and Parker (1985) and by Parker and Sutherland (1986). Suppose that there are p types of predator, differing in competitive ability and that there are n_{ij}

individuals of type j in patch i. Write $w_{ij}(n_{i1}, n_{i2}, \ldots, n_{ip})$ for the reward to a type-j individual in the ith patch, depending on the numbers of all the different types of competitors in the same patch. The conditions for an ideal free distribution, that one expects to hold at equilibrium, are that the reward to any type of individual must be the same in all patches in which it occurs and that this reward would not be exceeded in any patch where the type does not occur. This condition must hold separately for each of the types, but different types of individual may receive different rewards since they are not free to change type. In other words, the analogue of Equation 13 is

$$\begin{aligned} &w_{1j}(n_{11}, \ldots, n_{1p}) = w_{2j}(n_{21}, \ldots, n_{2p}) = \cdots = C_j \\ &\qquad\qquad\qquad\qquad\text{for all } i \text{ with } n_{ij} > 0 \\ &w_{ij}(n_{i1}, \ldots, n_{ip}) \le C_j \quad \text{if } n_{ip} = 0 \end{aligned} \tag{20}$$

These conditions must hold separately for each predator type, $j = 1, \ldots, p$.

I shall first consider a simple example of the continuous-input model with two types of predator, the second being twice as efficient as the first, and two patches, the rate of input of prey being 2 per minute in the first and 4 per minute in the second. The reward functions are

$$\begin{aligned} w_{11} &= 2/(n_{11} + 2n_{12}) \\ w_{12} &= 4/(n_{11} + 2n_{12}) \\ w_{21} &= 4/(n_{21} + 2n_{22}) \\ w_{22} &= 8/(n_{21} + 2n_{22}) \end{aligned} \tag{21}$$

Suppose that there are 60 type-1 and 30 type-2 predators. An ideal free distribution is obtained with each of the types present in the two patches according to the matching rule, the first type distributed 20:40 and the second 10:20. This mimics the matching rule applied to the total number of individuals regardless of type. However, there is also a range of alternate solutions that do not mimic the matching rule applied to the total number, since it makes no difference to the reward functions if x type-2 individuals are moved from one patch to the other as long as $2x$ type-1 individuals are moved in the other direction. For example, {30:30, 5:25} and {10:50, 15:15} are alternative solutions satisfying Equation 20 but not mimicking the matching rule for total numbers. Houston and McNamara (1988) have worked out the most likely solution to be observed in this situation, and they conclude that there is a weak tendency for more efficient animals to aggregate in the more profitable patch, so that the total number of animals in this patch is slightly less than the matching rule predicts. The empirical data are consistent with this prediction. As we shall see in Table 7.2, the tendency for more competitive animals to aggregate in the more profitable patch becomes much stronger under interference competition.

The existence of a range of ideal free solutions in the above example is due to the fact that the ratio of the efficiencies of the two types of predator is the same in both patches. Suppose now that they have the same efficiency in the first patch, while type 2 is twice as efficient as type 1 in the second patch as before. The reward functions are

$$\begin{aligned} w_{11} &= 2/(n_{11} + n_{12}) \\ w_{12} &= 2/(n_{11} + n_{12}) \\ w_{21} &= 4/(n_{21} + 2n_{22}) \\ w_{22} &= 8/(n_{21} + 2n_{22}) \end{aligned} \tag{22}$$

There is a unique ideal free solution with type-1 individuals distributed 40:20 between patches 1 and 2, while all 30 of the type-2 individuals are in patch 2. It is intuitively reasonable that there should be a tendency for the types to aggregate in the patch in which their relative effiency is higher.

I finally consider the interference model with differences in the degree of interference between individuals. I shall describe one of the models analyzed numerically by Parker and Sutherland (1986). There are three patches and 10 types of predator, with 100 individuals of each type. The reward function is

$$w_{ij} = i \times n_i^{-m_{ij}} \tag{23}$$

The first term, i, represents patch quality and means that patch 2 is twice as good as patch 1, and patch 3 three times as good. The second term represents interference; n_i is the total number of individuals in patch i,

$$n_i = \sum_{j=1}^{10} n_{ij} \tag{24}$$

and m_{ij} is the degree of interference experienced by a type-j individual in patch i, defined as

$$m_{ij} = 0.1\bar{j}_i/j \tag{25}$$

where $\bar{j}_i$ is the average type of predator in patch i

$$\bar{j}_i = \sum_{j=1}^{10} jn_{ij}/n_i \tag{26}$$

The idea is that individuals of low type number are less competitive and thus experience more interference. The degree of interference experienced by an individual is inversely proportional to its type number relative to the average type within the patch, and it has an average value of 0.1 within each patch.

Parker and Sutherland (1986) found the ideal free solution by an iterative numerical procedure that is very useful in this type of situation. Starting from some set of initial values, they allowed the individuals within each patch to replicate in proportion to their payoffs and then standardized the numbers so that there were 100 individuals of each type. In other words, they used the recurrence relations

$$n_{ij}(t+1) = \frac{100\, w_{ij} n_{ij}(t)}{\sum_{i=1}^{3} w_{ij} n_{ij}(t)} \tag{27}$$

Iteration was continued until convergence occurred. Parker and Sutherland (1986) describe the simulation as follows:

> A[n iterative] round could be envisaged as a round of reproduction in an evolutionary process, but is probably best considered as a round of sampling and dispersal, with individuals most likely to leave patches in which they experience low payoffs and least likely to leave patches where they experience high payoffs. Either way, the model corresponds qualitatively to a system in which strategies that are more successful are either repeated more often (learning and dispersal) or proliferate at an enhanced rate, until equilibrium is achieved.

The results are shown in Table 7.2. The weak competitors aggregate in the worst patch, the intermediate competitors in the intermediate patch, and the strong competitors in the best patch. The payoffs show that this is an ideal free solution; the weak competitors would do worse if they moved to a better patch. However, the *average* payoff is much lower in the bad than in the good patch, so that one might conclude, if one only had data on the total numbers in the three patches and their average payoffs, that they were not in agreement with ideal free theory. This may explain observations on the feeding behavior of Herring Gulls on a refuse dump (Monaghan, 1980). Adults fed in the main dump where the average intake, the number of gulls, and the number of aggressive encounters were all greater, whereas immature gulls tended to feed in a less profitable secondary area where there were fewer aggressive encounters. This could conform to an ideal free distribution if immature gulls are more susceptible to aggressive encounters than are adults. It is not necessary to postulate that immature birds are forced out of profitable areas by dominant adults.

Table 7.2. Ideal free solution of the competitive interference model with differences in competitive ability defined in Equation 23. Equilibrium numbers are shown in bold type, and their payoffs are shown in ordinary type.

	Patch type					
Predator type	1		2		3	
1	**100**	0.45	**0**	0.14	**0**	0.02
2	**100**	0.67	**0**	0.54	**0**	0.24
3	**0**	0.77	**100**	0.83	**0**	0.56
4	**0**	0.82	**100**	1.04	**0**	0.85
5	**0**	0.85	**100**	1.18	**0**	1.09
6	**0**	0.88	**80**	1.29	**20**	1.29
7	**0**	0.89	**0**	1.37	**100**	1.46
8	**0**	0.90	**0**	1.44	**100**	1.59
9	**0**	0.92	**0**	1.49	**100**	1.71
10	**0**	0.92	**0**	1.54	**100**	1.81
Average payoff:		0.56		1.08		1.62

From Parker and Sutherland, 1986.

Competition between males for females: Dungflies

The reproductive success of a male depends on the number of females he can inseminate. Suppose that there are a number of patches with f_i available unmated females in the ith patch. How should the males distribute themselves between the patches? If there are m_i males in the ith patch, each competing on equal terms for the females, the average success of a male in that patch is f_i/m_i. It will be to the advantage of any male to move to a patch with a higher success rate, until equilibrium is attained when males have the same success rate in all patches. At this point the distribution of males matches that of females, $m_i \propto f_i$. This is an obvious analogue of the matching rule in Equation 15 obtained under the continuous-input model of Equation 14, with competition between predators for prey being replaced by competition between males for females.

A classic example is provided by the study of Parker (1970) on the mating behavior of male dungflies. Female dungflies come to cowpats to lay their eggs, and as the cowpat ages it becomes less attractive to females so that their rate of arrival decreases. Males congregate at pats and compete with each other to mate with the newly arriving females. How long should a male stay at a cowpat, assuming as is usually the case that he arrives at a fresh cowpat? If most males leave quickly, it will pay to wait longer, because the decreased competition more than offsets the lower arrival rate of females; if most males stay a long time it will pay to leave quickly. Thus the negative frequency-dependent factor leading to an ideal free distribution is present, though the strategy of a male dungfly is the choice of a waiting time rather than a spatial patch.

Estimates can be made of the arrival rate of females (f_t) and of the number of males present (m_t) t minutes after a fresh cowpat has appeared. Consider a male who waits T minutes after arrival before moving on to find a new cowpat. To estimate his success one must take into account not only his average number of matings over that time, given by

$$M(T) = \sum_{t=1}^{T} m_t/f_t \qquad (28)$$

but the fact that it takes him about four minutes to find a new cowpat. Thus his success rate is given by

$$w(T) = M(T)/(T+4) \qquad (29)$$

The prediction from ideal-free-distribution theory is that males will adjust their stay times to give a distribution of times such that $W(T)$ is the same for all T. The estimated relationship is shown in Figure 7.2. Agreement with prediction is quite good, though there is a slight drop in the success rate of males staying a very short time. This means that no

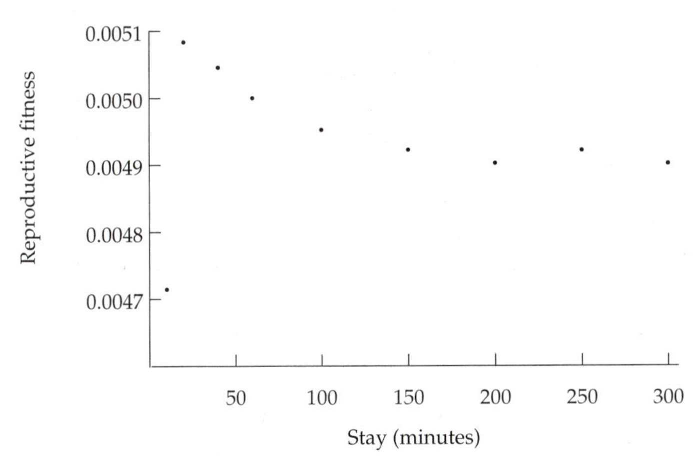

Figure 7.2. Reproductive fitness of male dungflies as a function of length of stay on cowpat. Source: Parker, 1970.

males should adopt this strategy, and in fact very few of them do. (See Curtsinger [1986] for a criticism of these conclusions and the reply by Parker and Maynard Smith [1987].) The mechanism by which males achieve the ideal free distribution is unknown, but it requires a mixture of strategies. Either some males must be long-stayers and some short-stayers, or they must all vary their strategy from time to time. In the first case it is likely that the difference between long-stayers and short-stayers is inherited, so that this would provide an example of the maintenance of genetic variability by frequency-dependent selection.

Competition between males for females: Protandry

> "Throughout the great class of insects the males almost always emerge from the pupal state before the other sex. ... The cause of this difference between the males and females in their periods ... of maturity is sufficiently obvious. Those males ... which in the spring were first ready to breed ... would leave the largest number of offspring; and these would inherit similar instincts and constitutions."
>
> — Charles Darwin, *The Descent of Man, and Selection in Relation to Sex* (1871)

In many insects there is a tendency for males to emerge before females, a phenomenon known as *protandry*.[2] For example, Figure 7.3 shows the distributions of male and female emergence dates in the checkerspot butterfly; males emerge on average about six days before females. Life-history characteristics predisposing a species with separate sexes toward protandry are a restricted breeding season and a tendency for females to mate once only, soon after they emerge from the pupa. The idea is that there is frequency-dependent selection on male mating success, depending on the number of females emerging on a particular day and on the number of males competing to mate with them, and that males will adjust their emergence times until the reproductive success of males emerging on different days equilibrates at the same value. This is an ideal free distribution, except that a distribution in space is replaced by one in time.

I shall now discuss a model for the evolution of protandry developed by Iwasa et al. (1983) and Bulmer (1983c). Let $m(t)$ and $f(t)$ be the rates of emergence of males and females at time t within the breeding season

[2] "Protandry" is derived from two Greek words meaning "male first." In species with separate sexes (the present context) it refers to males emerging before females. In hermaphrodite species it refers to male function preceeding female function; for example, the flowers of foxgloves function first as males and then as females (see Chapter 6), and some shrimp change sex from male to female (see Chapter 10).

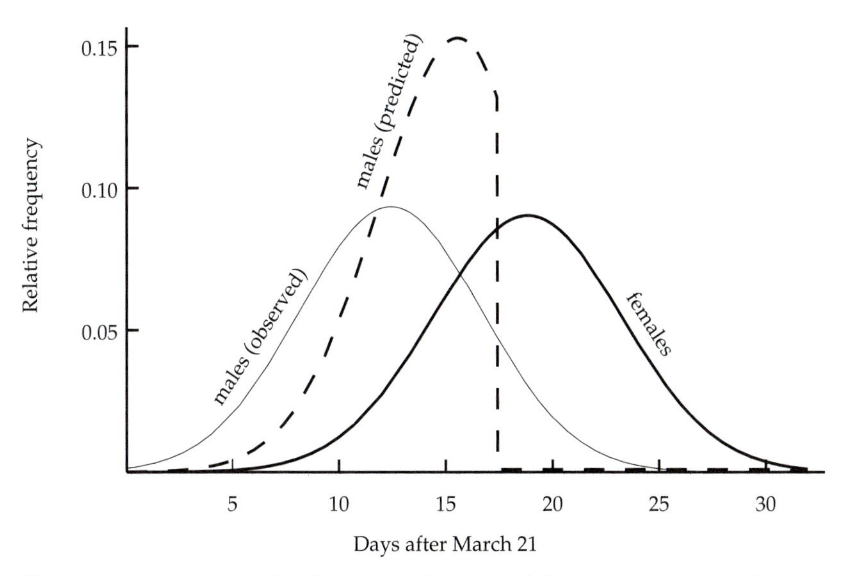

Figure 7.3. Observed distributions of male and female emergence dates in the checkerspot butterfly (solid curves), and predicted male distribution (dashed curve); observed distributions are normal distributions fitted to the observed data. Source: Iwasa et al., 1983.

($a < t < b$), each standardized so that the area under the curve is unity. We suppose that the female emergence curve, $f(t)$, is determined by factors such as the availability of the larval food supply, and we consider how the male emergence curve, $m(t)$, should respond. Suppose that males have constant mortality λ after emergence so that the probability of a male born at time t_1 being alive at some later time t_2 is $\exp[-\lambda(t_2 - t_1)]$. The number of males alive at time t, expressed as a proportion of of the total number of males emerging over the whole season, is

$$m^*(t) = \int_a^t \exp[-\lambda(t-u)]\, m(u)\, \mathrm{d}u \tag{30}$$

A male emerging at time t can expect to mate with

$$w(t) = \int_t^b \exp[-\lambda(u-t)] \frac{f(u)}{m^*(u)} \mathrm{d}u \tag{31}$$

females. (If F females and M males emerge over the whole season, there will be $Ff(u)$ newly emerged females on day u and $Mm^*(u)$ males competing for them. I assume an unbiased sex ratio so that $F = M$; I also assume that the females mate once only immediately after emergence and that they mate at random among the pool of males alive at that time.)

At equilibrium, under the ideal free distribution of emergence times, the reproductive success of males emerging at time t is

$$\begin{aligned} w(t) &= k \quad \text{when } m(t) > 0 \\ w(t) &\le k \quad \text{when } m(t) = 0 \end{aligned} \tag{32}$$

for some constant k; this constant is the expected reproductive success at emergence of any male and must therefore be 1. By equating dw/dt to zero, we find that

$$f(t) = \lambda m^*(t) \quad \text{whenever } m(t) > 0 \tag{33}$$

By differentiating this equation we find that

$$m(t) = f(t) + f'(t)/\lambda \quad \text{or} \quad 0 \tag{34}$$

As long as there are any females emerging early in the season it will be worthwhile for a few males to emerge to mate with them, but a time will come near the end of the season when the reproductive success of a newly emerging male may drop below the critical value making it worthwhile, partly because few females are emerging in comparison with the number of males alive and partly because there is little time left till the end of the season when female emergence stops entirely. Thus we might expect that there is a single switch point t^* at which males cease to emerge, so that

$$\begin{aligned} m(t) &= f(t) + f'(t)/\lambda \qquad t < t^* \\ m(t) &= 0 \qquad\qquad\qquad\quad t > t^* \end{aligned} \tag{35}$$

By integrating $m(t)$ up to t^*, we find that it satisfies

$$F(t^*) + f(t^*)/\lambda = 1 \tag{36}$$

where $F(t)$ is the cumulative distribution function of female emergence times. This equation can be solved numerically to obtain t^*, and one can then check whether the solution obtained satisfies Equation 32; if it does not, one would need to modify the assumption of a single switch point in Equation 35.

Parker and Courtney (1983) developed a very similar model to the above and showed that it predicted accurately the male emergence distribution of the orange tip butterfly in England from data on the female emergence distribution and male mortality. In particular, the male distribution had an earlier peak and a smaller variance than the female distribution, as predicted by theory. However, Iwasa et al. (1983) did not find as good agreement with prediction from data on the checkerspot butterfly in

California (Figure 7.3). The male distribution had an earlier peak as predicted (the phenomenon of protandry), but it did not have the reduction in variability or the abrupt threshold predicted. Bulmer (1983c) suggested that there is a genetic constraint that allows the male emergence distribution to adjust its mean but not its variance or its shape in response to selection. Recently, Iwasa and Haccou (1994) have extended the model to allow for stochastic variability in the female emergence distribution about which males have imperfect cues. This introduces an element of bet-hedging into the system that will select for greater variability in the male emergence distribution and might explain the failure to observe the reduction in variance predicted by the deterministic model.

FURTHER READING

Clarke and Partridge (1988) have edited a symposium reviewing frequency-dependent selection. Milinski and Parker (1991) review ideal free distributions.

EXERCISES

1. (a) Reproduce the plot of Δp against p in Figure 7.1a.
   (b) Write a program to simulate the recurrence relation in Equation 1 under the model in Table 7.1 with $s = 0.1$, writing $q = 1 - p$. Use the program to verify that there is a stable equilibrium with $p = 0.5$.

2. Suppose that the color polymorphism is determined by a locus with three alleles (A_1, A_2, and A_3) having frequencies p, q, and r, with A_1 dominant to A_2 dominant to A_3. (A_1A_1, A_1A_2, and A_1A_3 snails are red; A_2A_2 and A_2A_3 snails are yellow; A_3A_3 snails are green.) Suppose also that fitness declines linearly with frequency in each phenotypic class with coefficient $s = 0.1$. Extend the model in Table 7.1 to this situation and write a program to simulate the recurrence relations for p and q, writing $r = 1 - p - q$, and hence show that there is a stable equilibrium with all three phenotype frequencies equal.

3. Suppose that the three color morphs are determined asexually with the fitness defined in Equation 9. Verify the eigenvalues given in the text, and hence show that the equilibrium with all three morphs present at the same frequency is the unique stable equilibrium under negative frequency dependence.

4. Consider the host–parasite model in Equation 12 with $s = 0.1$, $t = 0.3$, and with (i) $m = 0$, (ii) $m = 0.003$, (iii) $m = 0.01$.
(a) Find the eigenvalues of the equilibria at {0.5, 0.5}, {0, 0}, and {0, 1} and hence determine their stability;
(b) simulate the behavior of the system for the three mutation rates.

5. Find the ideal free distributions for the reward functions in (a) Equation 21 and (b) Equation 22 when there are 60 predators of type 1 and 30 of type 2.

6. Write a program to verify that Table 7.2 gives the ideal free distribution for the model of Parker and Sutherland (1986) described in the text.

7. The average number of matings, $M(T)$ in Equation 28, for male dungflies staying different times T after arrival at a fresh cowpat are given in Table 7.3. The fitness of a male staying T minutes can be estimated as

$$\text{fitness} = \text{mating success}/(T + \tau)$$

where τ is the time taken to find a fresh cowpat after leaving the old one. Plot fitness against staying time for different values of τ (say, 0, 2, 4, 6, and 10 minutes) and see which value best fits the prediction of constant fitness. Direct observation suggests that τ is about four minutes.

Table 7.3. Relationship between mating success of male dungflies and the time they stay at a cowpat.

Male staying time (minutes)	Mating success
10	0.066
20	0.122
40	0.222
60	0.320
100	0.515
150	0.758
200	1.000
250	1.250
300	1.490

From Parker, 1970.

8. Verify the theory leading to Equations 33 and 34.

9. Iwasa et al. (1983) found the following numbers of male and female checkerspot butterflies emerging on days 1 through 31, day 1 being March 22, 1981:

   Males: {1, 1, 6, 1, 5, 4, 15, 18, 26, 25, 17, 33, 18, 36, 23, 21, 17, 8, 8, 6, 1, 1, 1, 0, 1, 0, 1, 0, 0, 0, 1};

   Females: {0, 0, 0, 0, 0, 0, 1, 2, 2, 1, 0, 4, 7, 5, 11, 14, 14, 17, 11, 12, 9, 5, 10, 14, 8, 3, 2, 1, 0, 0, 1}.

   Males had a daily mortality of 0.137. Find the mean and variance of these distributions. Show that these distributions are approximately normal. Assuming that the female distribution is normal, use Equation 36 to find the predicted truncation point for male emergence. Hence use Equation 35 to plot the predicted male emergence distribution and find its mean and variance by numerical integration.

CHAPTER 8

Evolutionary Game Theory

Maynard Smith and Price (1973) set out to answer the question of why conflicts between animals of the same species, in particular between males for mates or territory, are of "limited war" type, involving inefficient weapons or ritualized tactics that seldom cause serious injury to either contestant, rather than "total war." For example, in many snakes the males fight each other by wrestling without using their fangs. Prior to Maynard Smith and Price's paper, the accepted answer for the conventional nature of contests was that otherwise many individuals would be injured, that would militate against the survival of the species. But Maynard Smith and Price were not satisfied with this group-selectionist explanation, and they proposed an alternative explanation based on individual selection by applying in a biological context ideas from the theory of games, that was originally developed by economists and social scientists as a paradigm for rational behavior in human conflicts of interest. These ideas have been actively developed since then and have found applications in many areas of animal behavior. Before discussing them in detail I shall outline some of the concepts of classical game theory and contrast them with those of evolutionary game theory.

CLASSICAL AND EVOLUTIONARY GAME THEORY

An n-person game is a contest involving n players. I shall here only consider the simplest case of two-person games. The usual first step in the analysis is to reduce the game to normal form by writing down all the possible strategies for each of the players. A strategy for a player is a description of the decisions he will make at all the possible situations that can arise in the game; it will determine what he does at any point in the game whatever the other player does or whatever the outcome of chance events. A choice of strategies by each of the players completely determines the course and outcome of the game and the payoff made to each of them at the end.

Consider for example the game of simplified poker played between two players with a deck of just two cards, an Ace and a King. Each player puts two dollars into the pot and one of them (A) is nominated as the dealer and the other (B) as the recipient. A deals himself a card at random and after seeing it can either fold, in which case B wins, or raise by putting another two dollars into the pot. In the latter case, B can either fold, in which case he loses, or he can put another two dollars into the pot and demand to see the remaining card, winning if it is the Ace and losing if it is the King. The winner takes the contents of the pot.

The dealer has four strategies: (1) always raise, (2) raise with Ace and fold with King, (3) fold with Ace and raise with King, (4) always fold. The recipient has only two strategies: (1) stay in if A raises, (2) fold if A raises. If A is equally likely to deal himself an Ace or a King, the payoff matrix is

$$\begin{bmatrix} (0,0) & (2,-2) \\ (1,-1) & (0,0) \\ (-3,3) & (0,0) \\ (-2,2) & (-2,2) \end{bmatrix} \tag{1}$$

In this matrix A chooses a row and B a column. The pair of numbers in each cell are the payoffs to A and B resulting from the choice of a particular pair of strategies. (This is called the bimatrix representation of the game.) For example, if A chooses strategy 1 (always raise), while B chooses strategy 2 (fold if A raises), then A wins and B loses the two dollars that B put into the pot. It will be seen that B's payoff is always the negative of A's payoff; this arises because the winner gains the money that the loser has put into the pot. This property defines a *zero-sum game*. Write i and j for the strategies adopted by A and B, with $i = 1, \ldots, 4$ and $j = 1$ or 2, and write $E_A(i, j)$ and $E_B(i, j)$ for their respective payoffs. The zero-sum property is that

$$E_A(i,j) + E_B(i,j) = 0 \tag{2}$$

Many parlor games and military war games are zero-sum since the interests of the two players are directly opposed. A prey–predator interaction might be modeled as a zero-sum game, but biological games involving interactions between members of the same species are not usually zero-sum. Another important distinction between economic and evolutionary game theory is that in economic and social science applications the payoffs are expressed in money or more generally in the economic concept of utility while in biological applications the payoffs are changes in fitness to the two animals as a result of their interaction.

Having enumerated the strategies of the two players and written down the payoff matrix, the final step is to solve the game by finding the "best" strategies for the two players, which may be pure or mixed. (A mixed strategy, say $\mathbf{p} = \{p_1, p_2, \dots\}$, chooses different pure strategies with different probabilities, the ith strategy being chosen with probability p_i. The pure strategy i is a special case of a mixed strategy with $p_i = 1$ and $p_j = 0$ for $j \neq i$. We shall use the symbols $\mathbf{p}$ and $\mathbf{q}$ for alternative mixed strategies.) The concept of a solution has a different meaning in economic and evolutionary game theory. The social scientist wants to find the strategies that two "rational" players should adopt. The evolutionary biologist considers how natural selection will act on players adopting different strategies on the assumption that the "fittest" strategy will eventually evolve. This difference accounts for the greater success of game theory in evolutionary biology than in the social sciences because the concept of fitness is less nebulous than that of rationality. However, one concept important in both disciplines is that of a *Nash equilibrium* (Nash, 1951).

If A knows that B will play the (pure or mixed) strategy $\mathbf{p}_B$, he should play the strategy $\mathbf{p}_A$ that maximizes his payoff given that knowledge; this is called the "best reply" to $\mathbf{p}_B$. (It may happen that there are several equally good replies, in which case each of them is called a best reply.) Likewise if B knows that A will play the strategy $\mathbf{p}_A$, he should choose the strategy $\mathbf{p}_B$ that maximizes his payoff. A pair of strategies $\{\mathbf{p}_A, \mathbf{p}_B\}$, one for A and the other for B, is a Nash equilibrium if $\mathbf{p}_A$ is a best reply to $\mathbf{p}_B$ and vice versa:

$$\begin{aligned} E_A(\mathbf{p}_A, \mathbf{p}_B) &\geq E_A(\mathbf{q}_A, \mathbf{p}_B) \qquad \text{for all } \mathbf{q}_A \\ E_B(\mathbf{p}_A, \mathbf{p}_B) &\geq E_B(\mathbf{p}_A, \mathbf{q}_B) \qquad \text{for all } \mathbf{q}_B \end{aligned} \tag{3}$$

where

$$\begin{aligned} E_A(\mathbf{p}_A, \mathbf{p}_B) &= \sum_{i,j} p_{Ai} p_{Bj} E_A(i,j) \\ E_B(\mathbf{p}_A, \mathbf{p}_B) &= \sum_{i,j} p_{Ai} p_{Bj} E_B(i,j) \end{aligned} \tag{4}$$

If $\mathbf{p}_A$ is the unique best reply to $\mathbf{p}_B$, so that there are no alternative best replies, and likewise $\mathbf{p}_B$ is the unique best reply to $\mathbf{p}_A$, then $\{\mathbf{p}_A, \mathbf{p}_B\}$ is called a strict Nash equilibrium:

$$\begin{aligned} E_A(\mathbf{p}_A, \mathbf{p}_B) &> E_A(\mathbf{q}_A, \mathbf{p}_B) \qquad \text{for all } \mathbf{q}_A \neq \mathbf{p}_A \\ E_B(\mathbf{p}_A, \mathbf{p}_B) &> E_B(\mathbf{p}_A, \mathbf{q}_B) \qquad \text{for all } \mathbf{q}_B \neq \mathbf{p}_B \end{aligned} \tag{5}$$

Otherwise $\{\mathbf{p}_A, \mathbf{p}_B\}$ may be called a weak Nash equilibrium.

In a parlor game such as simplified poker, the rationale of the Nash equilibrium is that an individual who plays the game many times, both as dealer and recipient, against a number of opponents, will get to know how his opponents tend to play and will adapt his own play accordingly; if all players do this, an equilibrium reached by this procedure must be a Nash equilibrium. In an evolutionary context we suppose that many games are played between different pairs of animals with strategies determined by their genotypes so that at equilibrium the fittest strategy must predominate.

In simplified poker with payoff matrix (1) there is no Nash equilibrium if the players are restricted to pure strategies, that is to say, if A must always choose the same row and B the same column. To find an equilibrium we must introduce a mixed strategy in which a player plays different pure strategies with different probabilities. The Nash equilibrium using mixed strategies is

$$\begin{aligned} \mathbf{p}_A &= \{0.33, 0.67, 0, 0\} \\ \mathbf{p}_B &= \{0.67, 0.33\} \end{aligned} \tag{6}$$

(This means that A uses his strategies 1 and 2 with probabilities 0.33 and 0.67, respectively, while B uses his strategies 1 and 2 with probabilities 0.67 and 0.33.) If the players use this strategy mix, A will on average win and B will lose two-thirds of a dollar per game. (To make this a fair game we may suppose that the players initially toss a coin to decide which of them shall be the dealer.) If A deviates from this strategy, then B could restrict his own loss, and A's gain, to less than one third of a dollar by an appropriate change of strategy, and vice versa. It makes sense that A should never use his third or fourth strategies that are clearly suboptimal. It also makes sense in a game of poker that A should bluff some of the time, but not all the time, and likewise that B should call his bluff some of the time, but not all the time. Note also that if A plays $\mathbf{p}_A$, then B will lose two-thirds of a dollar whatever strategy mix he uses, and likewise that if B plays $\mathbf{p}_B$ then A will gain two-thirds of a dollar whatever strategy mix he uses, provided that it is confined to his first two pure strategies. Thus Equation 6 is only a weak Nash equilibrium. It is typical that a Nash equilibrium using mixed strategies is only weak, whereas a Nash equilibrium using pure strategies is usually strict.

The problem with the Nash equilibrium concept is that there may be many equilibria, some of which are less obviously appealing than others as solutions, particularly in large, nonzero-sum games. The social scientist therefore seeks additional criteria for rational behavior. The evolutionary biologist travels a different route, since he is concerned not with rationality but with how natural selection might mold organisms to maximize their fitness, and seeks to find stable equilibria. A strict Nash equilibrium is clearly stable, but it is necessary to find an additional criterion for the stability of a weak Nash equilibrium.

Simplified poker is an asymmetric game in which the two players have different roles (dealer and recipient), with different strategy sets and different payoffs. In symmetric games, on the other hand, there is no role asymmetry beteween the players, so that they have the same strategy set and their payoffs are symmetric; at equilibrium each of the two players will play the same (pure or mixed) strategy, since there is no distinction between them that would allow them to play different strategies, and the relevant Nash equilibrium is a symmetric Nash equilibrium. Write $E(\mathbf{p}, \mathbf{q})$ for the payoff to the row player in a symmetric game when he plays the (pure or mixed) strategy $\mathbf{p}$ and his opponent plays $\mathbf{q}$; by symmetry the payoff to the column player is $E(\mathbf{q}, \mathbf{p})$. A symmetric Nash equilibrium $\mathbf{p}$ is a strategy that is the best reply to itself:

$$E(\mathbf{p},\mathbf{p}) \geq E(\mathbf{q},\mathbf{p}) \qquad \text{for all } \mathbf{q} \tag{7}$$

A strict symmetric Nash equilibrium is the unique best reply to itself:

$$E(\mathbf{p},\mathbf{p}) > E(\mathbf{q},\mathbf{p}) \qquad \text{for all } \mathbf{q} \neq \mathbf{p} \tag{8}$$

Equation 7 provides a necessary condition for a stable equilibrium of a symmetric evolutionary game, and Equation 8 provides a sufficient condition.

In the child's game of stone–scissors–paper, for example, each player simultaneously announces a choice of one of these objects. If the players have chosen different objects the loser pays the winner one dollar, with the rule that stone blunts scissors, scissors cut paper, but paper wraps stone; if they have chosen the same object they each pay ten cents to a charity. The payoff matrix to the row player is shown in Table 8.1. Write $a_{ij} = E(i, j)$ for the ijth element of this matrix, the payoff to the row player when he plays the pure strategy i and his opponent plays strategy j; by symmetry, the payoff to his opponent is a_{ji}, so that it is unnecessary to specify the payoff matrix to the column player separately. The unique symmetric Nash equilibrium is $\mathbf{p} = \{0.33, 0.33, 0.33\}$, choose the three objects at random with equal probabilities. Since this is a mixed strategy it is only a weak Nash equilibrium.

In the rest of this chapter I shall review in turn the main properties of symmetric, asymmetric, and many-person evolutionary games. As an

Table 8.1. Payoff matrix to the row player in the game of stone–scissors–paper.

	Stone	Scissors	Paper
Stone	−0.1	1	−1
Scissors	−1	−0.1	1
Paper	1	−1	−0.1

introduction to the study of symmetric, two-person evolutionary games I shall consider the hawk–dove game which, though naive in its assumptions, has become one of the paradigms of evolutionary game theory.

THE HAWK–DOVE GAME

Imagine that two animals of the same species are contesting a resource of value V, such that the Darwinian fitness of an individual obtaining the resource would be increased by V. Suppose that individuals in such contests adopt one of two strategies:

Hawk: escalate and continue until injured or until opponent retreats

Dove: display; retreat at once if opponent escalates

If both opponents escalate it is assumed that one of them is eventually injured and retreats, and that injury reduces fitness by a cost C.

The payoffs to the row player in this contest, depending on the strategies adopted by the two players, can be written in the payoff matrix in Table 8.2. (A hawk playing another hawk has an even chance of gaining the resource or of injury; a hawk playing a dove always wins the resource while a dove playing a hawk never does; and a dove playing another dove has an even chance of gaining the resource without injury to either.) It is assumed that there is no difference between the two players that could

Table 8.2. Payoff matrix for the hawk–dove game.

	Hawk	Dove
Hawk	$\frac{1}{2}(V-C)$	V
Dove	0	$\frac{1}{2}V$

favor one of them in an escalated contest or that could be used as a criterion for adopting different strategies.

The nature of the game depends qualitatively on whether the value of the resource is greater or less than the cost of injury. I shall use the following numerical examples to illustrate this difference. When $V = 4$ and $C = 2$, the payoff matrix is

$$\begin{bmatrix} 1 & 4 \\ 0 & 2 \end{bmatrix} \tag{9}$$

whereas when $V = 2$ and $C = 4$, the payoff matrix is

$$\begin{bmatrix} -1 & 2 \\ 0 & 1 \end{bmatrix} \tag{10}$$

The elements of the payoff matrix represent the increase in Darwinian fitness resulting from the contest; absolute fitness is given by adding the baseline fitness w_0 to the relevant entry in the matrix. Suppose for simplicity that hawks and doves breed true (asexually) at rates equal to their fitness. A population of doves can be invaded by a rare hawk mutant, since the second column shows that hawks are at an advantage when playing against doves; thus dove is not a stable strategy. In a population of hawks (first column) hawk is also at an advantage to dove when $V > C$ [e.g., the payoff matrix (9)], so that the hawk strategy is stable and cannot be invaded by doves. Thus when the value of the resource exceeds the cost of injury, one would expect the population to evolve to consist entirely of hawks. However, when $V < C$ [e.g., the payoff matrix (10)], doves are at an advantage over hawks when the latter are common, so that neither the pure hawk nor the pure dove strategy is stable. At equilibrium one would expect to find a mixture of hawks and doves. The reason for this is that the relative fitnesses of hawks and doves depend on their population frequencies. The hawk strategy is at an overall advantage to the dove strategy when doves are common, but at a disadvantage when they are rare. This is the classic situation of negative frequency dependence discussed in Chapter 7.

Suppose that the frequency of hawkish behavior in the population at a given time is p; this is the probability that a player's opponent will act as a hawk rather than as a dove. A player's fitness can be computed from the payoff matrix in Table 8.2 as

$$\begin{aligned} w_{\text{hawk}} &= w_0 + \tfrac{1}{2}(V - C)p + V(1 - p) \\ w_{\text{dove}} &= w_0 + \tfrac{1}{2}(1 - p) \end{aligned} \tag{11}$$

depending on whether he acts like a hawk or a dove. This is plotted in Figure 8.1 with $w_0 = 10$ for the payoff matrices (9) and (10). In both cases fitness decreases linearly as the frequency of hawks increases. (This is true in general for the hawk–dove game since it is always better to have a dove as your opponent.) The hawk strategy has higher fitness than the dove strategy for any frequency of hawks (between 0 and 1) in Figure 8.1a, so that one expects doves to be eliminated from the population in

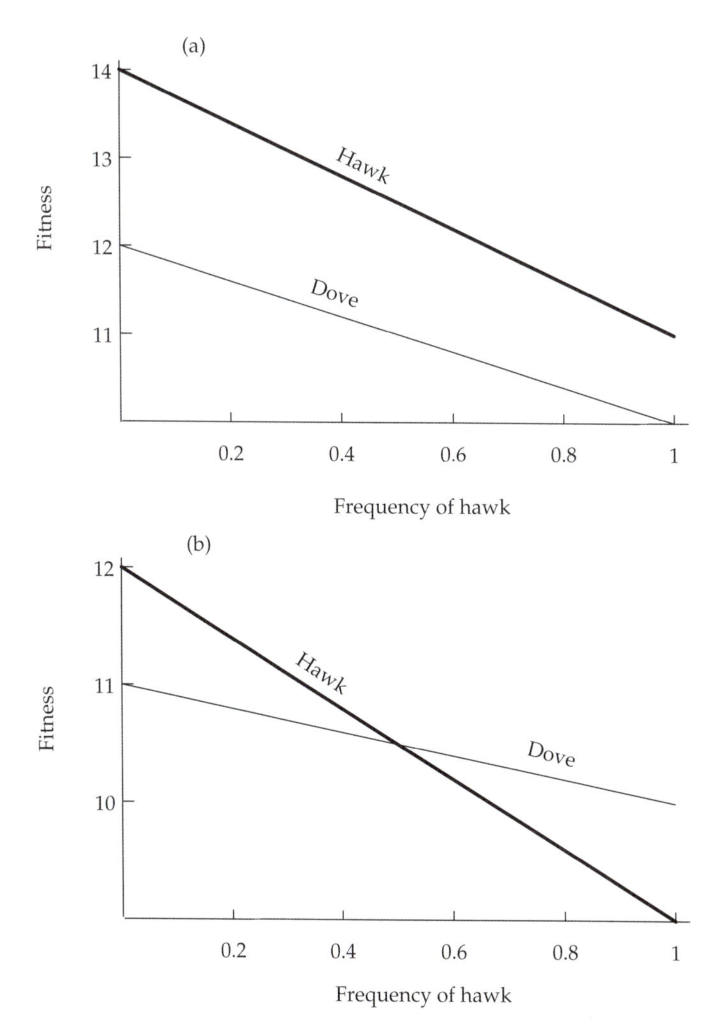

Figure 8.1. Fitness of hawk and dove plotted against the frequency of hawk. (a) Payoff matrix in (9); (b) payoff matrix in (10).

this situation. However, in Figure 8.1b, hawks are fitter than doves when $p < 0.5$, but they are less fit than doves when $p > 0.5$. Thus in this situation one would expect doves and hawks to coexist at the same frequency. In general for the hawk–dove game, one expects hawks to exclude doves when the value of the resource exceeds the cost of injury ($V > C$), but one expects hawks and doves to coexist with the frequency of hawks being V/C otherwise. (This value is obtained by equating the fitnesses of hawks and doves in Equation 11.) This oversimplified model cannot be expected to give a complete explanation of the avoidance of injurious tactics in animal conflict, but it at least suggests a reason why such tactics should be rare.

In this account I have assumed that individuals always play either hawk or dove, an equilibrium population for the model in Figure 8.1b containing an equal mixture of hawks and doves. This is the pure-strategy model of the next section. Under the mixed-strategy model, that is more natural in a game-theory context, all individuals play the same strategy, but it may be a mixture of the two pure strategies. In game-theoretic terms, playing hawk with probability p and dove with probability $q = 1 - p$ is the mixed strategy $\{p, q\}$. For the payoffs in (9), the pure hawk strategy $\{1, 0\}$ is the best response to any strategy of one's opponent and is there-

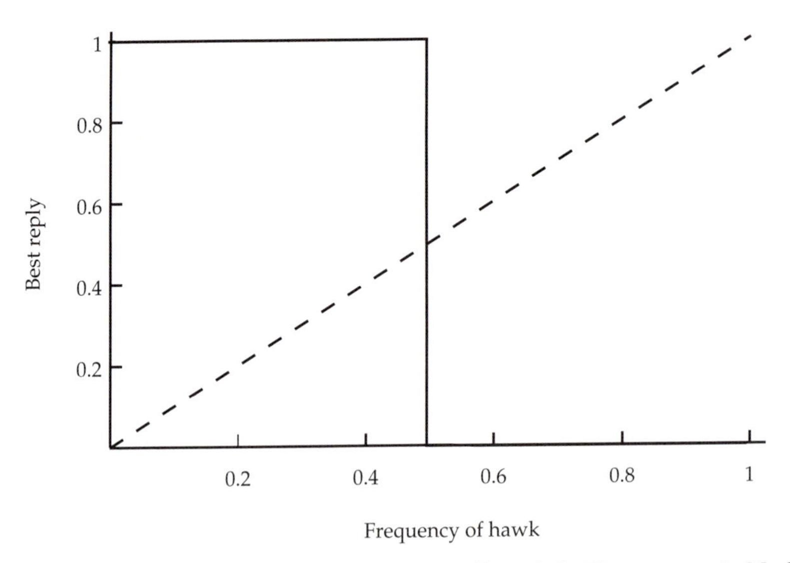

Figure 8.2. Best-reply diagram for the payoffs in (10). The symmetric Nash equilibrium is the intersection of the best-reply (solid) line with the diagonal (dashed) line.

fore a strict symmetric Nash equilibrium, as we have already seen. For the payoffs in (10), hawk is the best reply to any strategy with $p < 0.5$, dove is the best reply to any strategy with $p > 0.5$, but any strategy is an equally good reply to $p = 0.5$; hence {0.5, 0.5} is a weak symmetric Nash equilibrium, as illustrated in Figure 8.2. In both cases, there is only one symmetric Nash equilibrium.

SYMMETRIC, TWO-PERSON GAMES

We now consider the general properties of a symmetric, two-person evolutionary game. Suppose that there are n possible strategies, numbered 1, $\cdots$, n. Games (contests) occur between randomly chosen pairs of animals. There is an associated payoff matrix of order $n \times n$, with a_{ij} being the payoff (the increase in fitness) to an animal playing strategy i against an animal playing strategy j.

The evolution of this system can be investigated in two slightly different ways. First, suppose that an individual always plays the same pure strategy, and that this characteristic is inherited asexually by all of his offspring. Write n_i for the number of i-strategists at a particular time and p_i for their relative frequency. The fitness of an i-strategist is

$$w_i = w_0 + \sum_{j=1}^{n} a_{ij} p_j \tag{12}$$

The dynamics of the system in continuous time can be examined by the method of Taylor and Jonker (1978) described in Chapter 7 (see Equations 7.3–7.8). At equilibrium, either $p_i = 0$ or $w_i = \overline{w}$. Thus all strategies present at equilibrium with nonzero frequency (these strategies are called the *support* of the equilibrium) must have the same fitness. The intuitive reason is illustrated in Figure 8.1b; under frequency-dependent selection, strategies will adjust their frequencies to have the same fitness. This fact can be used to find all possible equilibria. The local stability of these equilibria can be investigated by finding the eigenvalues of the Jacobian matrix defined in Equation 7.8. A stable equilibrium under this model, $\hat{\mathbf{p}} = \{\hat{p}_1, \hat{p}_2, \ldots, \hat{p}_n\}$, is called an *evolutionarily stable state*. This kind of model leads to a polymorphism of pure strategies and may be called the "pure-strategy model"; the components of $\hat{\mathbf{p}}$ represent the frequencies with which individuals using the different pure strategies are present in the population at equilibrium.

The second approach, which fits more naturally into the context of game theory, supposes that an individual can choose a *mixed strategy*, playing different pure strategies with different probabilities. For example, an animal might have the strategy of being a hawk half the time and a dove the other half. This approach seeks a strategy (pure or mixed) that when it is common cannot be invaded by any rare mutant strategy. Such a strategy is called an *evolutionarily stable strategy* or ESS. The advantage of this approach, that we may call the "mixed-strategy model," is that it is possible to determine the stability of an equilibrium of a strategy without explicitly considering the dynamics of the situation. Consider the mixed strategies **p** and **q**, that play pure strategy i with probabilities p_i and q_i, respectively. The average payoff to a **p**-strategist playing against a **q**-strategist is

$$E(\mathbf{p},\mathbf{q}) = \sum_{i,j=1}^{n} p_i q_j a_{ij} \tag{13a}$$

or in matrix terminology

$$E(\mathbf{p},\mathbf{q}) = \mathbf{pAq} \tag{13b}$$

(Note that the payoff is a linear function of the probabilities used by either player; this property of linearity simplifies the theory, but it does not usually hold for "playing-the-field" models considered at the end of this chapter.) If **p** is an ESS, so that when it is common it cannot be invaded by any rare mutant **q**, then the fitness of **p** when it is common must exceed the fitness of **q**, which is assumed to be rare. If **q** has frequency x, then

$$\begin{aligned} w_{\mathbf{p}} &= w_0 + (1-x)E(\mathbf{p},\mathbf{p}) + xE(\mathbf{p},\mathbf{q}) \\ w_{\mathbf{q}} &= w_0 + (1-x)E(\mathbf{q},\mathbf{p}) + xE(\mathbf{q},\mathbf{q}) \end{aligned} \tag{14}$$

To have $w_{\mathbf{p}} > w_{\mathbf{q}}$ with x small, we must have

$$E(\mathbf{p},\mathbf{p}) \geq E(\mathbf{q},\mathbf{p}) \tag{15a}$$

If $E(\mathbf{p}, \mathbf{p}) = E(\mathbf{q}, \mathbf{p})$, we must also have

$$E(\mathbf{p},\mathbf{q}) > E(\mathbf{q},\mathbf{q}) \tag{15b}$$

The conditions for **p** to be an ESS are that these conditions hold for all $\mathbf{q} \neq \mathbf{p}$. The interpretation of Equation 15a holding for all **q** is that **p** is the best reply to itself; if you know that your opponent will play **p**, then you cannot do better than to play **p** yourself. In other words this condition requires that **p** should be a Nash equilibrium. The interpretation of Equation 15b is that if **q** is an alternative best reply to **p**, then **p** must do better against **q** than **q** does against itself; this ensures that a rare mutant using **q** cannot invade a population of **p**-strategists. When **p** is a strict Nash equilibrium it is automatically an ESS; the second condition in Equation 15b is only operative when **p** is a weak Nash equilibrium.[1]

To check that $p = 0.5$ is an ESS for the hawk–dove game with $V = 2$, $C = 4$, suppose that $\mathbf{p} = \{0.5, 0.5\}$ and $\mathbf{q} = \{q, 1-q\}$. Then, from Equation 10,

$$\begin{aligned} E(\mathbf{p},\mathbf{p}) &= 0.5 \\ E(\mathbf{q},\mathbf{p}) &= 0.5 \\ E(\mathbf{p},\mathbf{q}) &= 1.5 - 2q \\ E(\mathbf{q},\mathbf{q}) &= 1 - 2q^2 \end{aligned} \tag{16}$$

Since $E(\mathbf{p}, \mathbf{p}) = E(\mathbf{q}, \mathbf{p})$, the relevant condition for stability is that

$$E(\mathbf{p},\mathbf{q}) - E(\mathbf{q},\mathbf{q}) = \tfrac{1}{2}(1-2q)^2 > 0 \tag{17}$$

This condition is satisfied when $q \neq 0.5$.

It is no accident that $E(\mathbf{p}, \mathbf{p}) = E(\mathbf{q}, \mathbf{p})$ at the equilibrium for the hawk–dove game. Bishop and Cannings (1978) showed that all the pure strategies used in an evolutionarily stable mixed strategy **p** with non-zero probability (i.e., all the pure strategies in the support of **p**) must have the same fitness when playing against **p**. (See Theorem 1 in the appendix to this chapter.) This ensures that the equilibria under the mixed-strategy model are the same as those under the pure-strategy model, since the latter also implies that the fitnesses of all strategies used in the polymorphism are the same.

The various types of equilibria can be summarized as follows. (1) An *equilibrium* is a probability vector $\mathbf{p} = \{p_1, p_2, \ldots, p_n\}$ with either $p_i = 0$ or $w_i = \overline{w}$; the strategies i for which $p_i > 0$ (and hence necessarily $w_i = \overline{w}$) form the *support* of the equilibrium. This definition of an equilibrium is valid under either the pure-strategy or the mixed-strategy model. (2) An *evolutionarily stable state* is a stable equilibrium under the pure-strategy model in continuous time with p_i representing the relative frequency of individuals playing the pure strategy i (Equations 7.3–7.8). (3) A *Nash equi-*

[1] There are two ways in which the condition in Equation 15b may fail. If $E(\mathbf{p}, \mathbf{p}) = E(\mathbf{q}, \mathbf{p})$ and $E(\mathbf{p}, \mathbf{q}) < E(\mathbf{q}, \mathbf{q})$, Equation 14 shows that **q** has a positive selective advantage over **p**, so that the equilibrium is unstable. If $E(\mathbf{p}, \mathbf{p}) = E(\mathbf{q}, \mathbf{p})$ and $E(\mathbf{p}, \mathbf{q}) = E(\mathbf{q}, \mathbf{q})$, **p** and **q** are equally fit; **q** may increase in frequency through drift so that **p** is only neutrally stable and does not satisfy the strict criteria of an ESS. Some caution must be used in interpreting what is likely to happen under neutral stability; once **q** has gained ground in the population it may be that a third mutant will arise which makes **q** less fit than **p** and so stabilizes the equilibrium, or the opposite may happen. (See the discussions of the tit-for-tat strategy in the following section and of Selten's theorem in this chapter's appendix.)

librium $\mathbf{p}$ is an equilibrium interpreted under the mixed-strategy model, with p_i representing the relative frequency with which a $\mathbf{p}$-strategist plays the ith pure strategy, that is the best reply to itself in the sense that no other mixed strategy $\mathbf{q}$ has a better return against a $\mathbf{p}$-strategist (Equation 15a). (4) An *evolutionarily stable strategy* is a Nash equilibrium under the mixed-strategy model that is either a strict Nash equilibrium or satisfies the stability criterion in Equation 15b. This ensures that if there are any alternative strategies $\mathbf{q}$ that are alternative best replies to $\mathbf{p}$ (that will normally use the same pure strategies as $\mathbf{p}$ or a subset of them), they cannot invade a population of $\mathbf{p}$-strategists as rare mutants.

The dominance hierarchy of these types of equilibrium can be represented as follows:

$$\begin{aligned}\text{equilibrium} &\supseteq \text{Nash equilibrium}\\ &\supseteq \text{evolutionarily stable state}\\ &\supseteq \text{evolutionarily stable strategy}\end{aligned} \tag{18}$$

In other words, a Nash equilibrium is always an equilibrium, but an equilibrium need not be a Nash equilibrium, and so on. The pure dove strategy {0, 1} for matrix (9) is an equilibrium but not of course a Nash equilibrium. (It is the worst reply to itself.) The payoff matrix

$$\begin{bmatrix} 2 & 3 \\ 1 & 4 \end{bmatrix} \tag{19}$$

has three Nash equilibria, {1, 0}, {0.5, 0.5}, and {0, 1}. The first and third, the pure strategies, are strict Nash equilibria and are therefore evolutionarily stable both as states and strategies; one might expect to observe either one of them depending on initial conditions. The second is a weak Nash equilibrium, but it is not evolutionarily stable either as a state or a strategy, and one would therefore not expect to observe it; as a strategy it can in fact be invaded by any alternate strategy.

If there are only two pure strategies, an evolutionarily stable state is also an evolutionarily stable strategy, but the following payoff matrix with three pure strategies provides a counterexample:

$$\begin{bmatrix} 0 & 5 & -4 \\ -7 & 0 & 8 \\ -1 & 2 & 0 \end{bmatrix} \tag{20}$$

There are three Nash equilibria, {1, 0, 0}, {0.8, 0, 0.2}, and {0.33, 0.33, 0.33}; the first and third, but not the second, are evolutionarily stable states, but only the first is an evolutionarily stable strategy. The reason is that neither strategy 2 nor strategy 3 on its own can invade a population playing {0.33, 0.33, 0.33}, but a mutant that is equally likely to play strategy 2 or strategy 3 can invade. It is however rather rare for these two criteria of stability to disagree.

There is no guarantee that an ESS will be unique [see the payoff matrix (19)]. When there are two or more ESS's, initial conditions will determine which one is reached. It may also happen if there are more than two pure strategies that there is no ESS. An example is provided by the following payoff matrix

$$\begin{bmatrix} 0.1 & 1 & -1 \\ -1 & 0.1 & 1 \\ 1 & -1 & 0.1 \end{bmatrix} \tag{21}$$

This is the payoff matrix for the game of stone–scissors–paper if the players each receive 10 cents when they choose the same object. The population continually cycles round from 1 to 2 to 3 and back to 1 and so on.

When there are several strategies it is rather tedious to identify all possible equilibria and test them for stability.[2] There are two general rules that often simplify the task and enable the ESS's to be identified by inspection. The first rule finds all pure ESS strategies. If the *i*th diagonal element in the payoff matrix is strictly larger than all the other elements in the same column then this strategy is a strict Nash equilibrium and hence a pure ESS:

Rule 1: *If* $a_{ii} > a_{ji}$ *for* $j \neq i$, *strategy i is an ESS* (22)

If $a_{ii} \geq a_{ji}$ for $j \neq i$, it is also necessary that $a_{ji} > a_{jj}$ for those values of j for which $a_{ji} = a_{ii}$. (The derivation of this extension to the rule from Equation 17b is straightforward.) The second rule states that if **p** and **q** are ESS's, the support of **p** cannot be a subset of the support of **q** and vice versa:

Rule 2: *If* **p** *and* **q** *are ESS's*, $S(\mathbf{p}) \not\subset S(\mathbf{q})$ (23)

[$S(\mathbf{p})$ means the support of **p**, the set of all strategies appearing with positive probability in **p**. See the appendix to this chapter for a proof of this rule.]

These rules enable one to find the ESS for any game with only two pure strategies. The three generic possibilities are illustrated in the payoff matrices **A**, **B**, and **C**, reproduced from (9), (10), and (19):

[2] A program is provided in the *Mathematica* Notebooks to perform this task, using the method of Haigh (1975) to test for the stability of an evolutionarily stable strategy.

$$\mathbf{A} = \begin{bmatrix} 1 & 4 \\ 0 & 2 \end{bmatrix}$$
$$\mathbf{B} = \begin{bmatrix} -1 & 2 \\ 0 & 1 \end{bmatrix} \tag{24}$$
$$\mathbf{C} = \begin{bmatrix} 2 & 3 \\ 1 & 4 \end{bmatrix}$$

In **A**, {1, 0} is a pure ESS but {0, 1} is not by rule 1; by rule 2 there is no other ESS. In **B** there is no pure ESS; the mixed ESS $\mathbf{p}$ = {0.5, 0.5} can be found from the condition that $E(1, \mathbf{p}) = E(2, \mathbf{p})$ but must then be tested for stability from the criterion in Equation 15b. (In fact, if neither of the pure stategies is an ESS, the appropriate mixed strategy is always an ESS.) In **C** both {1, 0} and {0, 1} are pure ESS's by rule 1, and there is no mixed ESS by rule 2.

RECIPROCAL ALTRUISM AND THE PRISONER'S DILEMMA

Trivers (1971) suggested that altruism can evolve between two unrelated individuals if they agree to reciprocate some act that they cannot do themselves—you scratch my back today and I'll scratch yours tomorrow. If the benefit of having your back scratched (getting rid of parasites) is greater than the cost of scratching someone else's back (loss of time and energy) and if the agreement to reciprocate is honored, then both parties will gain. The problem, both in an evolutionary and a human context, is that it is open to cheating.

This question has been studied by a game-theoretic analysis of a game called the prisoner's dilemma, that has its origin in the following scenario. Two men are being held separately in prison. They are accused of having conspired together to commit a crime, that they have previously agreed together to deny, and a detective is trying to persuade each of them to implicate the other with the following inducement. If neither of them confesses they will both be set free because there is no evidence against them; if both confess, they will both be punished; but if one confesses and the other does not, the confessor is set free and in addition receives a reward while his partner is punished more severely than if he had confessed. The rational decision for each of them is to confess since each of them will be better off by confessing whatever the other one does. They therefore both confess and are punished. The dilemma is that they could both do better than this and be set free if they had stuck to their original agreement to deny the charge, but there seems to be no rational way of getting to this result.

This game serves as a paradigm for the evolution of reciprocal altruism, because it expresses the fact that each player in a "You scratch my back and I'll scratch yours" situation can do better by being selfish but that they would both do better if they both cooperate than if neither of them does. In the simplest case, we may suppose that each player incurs a cost c for cooperating and receives a benefit b (greater than c) from his opponent if the latter cooperates; the payoff matrix for the row player is shown in Table 8.3. If the game is only played once between any pair of animals, defect is the only ESS. However, if a pair of animals plays the game repeatedly, they may adopt more complex strategies, depending on the past history of their interaction, that may enable them to cooperate.

The political scientist Axelrod (1984) organized a competition in which a number of game theorists were invited to submit computer programs defining their strategy for playing the iterated prisoner's dilemma. Each strategy was played 200 times against each of the others, and the average payoff was calculated. The winning strategy was a very simple one called *tit for tat:* cooperate on the first move and then do what your opponent did on the previous move. Axelrod describes the reasons for the success of tit for tat as follows: "It is nice, forgiving and retaliatory. It is never the first to defect; it forgives an isolated defection after a single response; but it is always incited by a defection no matter how good the interaction has been so far."

Axelrod has applied this result to political problems such as the development of reciprocity (for example, vote-trading in the U.S. Senate). To pursue its evolutionary implications he collaborated with the evolutionary biologist W. D. Hamilton, and the following discussion of the problem is based on their conclusions (Axelrod and Hamilton, 1981 [reprinted as Chapter 5 of Axelrod, 1984]; Axelrod and Dion, 1988). First, as already seen, the only stable strategy in a single game is defection. The same conclusion holds for a game played a known fixed number of times, for the following reason. In the last round, defection is the only stable strategy; in the penultimate round, the same is true, since the strategy in the next round is determined; and so on, back to the first round. However, it is more realistic to

Table 8.3. Payoff matrix for the prisoner's dilemma.

	Defect	Cooperate
Defect	0	b
Cooperate	$-c$	$b-c$

Table 8.4. Payoff matrix for two strategies in the iterated prisoner's dilemma.

	AD	TFT
AD	0	b
TFT	$-c$	$(b-c)/(1-p)$

suppose that the game will be played an unknown number of times, with a fixed probability p that the two players will meet again. This game is known as the *iterated prisoner's dilemma.* It has been used extensively as a possible model for the evolution of cooperation. Theoretical analysis is complicated because of the large number of possible strategies, since a strategy may depend on all the previous moves of one's opponent. The analysis here will be restricted to three possible strategies, (1) always defect (AD), (2) tit for tat (TFT), and (3) always cooperate (AC).

Consider first only the first two strategies. The payoff of TFT playing against itself is

$$(b-c)+p(b-c)+p^2(b-c)+p^3(b-c)+\cdots=(b-c)/(1-p) \tag{25}$$

The other payoffs are easily calculated, and the payoff matrix is shown in Table 8.4. AD is always an ESS; TFT is also an ESS provided that

$$(b-c)/(1-p)>b \tag{26a}$$

or

$$pb>c \tag{26b}$$

(This is the same as the criterion for the evolution of altruism under kin selection discussed in Chapter 9, with p, the probability that the two players will meet again, replacing r, the coefficient of relatedness.)

Thus TFT is an ESS when $pb > c$, if the only other permissible strategy is AD, but AD is also an ESS since the best response to a player who always defects is defection. If noncooperation is the primitive condition, it is difficult to see how reciprocal altruism can evolve from it. Axelrod and Hamilton (1981) suggest that cooperative behavior might originate as altruism between relatives selected by kin selection and then spread to encompass nonrelatives, or it might spread from a small cluster of cooperative individuals, but the transition from noncooperation to reciprocal altruism is clearly difficult.

We now introduce the third strategy, AC. The payoff matrix is shown in Table 8.5. (Note that there is no difference between TFT and AC in games between these two strategies; the superiority of TFT only manifests

Table 8.5. Payoff matrix for three strategies in the repeated prisoner's dilemma.

$$\begin{array}{c} \\ \text{AD} \\ \text{TFT} \\ \text{AC} \end{array} \begin{array}{c} \begin{array}{ccc} \text{AD} & \text{TFT} & \text{AC} \end{array} \\ \begin{bmatrix} 0 & b & b/(1-p) \\ -c & (b-c)/(1-p) & (b-c)/(1-p) \\ -c/(1-p) & (b-c)/(1-p) & (b-c)/(1-p) \end{bmatrix} \end{array}$$

itself against AD.) TFT is a weak Nash equilibrium if $pb \geq c$ but it is not an ESS because the payoff for AD when playing either TFT or itself is the same as the payoff for TFT against itself. In a pure TFT population, AC has the same fitness as TFT and can therefore increase in frequency by drift; once this has happened AD can begin to spread because of its superiority over AC. Another weakness of the TFT strategy is that it is vulnerable to occasional mistakes; in an interaction between two TFT players, if one of them defects by mistake, they are subsequently locked in to an alternating series of cooperation and defection.

These drawbacks have stimulated theoretical work to find alternative strategies that are more robust than TFT. Nowak and Sigmund (1993) have shown that the strategy *Pavlov* outperforms TFT because it can correct occasional mistakes and exploit unconditional cooperators. Pavlov adopts the rule: repeat what you did in the previous encounter if you were successful (gaining more than zero), do the opposite if you were unsuccessful (gaining zero or less). Thus Pavlov cooperates on the current move if both players cooperated or if they both defected on the previous move and defects if one cooperated and the other defected.

A good illustration of reciprocal altruism is provided by reciprocal food-sharing in vampire bats (Wilkinson, 1984). Individuals who failed to obtain a blood meal during the night begged for blood from well-fed individuals in the daytime roosts and were often given it. It seems likely that a small amount of blood increases the survival chances of the recipient considerably more than it reduces those of the donor, so that this condition for the evolution of altruism is met. However, regurgitation only occurs between close relatives or between unrelated individuals who were frequent roost-mates. It is plausible that this behavior arose by kin selection, "roost-mate" being used as a cue for "relative." Now that the behavior has spread it is a stable example of reciprocal altruism, and there is no selection pressure to use a more reliable cue to relatedness. There are two barriers to the spread of this behavior between individuals who are not frequent roost-mates. First, the chances that they will meet again are small so that reciprocal food sharing between them, even if it were common, would not be a stable ESS. Secondly, even apart from this, non-roost-mates are currently at the stable ESS of noncooperation from which there is no way of escaping. Two examples of cooperative behavior in which experimental manipulation suggests that animals employ a TFT strategy are the interaction between parent and nonbreeding Tree Swallows (Lombardo, 1985) and the behavior of sticklebacks in the presence of a predator (Milinski, 1987), but it is not easy to determine whether animals are using a strategy like TFT or one like Pavlov; Milinski (1993) suggests that sticklebacks may be using Pavlov.

ASYMMETRIC GAMES

The games analyzed so far are symmetric in the sense that the two players are identical; their rewards are the same, their chances of winning an escalated contest between them are the same (at least as far as they can tell), and there is no other obvious difference between them that they could use as a cue to determine their strategies. Thus any feasible solution must be a symmetric Nash equilibrium in which both players use the same strategy; that is to say, the equilibrium strategy must be the best reply to itself.

Most pairwise contests in real life, however, are asymmetric in the sense that there are role differences between the contestants; they may be contests between a male and a female, between a larger and a smaller male, or between the owner of a territory and an intruder. The roles of the two players may affect the likely outcome of a contest or the payoffs to the players. Even if they do not affect the payoffs, role differences known to the players at the start of the game can, and often should, be used as a basis for determining their strategies, leading to the possibility of an asymmetric Nash equilibrium; in the hawk–dove game, for example, the player in one role might always plays hawk, and the player in the other role might always play dove.

To deal with asymmetric games by the methodology developed above for symmetric games it is often convenient to symmetrize them by introducing the idea of a conditional strategy. Consider a contest between two animals in different roles A and B (for example, owner and intruder, large and small, male and female). We suppose that both animals know both their own role and that of their opponent. Suppose that m strategies are open in role A and n strategies in role B. (The strategy set might be the same for both roles, for example, if they are size differences, but it might differ if they are male and female.) Write $E_A(i, j)$ for the payoff to the A animal if he adopts strategy i and his opponent adopts strategy j and $E_B(i, j)$

for the payoff to the B animal. (Compare the definitions above Equation 2.) We now introduce the set of mn conditional strategies, i/j: "If I am in a contest between an A animal and a B animal, adopt strategy i if I am the A animal and strategy j if I am the B animal." The payoff to an i/j strategist playing a k/l strategist is

$$E(i/j,k/l) = \tfrac{1}{2} E_{\mathrm{A}}(i,l) + \tfrac{1}{2} E_{\mathrm{B}}(k,j) \tag{27}$$

since the i/j strategist is equally likely to be in the A or the B role. This formula defines the payoff matrix for the equivalent symmetric game with mn pure strategies that can be analyzed by the methods just described for symmetric two-person games.

This analysis is based on the assumption that animals have complete information about their own role and about that of their opponent, so that, for example, an animal never thinks that its opponent is larger or smaller than it really is; we are also assuming a game between animals in different roles, excluding for example a contest between two animals of the same size. A general feature of models with asymmetric roles with complete information is that an ESS must be a pure strategy, though it is possible for an evolutionarily stable state to be polymorphic. (For a derivation of this result and a critique of its implications see the discussion of Selten's theorem in the appendix to this chapter.) This makes it easy to identify all the ESS's since the diagonal element in the symmetrized payoff matrix must be greater than the other elements in the same column. It is in fact easy to identify all pure-strategy ESS's without symmetrizing the game since they are the (possibly asymmetric) pure-strategy Nash equilibria in the original game, but the technique of symmetrization is useful in unifying the theory of symmetric and asymmetric games, in determining whether or not there exist any polymorphic evolutionarily stable states, and in analyzing asymmetric games with incomplete information that may have mixed-strategy ESS's (though the extension needed to deal with incomplete information will not be discussed here). I shall now describe three examples of the application of this methodology.

The owner–intruder asymmetry

Consider a contest between the owner of a territory and an intruder, and suppose for simplicity that this role makes no difference to the value of the territory or to the chances of winning an escalated contest, so that the payoff matrix is given by Table 8.2 whether the row player is the owner or the intruder. It is clear that for the owner to play hawk and for the intruder to play dove is a strict asymmetric Nash equilibrium when $V < C$, since dove is the best response to hawk and vice versa; for the same reason it is a strict Nash equilibrium for the owner to play dove and for the intruder to play hawk. When $V > C$, hawk is the best reply to itself, and this is the only Nash equilibrium.

To symmetrize the game, define four conditional strategies: (1) hawk/hawk (H/H), (2) hawk/dove (H/D), (3) dove/hawk (D/H), and (4) dove/dove (D/D). For example H/D means "Play hawk when you are the owner and dove when you are the intruder." Maynard Smith (1982) calls this the "Bourgeois" strategy, and we may by analogy call D/H the anti-Bourgeois strategy. We can now calculate the payoff matrix for these four strategies from Equation 27 using the values for $E(i, j)$ in Table 8.2 together with the relationships

$$\begin{aligned} E_{\mathrm{A}}(i,l) &= E(i,l) \\ E_{\mathrm{B}}(k,j) &= E(j,k) \end{aligned} \tag{28}$$

from the payoff symmetry of the original hawk–dove game. This payoff matrix is shown in Table 8.6. (The entries in this matrix can also be calculated by a simple direct argument.)

With $V = 4$ and $C = 2$, the payoff matrix is

$$\begin{bmatrix} 1 & 2.5 & 2.5 & 4 \\ 0.5 & 2 & 1.5 & 3 \\ 0.5 & 1.5 & 2 & 3 \\ 0 & 1 & 1 & 2 \end{bmatrix} \tag{29}$$

and with $V = 2$ and $C = 4$, it is

$$\begin{bmatrix} -1 & 0.5 & 0.5 & 2 \\ -0.5 & 1 & 0 & 1.5 \\ -0.5 & 0 & 1 & 1.5 \\ 0 & 0.5 & 0.5 & 1 \end{bmatrix} \tag{30}$$

Table 8.6. Symmetrized payoff matrix for the hawk–dove game with an owner-intruder asymmetry.

	H/H	H/D	D/H	D/D
H/H	$\frac{1}{2}(V-C)$	$\frac{1}{4}(3V-C)$	$\frac{1}{4}(3V-C)$	V
H/D	$\frac{1}{4}(V-C)$	$\frac{1}{2}V$	$\frac{1}{4}(2V-C)$	$\frac{3}{4}V$
D/H	$\frac{1}{4}(V-C)$	$\frac{1}{4}(2V-C)$	$\frac{1}{2}V$	$\frac{3}{4}V$
D/D	0	$\frac{1}{4}V$	$\frac{1}{4}V$	$\frac{1}{2}V$

When $V > C$ [e.g., the payoff matrix (29)] there is a unique ESS with both owner and intruder playing hawk. When $V < C$ [e.g., the payoff matrix (30)] both the Bourgeois and the anti-Bourgeois strategy are ESS's since each is the unique best reply to itself (a Nash equilibrium with no equally good alternative reply); and it can be verified that they are the only ESS's. In particular, the original ESS of playing hawk or dove with probability 0.5 regardless of role in the game with payoffs in (30) is no longer an ESS in the new setup though it is a Nash equilibrium. Thus the asymmetry of ownership can be used as a convention to settle the contest provided that $V < C$.

An example of the Bourgeois strategy is provided by the study of Davies (1978) on territorial behavior in the speckled wood butterfly. Males defend patches of sunlight on the floor of woodland, that gives them mating opportunities. When two males meet in a sunspot, they spiral up into the canopy in a contest, but the resident always wins and returns to the sunspot. The anti-Bourgeois strategy is also an ESS when $V < C$, but this leads to an unstable situation with frequent changes of ownership. A possible example occurs in the social spider *Oecibus civitas* (Burgess, 1976).

These results seem to provide a general explanation for the fact that, in a contest between a territory-holder and an intruder, the territory-holder usually wins without appreciable injury to either party. The explanation is that the contestants are using ownership as an arbitrary asymmetry to settle the dispute in favor of the owner, in accordance with the finding that the Bourgeois strategy is an ESS. But Grafen (1987) has demonstrated a problem with this interpretation, that V and C will depend on the strategy for settling disputes adopted by the population. In a population playing the Bourgeois strategy, that respects ownership, the value of a territory V will be high and the cost of injury C will be low, so that it is unlikely that the condition $V < C$ for maintaining Bourgeois as an ESS will be satisfied. The value of a territory will be high because a territory-owner will hold it indefinitely. The cost of losing a fight will be low because it is the difference in future reproductive success between playing hawk and losing compared with playing dove; if a non-territory-holder requires a territory to breed and has little chance of gaining one except by ousting an owner, his expected reproductive success will be very small if he never tries to acquire one, and he will have little to lose by doing so. These arguments do not apply to the speckled wood butterfly, since a patch of sunlight is by its nature temporary, but they severely limit the applicability of the model in many situations.

Size asymmetry

In the above example, ownership was used as a purely conventional role difference to determine strategy, but role differences will usually affect either the value of the resource or the probability of winning an escalated contest. Suppose that the contest is between a large male and a small male, the large male having a chance $p > 0.5$ of winning an escalated contest and a probability $q = 1 - p$ of losing.

The bimatrix representation of the payoffs is shown in Table 8.7. The (possibly asymmetric) pure-strategy solutions to the contest can be found directly from this table without symmetrizing the game. The pure strategy "Play hawk whether you are large or small" is an ESS when (a) $pV - qC > V$ and (b) $qV - pC > V$, since the first condition is the criterion for hawk to be the best response by a large animal when a small animal plays hawk and the second condition is the criterion for hawk to be the best response by a small animal when a large animal plays hawk. The second condition is more stringent, so that the condition for hawk/hawk to be an ESS is that $V/C > p/q$; in this case, it is the unique ESS. In the same way it is easy to verify that hawk/dove is the unique ESS if $q/p < V/C < p/q$; and hawk/dove and dove/hawk are both ESS's if $V/C < q/p$. If the large male has a high probability of winning an escalated contest it is likely that $q/p < V/C < p/q$, confirming the intuitive expectation that contests should be won by the larger animal without fighting.

Parental care

Maynard Smith (1977) considers ESS models for the evolution of parental care. Should both parents, only one, or neither care for their offspring? If it is one parent, which one should it be, the mother (the duck strategy) or the father (the stickleback strategy)? Maynard Smith tries to find a pair of strategies, say α for males and β for females, such that it would not pay a male to diverge from α so long as females adopt β and it would not pay a

Table 8.7. Bimatrix representation of the hawk–dove game with size asymmetry.

$$\text{Large}\begin{cases}\text{Hawk} \\ \text{Dove}\end{cases}\overset{\overbrace{\text{Hawk} \qquad\qquad\qquad \text{Dove}}^{\text{Small}}}{\begin{bmatrix}(pV - qC, qV - pC) & (V, 0) \\ (0, V) & (\frac{1}{2}V, \frac{1}{2}V)\end{bmatrix}}$$

female to diverge from β so long as males adopt α. This is a (possibly asymmetric) Nash equilibrium in pure strategies.

Maynard Smith considers two models with discrete breeding seasons. In the first model reproductive success is limited mainly by parental care. Suppose the number of surviving young is V_2 if both parents care for them and V_1 if only one of them does; if both parents desert, no young survive. A male who deserts his mate after she has laid her eggs has a chance p of mating a second female, and his fitness is then $V_1 + pV_2$ if he does not desert the second female; his fitness if he does not desert his first mate is V_2. Desertion will be favored if

$$p > (V_2 - V_1)/V_2 \qquad (31)$$

A female can try to ensure the faithfulness of her mate by lowering p, the chance that he will find a second mate; she can do this by declining to mate with him until they have been paired for some time, by which time his chances of finding a second mate are small. Note also that there is a second ESS in which the female deserts, leaving the brood to be cared for by the male. In this case p is to be interpreted as the chance that a female will find a second male to mate with her. Which of these two alternative ESS's evolves depends on the preadaptation to them. In mammals the female lactates and so cannot desert. In birds and other animals with internal fertilization, the male is in a position to desert before the female, so that the duck strategy is preferred. In sticklebacks the male defends a territory in which the female lays her eggs, that are then fertilized by the male; this situation predisposes to female rather than male desertion.

In the second model it is assumed that a female can only care for the brood at the expense of laying fewer eggs. Let P_0, P_1, and P_2 be, respectively, the probabilities of survival of an egg guarded by 0, 1, and 2 parents. In each sex there are two strategies: guard and desert. A male who deserts has a chance p of mating again. A female who deserts lays W eggs, and one who guards lays w eggs, with $W > w$. The payoff matrix to the two sexes is shown in Table 8.8. Four pure-strategy combinations are possible: (1) both desert; (2) male guards and female deserts (the stickleback strategy); (3) male deserts and female guards (the duck strategy); and (4) both guard. The conditions for the strategy {α, β} to be an ESS is that β must be the female's best response to the male strategy α and vice versa, that can be determined from Table 8.8. For example, the conditions for (1) to be an ESS are that (a) $P_0(1+p) > P_1$ (otherwise the male guards) and (b) $WP_0 > wP_1$ (otherwise the female guards). Combinations (2) and (3) may be alternative ESS's, in which case preadaptation will determine which of them evolves, as discussed above.

Table 8.8. Bimatrix representation for a model of parental care.

		Female: Guards	Female: Deserts
Male	Guards	(wP_2, wP_2)	(WP_1, WP_1)
	Deserts	$[wP_1(1+p), wP_1]$	$[WP_0(1+p), WP_0]$

MANY-PERSON GAMES WITH CONTINUOUS STRATEGIES

We have so far considered games in which each player must choose between a small number of pure strategies at each play (for example, hawk or dove). It is sometimes more natural to consider the strategy set as a continuous variable, such as the proportion of time or resources invested in one activity rather than another. At the same time as doing this I shall extend the model to consider not only games between two players, as we have done so far in this chapter, but games involving any number of players. Evolutionary games with many players are usually of the type called "playing the field," in which the payoff to an individual playing a particular strategy depends not on the strategy adopted by an individual opponent, but on some average property of the population or some subgroup of it. Most playing-the-field games involve continuous strategies, so that it is natural to consider these two extensions of evolutionary game theory together.

Consider for example the following rather crude model of vigilance against predators (Parker and Hammerstein, 1985). Birds forage together in groups of size n, and each bird allocates a proportion of its time to vigilance (watching for predators) and the remainder to feeding. Assume that a predator attack will occur once during the course of a day and that one of the birds will be killed if none of them is vigilant at that time, whereas all of them will escape if at least one of them is being vigilant. The cost of vigilance is that less time is spent feeding, and we assume that the fitness of a bird spending a proportion of its time v in vigilance is $1 - v^2$ if it survives predation (the squared term representing diminishing returns on investment in foraging). If the ith bird spends time v_i in vigilance, the chance that a particular bird suffers predation is

$$(1-v_1)(1-v_2) \cdots (1-v_n)/n \qquad (32)$$

(This is the probability of an attack occurring when all the birds are feeding, divided by the number of birds in the group, since only one of them is killed in a successful attack.) The payoff to the first bird in the group is

$$w = [1 - (1 - v_1)(1 - v_2) \cdots (1 - v_n)/n](1 - v_1^2) \tag{33}$$

(the probability of survival multiplied by the fitness accrued from foraging).

We now want to find the optimum vigilance. Suppose that it is the pure strategy v^*, that is to say a bird always spends a fraction v^* of its time being vigilant. If nearly all birds adopt a strategy v, a rare mutant bird adopting a different strategy u will find itself in a group with $n - 1$ birds using the v strategy and will have fitness

$$w(u,v) = [1 - (1 - u)(1 - v)^{n-1}/n](1 - u^2) \tag{34}$$

The optimal response u to the population strategy v can be calculated by solving

$$\left. \frac{\partial w}{\partial u} \right|_{u=v} = 0 \tag{35}$$

If v is the optimal strategy v^* then $u = v^*$ must be the best reply to v, so that

$$\left. \frac{\partial w}{\partial u} \right|_{u=v=v^*} = 0 \tag{36}$$

The optimal vigilance can be calculated from this equation. It is also necessary to check that it maximizes (rather than minimizes) the fitness by verifying that

$$\left. \frac{\partial^2 w}{\partial u^2} \right|_{u=v=v^*} \leq 0 \tag{37}$$

Equations 36 and 37 ensure that v^* is a Nash equilibrium; if the strict inequality holds in Equation 37, it is a strict Nash equilibrium.

Consider as an example the model with $n = 4$. It is straightforward to find the equilibrium from Equation 36 to be $v^* = 0.105$ and to verify that the second derivative is negative so that it is a strict Nash equilibrium. Figure 8.3 shows the best-reply diagram obtained by calculating the optimal response u to different levels of population vigilance v. It will be seen that $u = 0.105$ is the unique best reply to the population strategy $v = 0.105$, and that there are no alternative best replies; from continuity considerations, their existence would have been shown by a vertical line as in Figure 8.2. (Note however that these two best-reply diagrams are rather different. Figure 8.2 shows the optimal probability in a mixed strategy,

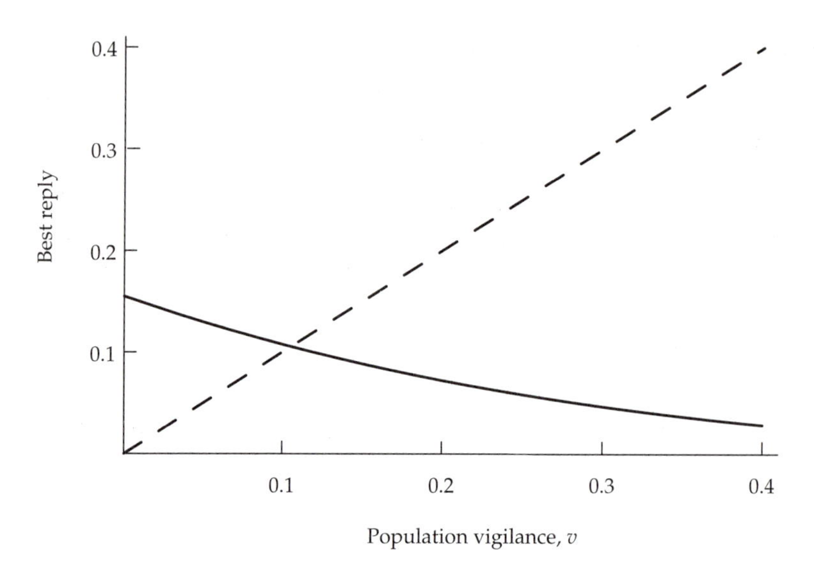

Figure 8.3. Best-reply diagram for vigilance model with n = 4.

whereas Figure 8.3 shows the optimal proportion of vigilance assumed to be a pure strategy chosen from a continuous range of values.) Thus v = 0.105 is a strict Nash equilibrium and is therefore an ESS; this population strategy is uninvasible, since any other mutant strategy is strictly less fit.

Another type of stability is also important in continuous-strategy models. Suppose that the population strategy v moves a small distance from its optimal level v^*. For stability one wants mutants that move it back toward the optimum to be favored and those that move it further away from the optimum to be disfavored. This type of stability is called *continuous stability* (Eshel, 1983). The algebraic condition for continuous stability is that

$$\left[\frac{\partial^2 w}{\partial u^2} + \frac{\partial^2 w}{\partial u \partial v}\right]_{u=v=v^*} < 0 \tag{38}$$

(See Theorem 4 in the appendix to this chapter.) It is straightforward to verify that the equilibrium in the vigilance model is continuously stable by this criterion. This can alternatively be inferred from the fact that the best-reply function is a decreasing function of the population vigilance in Figure 8.3 so that if $v < v^*$ the best reply exceeds v^* and vice versa.

This vigilance model has the expected prediction that individual vigilance v should decrease with group size n. This is due both to Hamilton's (1971) selfish-herd effect (an individual's chance of being captured is diluted by the presence of other individuals) and in addition to the fact that each individual benefits from the vigilance of other group members. The combination of these two factors is in fact so great that, under this model, the total vigilance nv decreases with group size. This is contrary to field data that suggest that individual vigilance decreases but total vigilance increases with group size. The assumption that the probability of attack is independent of group size may be too extreme. Alternatively, animals may be able to achieve a cooperative solution through reciprocal altruism, that is nearer to the interests of the group than of the individual.

FURTHER READING

Maynard Smith (1982) is the best account of evolutionary game theory. Axelrod (1984), updated by Axelrod and Dion (1988), and Poundstone (1992) are readable accounts of the prisoner's dilemma.

EXERCISES

1. For the game of simplified poker, verify that the matrix (1) is the payoff matrix, and show that Equation 6 gives the Nash equilibrium for this game.

2. For the hawk–dove game with $V = 2$ and $C = 4$, show that $p = 0.5$ is a Nash equilibrium, an evolutionarily stable state, and an evolutionarily stable strategy.

3. For the payoff matrix (19), show that $\{0.5, 0.5\}$ is a weak Nash equilibrium but that it is not an evolutionarily stable state and that as a strategy it can be invaded by any alternate strategy. Draw the best-reply diagram.

4. For the payoff matrix (20), show that $\mathbf{p} = \{0.33, 0.33, 0.33\}$ is an evolutionarily stable state; verify that it is not an evolutionarily stable strategy by showing that $E(\mathbf{p}, \mathbf{q}) < E(\mathbf{q}, \mathbf{q})$ for $\mathbf{q} = \{0, 0.5, 0.5\}$.

5. For the game with payoff matrix (21), show that $\{0.33, 0.33, 0.33\}$ is a Nash equilibrium but not an evolutionarily stable state or strategy. Write a computer program to simulate the behavior of the system. (Use the continuous-time model defined in Equations 7.5 and 8.14 if you have access to a program for numerical solution of differential

equations; otherwise use the corresponding discrete-time model. Take $w_0 = 20$ to make the fitnesses positive.)

6. If you have the *Mathematica* solutions, use the program there to find the Nash equilibria and evolutionarily stable states and strategies for the games with payoff matrices (9), (10), (19), (20), and (21).

7. Consider the iterated prisoner's dilemma with only TFT and AD present. Verify the payoff matrix in Table 8.4. Evaluate this matrix with $b = 4$, $c = 2$, and $p = 0.75$. Find the Nash equilibria and the evolutionarily stable states and strategies. When would you expect TFT to become the dominant strategy?

8. Verify the payoff matrix for the hawk–dove game with an owner–intruder role asymmetry in Table 8.6. For the game with $V = 2$ and $C = 4$ show that $\{0.5, 0, 0, 0.5\}$ is a Nash equilibrium but that it is not evolutionarily stable because it can be invaded by Bourgeois.

9. Verify the results in the text for the hawk–dove game with size asymmetry.

10. For the model of parental care in Table 8.8, find parameter values with the following evolutionarily stable strategies: (a) both desert; (b) stickleback only; (c) duck only; (d) both duck and stickleback; (e) both guard.

11. Find the equilibrium value for the vigilance model with $n = 2$ and verify that it is both a Nash equilibrium (Equation 37) and continuously stable (Equation 38). Draw the best-reply diagram and hence verify that it is a strong Nash equilibrium.

APPENDIX 8.1. THEOREMS ABOUT EVOLUTIONARILY STABLE STRATEGIES

Theorem 1: *If* $\mathbf{p}$ *is a mixed ESS in which the pure strategies i, j, and k, say, are used with nonzero probabilities, then* $E(i, \mathbf{p}) = E(j, \mathbf{p}) = E(k, \mathbf{p}) = E(\mathbf{p}, \mathbf{p})$ (Bishop and Cannings, 1978).

Proof: Suppose to the contrary that $E(i, \mathbf{p}) < E(\mathbf{p}, \mathbf{p})$. Define $\mathbf{q}$ as the strategy used by $\mathbf{p}$ conditional on not using i. Then

$$E(\mathbf{p},\mathbf{p}) = p_i E(i,\mathbf{p}) + (1-p_i)E(\mathbf{q},\mathbf{p})$$
$$< p_i E(\mathbf{p},\mathbf{p}) + (1-p_i)E(\mathbf{q},\mathbf{p}) \tag{39}$$

Hence $E(\mathbf{p}, \mathbf{p}) < E(\mathbf{q}, \mathbf{p})$, contrary to the assumption that $\mathbf{p}$ is an ESS. Thus $E(\mathbf{p}, \mathbf{p}) \leq E(i, \mathbf{p})$. But $E(\mathbf{p}, \mathbf{p})$ cannot be strictly less than $E(i, \mathbf{p})$ since $\mathbf{p}$ is an ESS. Hence $E(i, \mathbf{p}) = E(\mathbf{p}, \mathbf{p})$.

Theorem 2: *Suppose* $\mathbf{p}$ *and* $\mathbf{q}$ *are ESS's. Then* $S(\mathbf{p}) \not\subset S(\mathbf{q})$, *where* $S(\mathbf{p})$ *means the support of* $\mathbf{p}$, *the set of all strategies appearing with positive probability in* $\mathbf{p}$.

Proof: Suppose that $\mathbf{q}$ is an ESS and that $S(\mathbf{p}) \subset S(\mathbf{q})$. From Theorem 1, $E(\mathbf{q}, \mathbf{q}) = E(j, \mathbf{q})$ for all j in $S(\mathbf{q})$ and hence for all j in $S(\mathbf{p})$. Hence,

$$E(\mathbf{p},\mathbf{q}) = \sum_j p_j E(j,\mathbf{q}) = E(\mathbf{q},\mathbf{q}) \tag{40}$$

Since $\mathbf{q}$ is an ESS we must have $E(\mathbf{q}, \mathbf{p}) > E(\mathbf{p}, \mathbf{p})$, whence $\mathbf{p}$ cannot be an ESS.

Theorem 3: *In an asymmetric game with two roles,* A *and* B, *known to both players, an ESS must be a pure strategy* (Selten, 1980).

Proof: Suppose that $\alpha = \{\mathbf{p}_A, \mathbf{p}_B\}$ is a Nash equilibrium in which $\mathbf{p}_A$, the strategy played in the A role is a mixed strategy with the pure strategy i in its support. Consider the alternative pair of strategies $\beta = \{i, \mathbf{p}_B\}$ in which the animal in the A role always plays the pure strategy i. In the symmetrized game,

$$\begin{aligned}
E(\alpha,\alpha) &= \tfrac{1}{2}E_A(\mathbf{p},\mathbf{q}) + \tfrac{1}{2}E_B(\mathbf{p},\mathbf{q}) \\
E(\beta,\alpha) &= \tfrac{1}{2}E_A(i,\mathbf{q}) + \tfrac{1}{2}E_B(\mathbf{p},\mathbf{q}) \\
E(\alpha,\beta) &= \tfrac{1}{2}E_A(\mathbf{p},\mathbf{q}) + \tfrac{1}{2}E_B(i,\mathbf{q}) \\
E(\beta,\beta) &= \tfrac{1}{2}E_A(i,\mathbf{q}) + \tfrac{1}{2}E_B(i,\mathbf{q})
\end{aligned} \tag{41}$$

Since

$$E_A(i,\mathbf{q}) = E_A(\mathbf{p},\mathbf{q}) \tag{42}$$

it follows that

$$\begin{aligned}
E(\alpha,\alpha) &= E(\beta,\alpha) \\
E(\alpha,\beta) &= E(\beta,\beta)
\end{aligned} \tag{43}$$

Hence α is not evolutionarily stable, but only neutrally stable, since β can increase in frequency by drift (Equation 15b).

This conclusion may however be too far-reaching unless there is some alternate strategy that is positively selected, as is the case in many biological examples such as the hawk–dove game with an owner–intruder asymmetry. Selten's theorem suggests that the mixed Nash equilibrium solution in Equation 6 to the game of simplified poker is not evolutionarily stable, despite the universal acceptance among game theorists that it is the unique stable solution to the game. The problem is that the above analysis does not take into account that, once some A-role players start using strategy i, there is selection on the B-role players to change their strategy from $\mathbf{q}$ to something else, that may (and in the case of simplified poker probably does) select for A-role players to revert to $\mathbf{p}$, thus stabilizing the equilibrium. This type of consideration is taken into account in the concept of continuous stability discussed below, and the failure of the ESS approach to address it is the main weakness of this approach.

Theorem 4: *In a continuous-strategy game, the condition for the equilibrium value* v^* *to be continuously stable is that*

$$\left[\frac{\partial^2 w}{\partial u^2}+\frac{\partial^2 w}{\partial u \partial v}\right]_{u=v=v^*} < 0 \tag{44}$$

where $w(u, v)$ *is the fitness of a rare* u*-strategist in a population of* v*-strategists* (Eshel, 1983; Taylor, 1989).

Proof: Suppose that the population value v is displaced from its optimum v^* by a small amount ε and consider a mutant with value u that is displaced from the population value by a small amount δ. Thus the mutant will have value $u = v^* + \delta + \varepsilon$, and its fitness will be

$$\begin{aligned} w(u,v) &= w(v^*+\delta+\varepsilon, v^*+\varepsilon) \\ &\cong w(v^*+\varepsilon, v^*+\varepsilon)+\delta\left.\frac{\partial w}{\partial u}\right|_{u=v=v^*+\varepsilon} \\ &\cong w(v^*+\varepsilon, v^*+\varepsilon)+\delta\left\{\frac{\partial w}{\partial u}+\varepsilon\left[\frac{\partial^2 w}{\partial^2 u}+\frac{\partial^2 w}{\partial u \partial v}\right]\right\}_{u=v=v^*} \\ &= \text{wild-type fitness}+\delta\varepsilon\left[\frac{\partial^2 w}{\partial^2 u}+\frac{\partial^2 w}{\partial u \partial v}\right]_{u=v=v^*} \end{aligned} \tag{45}$$

The difference between mutant and wild-type fitness is therefore

$$\delta\varepsilon\left[\frac{\partial^2 w}{\partial^2 u}+\frac{\partial^2 w}{\partial u \partial v}\right]_{u=v=v^*} \tag{46}$$

For stability we require that selection should favor the mutant if δ and ε are of opposite sign (so that a mutation moving the population nearer to the equilibrium will increase in frequency) and should disfavor the mutant if δ and ε have the same sign (so that a mutation moving the population further away from the equilibrium will be eliminated). Thus this type of stability, which is called *continuous stability*, requires that the inequality (44) be satisfied. This result can be applied to any model with a continuous strategy set. As long as δ and ε are small so that (46) approximates the difference in fitness between wild-type and mutant, demonstration of this property for one pair of small values of δ and ε suffices to demonstrate it for all such values. This validates the argument in the text from the form of the best-reply diagram.

CHAPTER 9

Kin Selection and Inclusive Fitness

"I ... will confine myself to one special difficulty, that at first appeared to me insuperable, and actually fatal to my theory. I allude to the neuters or sterile females in insect-communities: for these neuters often differ widely in instinct and in structure from both the males and fertile females, and yet, from being sterile, they cannot propagate their kind. ...

"I will here take only a single case, that of working or sterile ants. ... The great difficulty lies in the working ants differing widely from both the males and the fertile females in structure ... and in instinct. With the working ant we have an insect differing greatly from its parents, yet absolutely sterile; so that it could never have transmitted successively acquired modifications of structure or instinct to its progeny. It may well be asked how is it possible to reconcile this case with the theory of natural selection? ...

"This difficulty, though appearing insuperable, is lessened, or, as I believe, disappears, when it is remembered that selection may be applied to the family, as well as to the individual, and may thus gain the desired end. Thus, a well-flavoured vegetable is cooked, and the individual is destroyed; but the horticulturalist sows seeds of the same stock, and confidently expects to get nearly the same variety; breeders of cattle wish the flesh and fat to be well marbled together; the animal has been slaughtered, but the breeder goes with confidence to the same family. I have such faith in the power of selection, that I do not doubt that a breed of cattle, always yielding oxen with extraordinarily long horns, could be slowly formed by carefully watching which individual bulls and cows, when matched, produced oxen with the longest horns; and yet no ox could ever have propagated its kind. Thus I believe it has been with social insects: a slight modification of structure, or instinct, correlated with the sterile condition of certain members of the community, has been advantageous to the community: consequently the fertile males and females of the same community flourished, and transmitted to their fertile offspring a tendency to produce sterile members having the same modification. And I believe that this process has been repeated, until that prodigious amount of difference between the fertile and sterile females of the same species has been produced, that we see in many social insects."

—Charles Darwin, *On the Origin of Species* (1859)

We saw in Chapter 5 that, in a stationary population, an organism is selected to maximize its lifetime reproductive success. How on this basis can we explain the evolution of sterile worker ants that leave no offspring? Darwin's solution to this problem relies typically on his knowledge of animal and plant breeding; it depends on what breeders call family selection, that is to say, breeding from the families of individuals with desirable characteristics. This idea contains the germ of the modern theory of kin selection, but he was unable to make it more precise because he did not know how heredity works. The general theory of kin selection, whereby natural selection can favor individuals who sacrifice their own fitness but enhance that of their relatives, has been developed by Hamilton (1964, 1972) and other authors and is today an influential concept in evolutionary biology.

Kin selection theory can be modeled by constructing a population genetic model to investigate the conditions under which altruism will evolve. This method is rigorous and is guaranteed to give the correct answer (if the genetics of the trait are known) but is often cumbersome. An alternative is to use the concept of inclusive fitness, leading to *Hamilton's rule*. Suppose that an altruist performs an act that decreases his own fitness by an amount c (the cost to the donor) but increases that of a relative by an amount b (the benefit to the recipient). Hamilton's rule states that altruism will evolve when

$$rb > c \tag{1}$$

where r is the coefficient of relatedness between donor and recipient. Inclusive fitness and Hamilton's rule are easier to use in practice since they bypass the genetics of the trait, but they are only approximately correct.

In this chapter I shall describe population genetic models before discussing the concepts of relatedness and inclusive fitness; I shall then apply these concepts to models of optimal dispersal of offspring and of kin selection in viscous populations with limited dispersal.

POPULATION GENETIC MODELS OF KIN SELECTION

Consider two genotypes, altruist and selfish. Altruists perform some act that benefits their relatives by an amount b at a cost c to themselves (b and c being measured in units of fitness). Altruistic individuals pay the cost c, but they are also likely to have altruistic relatives from whom they receive the benefit b; selfish individuals do not pay the cost c but are less likely to have altruistic relatives from whom they receive the benefit b. The overall fitness of a particular genotype, taking into account both the cost of any altruistic acts its bearer performs and the likely benefits he receives from his relatives, is the *neighbor-modulated fitness* of that genotype. Natural selection will favor the genotype with the highest neighbor-modulated fitness. This is usually determined by constructing a population genetic model and finding the rate of increase (or decrease) of the gene for altruism. In this section I shall illustrate how this can be done by considering a general model of altruism between siblings and two specific models for the evolution of siblicide (killing one's siblings) in birds and parasitic wasps. (Siblicide is of course selfish, so that refraining from it is the altruistic act.) Hamilton's rule holds exactly for the general model and its application to siblicide in birds but cannot be applied to the model of siblicide in parasitic wasps. We shall discuss later the reasons for this difference, that can be taken as a paradigm of the circumstances under which Hamilton's rule can or cannot be applied.

A model of altruism between siblings

Maynard Smith (1989) discusses the following model of altruism between siblings. Imagine a species of bird with a clutch size of two. The birds are monogamous, so that members of a clutch are full sibs. In typical members of the population (genotype aa), the probabilities of survival of the older and younger sib are S and s respectively ($s < S$). A rare dominant gene A (for altruism) has the following effects. If present in the older sib it causes that sib to be less greedy for food, so that its survival probability becomes $S - c$ and that of its sib becomes $s + b$; it has no effect if present in the younger sib.

Assume random mating so that the AA genotype has negligible frequency. Write x for the relative frequency of *Aa* individuals, which is presumed to be small. To order x, only two types of mating need be considered: $aa \times aa$ [probability $(1 - x)^2 \cong 1$], giving $(S + s)$ *aa* offspring; and $Aa \times aa$ or $aa \times Aa$ (probability $\cong 2x$), giving the offspring types shown in Table 9.1. The relative frequency of *Aa* individuals in the next generation is found by dividing the number of *Aa* survivors by the total number of survivors per mating. The linearized recurrence relationship for x is

$$\begin{aligned} x' &= \frac{1}{4}\left[\frac{s+(S-c)+(S-c+s+b)}{S+s}\right]2x \\ &= \left[1+\frac{(\frac{1}{2}b-c)}{S+s}\right]x \end{aligned} \qquad (2)$$

where x' is the value of x in the next generation. The gene for altruism will increase in frequency if $0.5b > c$, which is Hamilton's rule since the coefficient of relatedness between full sibs is 0.5. The intuitive reason for this result can be seen from the table of offspring types. The *Aa* genotype is equally likely to be found in the elder and the younger sib; in the former it always suffers the cost c while in the latter it has a 50% chance of enjoying the benefit b. A similar analysis shows that the altruistic gene, when common, can resist invasion by a rare selfish mutant under the same conditions.

The evolution of siblicide in birds

The above model can be applied directly to the evolution of siblicide in birds (Stinson, 1979; Godfray and Harper, 1990). Many raptors (eagles and hawks) lay two eggs that hatch asynchronously, so that the elder chick is larger than the younger one. If both eggs hatch, the younger chick is always killed by its older sibling, a practice called siblicide. We may identi-

Table 9.1. Offspring from $Aa \times aa$ matings under a model of altruism between sibs.

	Offspring types (elder sib first)			
	aa, *aa*	*aa*, *Aa*	*Aa*, *aa*	*Aa*, *Aa*
Frequency:	$\frac{1}{4}$	$\frac{1}{4}$	$\frac{1}{4}$	$\frac{1}{4}$
Aa survivors:	0	s	$S - c$	$S - c + s + b$

fy siblicide as the selfish trait and refraining from siblicide as the altruistic trait, from the viewpoint of the older chick. The cost c of refraining from siblicide to the older chick is the decrease in its chance of survival through having to share the food brought to the nest by the parents; the benefit b to the younger chick is its chance of survival in these circumstances. Siblicide will evolve when $c > 0.5b$, that is to say, when the decrease in the survival probability of the older chick if it has to share parental resources exceeds half the survival probability of the younger chick.

It has been assumed that the elder chick can determine whether or not to kill its younger sib. The parental interest is to maximize the number of surviving young, so that they should try to suppress siblicide when $c < b$. When $0.5b < c < b$ there is a conflict of interest between the elder chick and the parents, the former being selected to practice siblicide and the latter to suppress it. Which party will win such a conflict of interest will depend on the means available to each of them to gain the upper hand. It may be that the possession of talons by raptor chicks, coupled with asynchronous hatching, puts the elder chick in a strong position to get its way. It may also be asked why the parents should lay two eggs if they know that only one of the chicks will survive. The simplest explanation is the insurance-egg hypothesis (Anderson, 1990), that the second egg will be useful if the first egg fails to hatch.

The evolution of siblicide in parasitic wasps

Siblicide is rare in birds but is quite common in parasitic wasps. However, a different model must be used taking into account that the larvae are all of the same age and also the fact that wasps are haplodiploid, males being haploid and females diploid. I follow here the account of Godfray (1987).

Consider first the spread of a rare gene, A, for fighting. Females will be either Aa (with probability x) or aa (with probability $1 - x$), where x is sufficiently small that the frequency of AA females is negligible; males will be either A (with probability y) or a (with probability $1 - y$), where y is small. (The frequencies of mutant males and females must be accounted separately because of the asymmetry of haplodiploid inheritance.) Suppose that a mated female lays a clutch of n eggs. If none of them has the killer genotype they all have the same chance s of surviving to breed next year. If one or more of them is a killer, all the nonkillers and all but one of the killers die; the remaining killer, chosen at random, has the chance S of surviving to breed next year ($S > s$).

Three types of mating must be considered: $aa \times a$ [probability $(1 - x) \times (1 - y) \cong 1$], $Aa \times a$ (probability $\cong x$), and $aa \times A$ (probability $\cong y$). Assuming an equal sex ratio among the offspring, the first type of mating will produce on average $\frac{1}{2}ns$ *aa* and $\frac{1}{2}ns$ *a* offspring in the next generation. The second type of mating will produce *Aa*, *aa*, *A*, and *a* eggs with equal probability. The chance that the clutch will contain at least one mutant genotype is $1 - 0.5^n$, in which case the winner is equally likely to be an *Aa* female or an *A* male with a chance *S* of surviving to reproductive age. The third type of mating will produce *Aa* females and *a* males with equal probability. The chance that the clutch will contain at least one mutant *Aa* genotype is $1 - 0.5^n$, in which case the winner is an *Aa* female. The numbers of *Aa* females and *A* males next year (per mating this year) will be

$$\begin{aligned} Aa&: \ \tfrac{1}{2}\left[1-0.5^n\right]Sx+\left[1-0.5^n\right]Sy \\ A&: \ \tfrac{1}{2}\left[1-0.5^n\right]Sx \end{aligned} \tag{3}$$

These numbers are now divided by the total numbers of females and males, respectively, per mating to convert them to relative frequencies; both these numbers are $\frac{1}{2}ns$ since nearly all matings are of type $aa \times a$. The linearized recurrence relations for x and y, both assumed to be small, are

$$\begin{bmatrix} x' \\ y' \end{bmatrix} = \begin{bmatrix} C & 2C \\ C & 0 \end{bmatrix} \begin{bmatrix} x \\ y \end{bmatrix} \tag{4}$$

where

$$C = \left[1-0.5^n\right]S/ns \tag{5}$$

The dominant eigenvalue of the transition matrix is 2*C*, so that the condition for the killer genotype to spread is

$$2C > 1 \tag{6}$$

For example, if $n = 2$ the condition is

$$S > \tfrac{4}{3}s \tag{7}$$

To see under what circumstances siblicide is likely to arise one must make some further assumptions. Godfray assumed that survival rate declines exponentially with clutch size so that

$$s = S\exp[-h(n-1)] \tag{8}$$

where h is is a constant determining the strength of the fitness reduction as clutch size increases. He also supposed that the female lays a clutch size that maximizes the number of surviving offspring, *ns*, leading to the optimal clutch size (often called the "Lack solution" following the argument in Chapter 1) so that $n = 1/h$. Substituting $1/n$ for h in Equation 8 leads to a condition for the killer genotype to spread only involving n, which can be solved numerically. The critical value of n below which a gene for fighting will spread under these assumptions is 3.95, suggesting that a parental strategy of producing a clutch of two or three eggs will be evolutionarily unstable, resulting in the spread of the gene for fighting. Empirical observations provide strong circumstantial evidence for this model. The distribution of clutch size in different species of parasitic wasp is strongly bimodal with many solitary species, few species with clutches of two or three, and many species with clutches of between four and 30. The larvae of all solitary wasps (that develop alone in a host) are equipped with fighting mandibles, suggesting that they have evolved fighting behavior that has constrained their parents to lay only a single egg in a clutch, while no gregarious species has fighting mandibles.

Suppose now that the gene for fighting has spread through the population and that the environment becomes more favorable to larger broods. Under what conditions will a gene for tolerance spread? I now suppose that the common genotype, *aa* (in females) and *a* (in males), causes fighting and that there is a rare genotype *Aa* (with frequency x in females) and *A* (with frequency y in males) causing tolerance. The gene for tolerance will have no effect in single clutches, and I suppose that females lay clutches of two eggs (perhaps as insurance in case one of them does not hatch). Consider the three types of matings. First, $aa \times a$ matings [probability $(1 - x)(1 - y) \cong 1$] will produce on average $\frac{1}{2}S$ *aa* and $\frac{1}{2}S$ *a* offspring in the next generation. Second, $Aa \times a$ matings (probability $\cong x$) will produce two tolerant offspring with probability $\frac{1}{4}$, each with an equal chance of being male or female; thus they will on average produce $\frac{1}{4}s$ *Aa* and $\frac{1}{4}s$ *A* offspring in the next generation, plus some fighting offspring that we need not consider. Third, $aa \times A$ matings (probability $\cong y$) will produce two tolerant female offspring with probability $\frac{1}{4}$, and will thus produce $\frac{1}{2}s$ *Aa* offspring in the next generation, plus some fighting offspring that we need not consider. The numbers of *Aa* females and *A* males next year (per mating this year) will be

$$\begin{aligned} Aa&: \ \tfrac{1}{4}sx + \tfrac{1}{2}sy \\ A&: \ \tfrac{1}{4}sx \end{aligned} \qquad (9)$$

These numbers are now divided by the average numbers of surviving females and males respectively, which are in both cases $\frac{1}{2}S$ since nearly all matings are of type $aa \times a$, to convert them to relative frequencies. The linearized recurrence relations for x and y, both assumed to be small, are given by Equation 4 with $C = 0.5$. The condition for the gene for tolerance to spread is that $s > S$, which means that an individual must have a higher survival probability as one of a pair than when alone, which is very unlikely to be the case. Once fighting behavior has evolved there is no way in which the population can return to tolerance. Bull and Charnov (1985) have called this type of situation "irreversible evolution."

COEFFICIENTS OF CONSANGUINITY AND RELATEDNESS

The rest of this chapter will be devoted to the inclusive-fitness approach and its applications. This approach depends on the measurement of the relatedness between two individuals, that will be explained in this section.

Consider two homologous genes at the same locus; they may belong to different individuals or they may be the two genes at that locus in the same individual. Two such genes are said to be *identical by descent* if they are copies of the same gene in a common ancestor in the recent past. Such genes must be identical in state, barring mutation, but genes that are identical in state need not, of course, be identical by descent. Unless otherwise stated "identical" in this section means identical by descent. Relatives resemble each other because they share identical genes. Two useful measures of genetic resemblance are the coefficients of consanguinity and relatedness that will now be discussed in turn.

The coefficient of consanguinity

The *coefficient of consanguinity* F_{JK} between two individuals, J and K, is defined as the probability that two homologous genes drawn randomly, one from J and the other from K, are identical. To illustrate how this coefficient can be calculated suppose that J and K are full sibs in a diploid outbred population (see Figure 9.1). Write (1, 2) and (3, 4) for the parental genotypes (the use of distinct numbers indicating genes that are not identical by descent since there is no inbreeding) and suppose that J has genotype (1, 3); K is equally likely to inherit 1 or 2 from his mother (x) and 3 or 4 independently from his father (y). If we choose one gene at random from J and another from K, the chance that they are identical is $\frac{1}{4}$; for example, if we chose 1 from J, we would have to choose the maternal rather than the paternal gene from K and this gene would have to be 1 rather than 2. Hence $F_{JK} = \frac{1}{4}$.

Inbreeding in either of the parents in this pedigree would increase the coefficient of consanguinity between their children, as would consanguinity between the parents. For an inbred diploid individual, I, there is a positive probability that two homologous genes are identical; this probability is called the *coefficient of inbreeding*, denoted by F_I. Suppose that the moth-

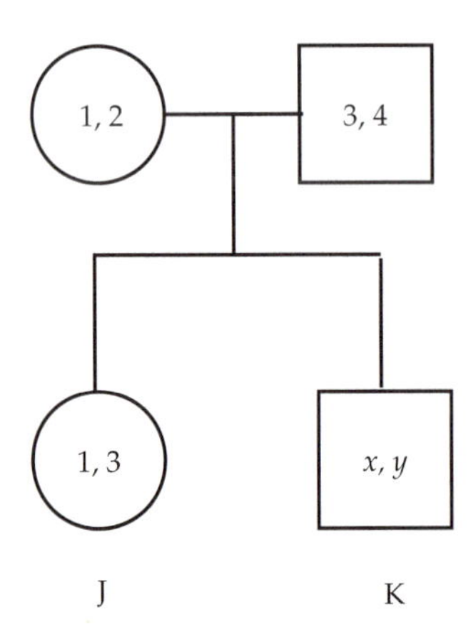

Figure 9.1. Calculation of the relatedness between sister and brother in an outbred diploid population.

er, M, in Figure 9.1 is inbred with $F_M = 0.2$, which means that there is a 20% chance of the genes labeled 1 and 2 being identical, while the father is not inbred and not related to the mother. A straightforward calculation shows that $F_{JK} = 0.275$.

The coefficient of consanguinity is defined in the same way in haplodiploids; the calculation of this coefficient between brother and sister in outbred haplodiploids is illustrated in Figure 9.2. The single gene x in the brother is derived from his mother and is therefore equally likely to be 1 or 2. If we choose one gene at random from J and another from K, the chance that they are identical is $\frac{1}{4}$ since we must choose 1 from J and x must be 1. Hence $F_{JK} = \frac{1}{4}$ as before. However, the coefficient of consanguinity often depends on the sex of the relatives in haplodiploids, because of the genetic asymmetry of the two sexes. For example, it is always $\frac{1}{4}$ between full sibs in diploids regardless of their sex, whereas it is $\frac{3}{8}$ between sisters and $\frac{1}{2}$ between brothers in haplodiploids.

We can also calculate F_{JJ}, the coefficient of consanguinity between an individual and himself (or herself), as the probability that two genes at the same locus drawn randomly *with replacement* are identical. For an outbred diploid individual $F_{JJ} = \frac{1}{2}$ since we have to choose the same gene each time. For an inbred diploid individual there is a positive probability that two homologous genes are identical; in this case

$$F_{JJ} = \frac{1}{2}(1 + F_J) \tag{10}$$

since the probability of identity is 1 if we choose the same gene each time

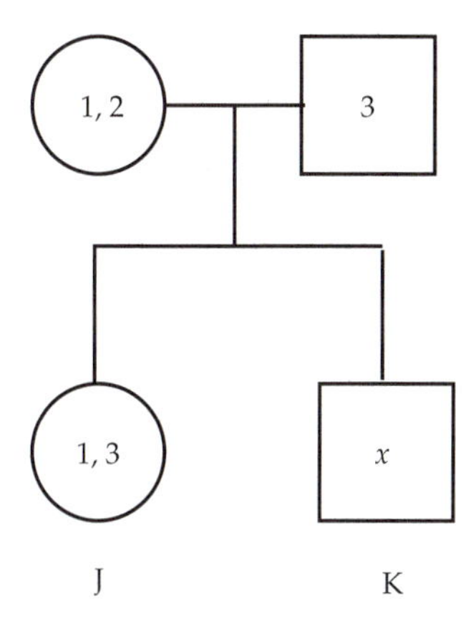

Figure 9.2. Calculation of the relatedness between sister and brother in outbred haplodiploids.

and F_J if we choose different genes. For a haploid individual $F_{JJ} = 1$ since we are bound to choose the same gene each time. (Note that the coefficient of inbreeding in a diploid individual is the probability that two genes at the same locus drawn randomly *without replacement* are identical; this concept is undefined in a haploid individual.)

The coefficient of relatedness

The coefficient of consanguinity is only a tool to be used in the construction of the coefficient of relatedness. There are several ways in which the coefficient relatedness, r_{JK}, can be defined. The most useful general definition in the context of inclusive fitness is

$$r_{JK} = F_{JK}/F_{JJ} \quad (11)$$

(I am grateful to P. D. Taylor for pointing out to me the generality and intuitive appeal of this definition.) This leads to the results

$$r_{JK} = 2F_{JK} \quad \text{if J is diploid and outbred} \quad (12a)$$

$$r_{JK} = 2F_{JK}/(1+F_J) \quad \text{if J is diploid and inbred} \quad (12b)$$

$$r_{JK} = F_{JK} \quad \text{if J is haploid} \quad (12c)$$

Whereas F_{JK} is symmetrical between J and K ($F_{KJ} = F_{JK}$), r_{JK} is not; if $F_{JJ} \neq F_{KK}$ (which will happen if J and K differ in ploidy or degree of inbreeding), then $r_{KJ} \neq r_{JK}$. Tables 9.2 and 9.3 show some values of the coefficient

Table 9.2. The coefficient of relatedness in outbred diploids.

Relationship	r_{JK}	Relationship	r_{JK}
Identical twins	1	Parent–offspring	$\frac{1}{2}$
Full sibs	$\frac{1}{2}$	Half sibs	$\frac{1}{4}$
Grandparent–grandchild	$\frac{1}{4}$	Uncle–nephew	$\frac{1}{4}$
Single first cousins	$\frac{1}{8}$	Double first cousins	$\frac{1}{4}$

Table 9.3. The coefficient of relatedness in outbred haplodiploids.

Relationship	r_{JK}	Relationship	r_{JK}
Mother–daughter	$\frac{1}{2}$	Daughter–mother	$\frac{1}{2}$
Mother–son	1	Son–mother	$\frac{1}{2}$
Father–daughter	$\frac{1}{2}$	Daughter–father	1
"Father"–son	0	Son–"father"	0
Sister–sister	$\frac{3}{4}$	Brother–brother	$\frac{1}{2}$
Sister–brother	$\frac{1}{2}$	Brother–sister	$\frac{1}{4}$

of relatedness in outbred diploids and haplodiploids, respectively.

The coefficient of relatedness defined in Equation 11 can be interpreted as the regression coefficient of the genotype of K on that of J. Consider a neutral locus with two alleles, A and a, with frequencies p and $q = 1 - p$ in an outbred diploid population. Under random mating the frequencies of the three genotypes, AA, Aa, and aa are p^2, $2pq$, and q^2 by the Hardy–Weinberg law. Define X as a variable taking the values 1, 0.5, and 0 for the genotypes AA, Aa, and aa respectively; it is the proportion of A genes in an individual's genotype. X is a random variable with the Hardy–Weinberg probabilities (i.e., taking the values 1, 0.5, and 0 with probabilities p^2, $2pq$, and q^2), and has mean and variance

$$\begin{aligned} E(X) &= p \\ \mathrm{Var}(X) &= \tfrac{1}{2}pq \end{aligned} \tag{13}$$

The values of X in two relatives, J and K, are correlated. It is shown in Appendix 9.1 that the regression of X_K on X_J, that is to say the expected value of X_K given X_J, is

$$E(X_K \mid X_J) = p + r_{JK}(X_J - p) \tag{14}$$

where r_{JK} is the coefficient of relatedness in Equation 12a. Thus the regression is linear with slope r_{JK}.

In an outbred haplodiploid population we consider a neutral locus with two alleles, A and a, as before and we define X as the variable taking values 1, 0.5, and 0 for the genotypes AA, Aa, and aa in females; in males we define X as taking values 1 and 0 for the genotypes A and a, so that it is the proportion of A genes in the individual's genotype in both cases. With this definition of X the regression of X_K on X_J is given by Equation 14 with r_{JK} defined in Equation 12a or 12c according to whether J is diploid or haploid.

The regression in Equation 14 also holds under inbreeding, provided it is interpreted as the best linear approximation to the true regression, since the true regression cannot be guaranteed to be linear, with r_{JK} defined in Equation 12b or 12c depending on whether J is diploid or haploid.

The coefficient of relatedness is the fundamental quantity in inclusive-fitness arguments, occurring in the formulation of Hamilton's rule, because it is the regression coefficient in Equation 14. To determine whether altruistic acts in natural populations can be explained by kin-selection theory using Hamilton's rule it is necessary to estimate the average coefficient of relatedness between interacting pairs of individuals. This can be done either from pedigrees or from genetic markers such as protein polymorphisms or DNA fingerprints (Queller and Goodnight, 1989).

INCLUSIVE FITNESS AND HAMILTON'S RULE

The purpose of having children (from a biological viewpoint) is to project one's genes into future generations. But one's relatives share some of one's own genes, so that the same purpose can be accomplished by helping them to project their genes into the future, after allowing for the fraction of one's own genes they are likely to share. This is the idea underlying the concept of inclusive fitness and Hamilton's rule. In this section I shall derive the rule under simplifying assumptions, first in a diploid and then in a haplodiploid organism, giving a few illustrations of its use, and I shall finally discuss the limitations on its validity.

Diploid organisms

Suppose that A is a gene causing J to be altruistic to K. To be precise, suppose that J is altruistic with probability $\pi(X_J)$ depending on the genotype of J, with $\pi(0) \leq \pi(0.5) \leq \pi(1)$, and that when J is altruistic he suffers a personal loss of fitness c while K gains b. [Two obvious submodels would be $\pi(0) = 0$, $\pi(0.5) = 0.5$, $\pi(1) = 1$ (no dominance) and $\pi(0) = 0$, $\pi(0.5) = \pi(1) = 1$ (complete dominance).] When J is selfish he contributes X_J A gametes to the next generation; when he is altruistic he contributes $(1 - c)X_J$ A gametes directly and $bE(X_K | X_J)$ indirectly through helping K. The total gametes (A or a) contributed through the actions of A are 1 when he is selfish and $1 - c + b$ when he is altruistic. The frequency of A in the next generation is the ratio of the number of A gametes to the total number of gametes

$$p' = \frac{E\{[1-\pi(X_J)]X_J + \pi(X_J)[(1-c)X_J + bE(X_K|X_J)]\}}{E\{[1-\pi(X_J)] + \pi(X_J)[1-c+b]\}} \tag{15}$$

the outer expectations being taken over the distribution of X_J. After substitution from Equation 14 and some straightforward algebraic simplification, it will be found that the change in the gene frequency is

$$\begin{aligned} \Delta p \equiv p' - p &= \frac{(r_{JK}b - c)E[\pi(X_J)(X_J - p)]}{1+(b-c)E[\pi(X_J)]} \\ &= \frac{(r_{JK}b - c)\mathrm{Cov}[\pi(X_J), X_J]}{1+(b-c)E[\pi(X_J)]} \end{aligned} \tag{16}$$

The fact that A is the gene *for* altruism embodied in the stipulation that $\pi(0) \leq \pi(0.5) \leq \pi(1)$ (with at least one of the inequalities holding strictly) ensures that $\pi(X_J)$ and X_J are positively correlated. Hence the gene for altruism will increase in frequency provided that $r_{JK}b > c$. This is Hamilton's rule. Equivalently we may define $r_{JK}b - c$ as the *inclusive fitness* of the individual J performing the altruistic act and assert that altruism will be favored by natural selection if J's inclusive fitness is positive. This accords with the definition of fitness as a quantity that is maximized by selection.

Several examples of the importance of kin selection are provided by cooperative breeding in birds and mammals in which related adult individuals in addition to the genetic parents regularly aid in the rearing of young. I shall here discuss data on the White-fronted Bee-eater (Emlen, 1990) and the Florida Scrub Jay (Woolfenden and Fitzpatrick, 1984).

White-fronted Bee-eaters nest in large colonies, and nesting pairs are often aided in rearing their young by one or more helpers. The data in Table 9.4 show that breeding success is greatly increased by the help received, and Emlen (1990) estimates that each helper increases the number of fledglings by about 0.44. There is also a strong tendency for help to be directed toward close relatives, and in consequence the average coefficient of relatedness between a helper and the young being helped is 0.33. There is thus good reason to suppose that helping at the nest in this species has evolved through kin selection. However, a simple-minded calculation suggests that the benefit of helping is not large enough to warrant this conclusion. Suppose that a bird has the following choice: either breed without help from others and have 0.41 young related to it by 0.5 or help a relative to have an additional 0.44 offspring related to it by 0.33. Since $0.41 \times 0.5 = 0.205 > 0.44 \times 0.33 = 0.145$ the bird should breed rather than help a relative. The problem with this calculation is that the class of birds that became helpers in a particular year would probably have had substantially less than 0.41 young if they had chosen not to help. Many of them are unmated birds, and it is unknown how many of them are involuntarily unmated and do not have the opportunity to breed. Another class of helpers are breeding birds whose nests have failed; their expected success if they try to breed again is probably small. The final class of helpers are mated birds who choose not to breed in a particular year. It is plausible that they choose not to breed because they are in worse condition, so that their expected success is lower than that of birds who choose to breed. The social structure implies that one bird in a mated pair is native to the breeding area and has close relatives in it, while the other is an immigrant and has no relatives nearby. When a mated pair does not breed, the native bird usually becomes a helper, while the immigrant rarely does so; this is strong evidence of the implication of kin selection.

In the Florida Scrub Jay (Woolfenden and Fitzpatrick, 1984) the social unit consists of a monogamous breeding pair together with some of the young of previous years who assist in defense of the territory and in feeding and guarding the nestlings. Removal experiments have shown that the help of the young substantially increases the reproduc-

Table 9.4. Fledging success in White-fronted Bee-eaters.

Group size	2	3	4	5	6
Young fledged	0.41	0.83	1.24	1.76	2.70

After Emlen, 1990.

tive success of their parents, each helper increasing the number of young fledged by 0.33; the average relatedness of a helper to the nestlings is 0.43. However, the average number of young reared by first-time breeders is 1.24. Since $1.24 \times 0.5 = 0.62$ is much greater than $0.33 \times 0.43 = 0.14$, it would clearly be better for a young bird to breed if it could, rather than to stay at home to help, even though there are other advantages to staying at home such as increasing its parents' chances of survival and learning how to become an efficient parent. The main difference from the previous example is that in bee-eaters breeding sites are readily available so that any bird that wants to do so can breed (if it has a mate), while in the scrub jay there is a scarcity of available breeding territories (the species occupies isolated islands of relict oak scrub habitat). Thus it would be a mistake to interpret helpers in the scrub jay as birds that have chosen to stay at home to help their parents rather than to breed themselves; instead, they have chosen to stay at home to help rather than to do nothing. This is favored by kin selection since $0.14 > 0$. The optimal strategy for a young scrub jay is to stay behind to help its parents until a breeding territory becomes available.

Other examples of cooperative breeding in which kin selection plays a role are provided by the Acorn Woodpecker (Koenig and Mumme, 1987) and in mammals by lions (Packer, 1986) and by the naked mole rat (Sherman et al., 1990), the only eusocial mammal.

Haplodiploid organisms

We turn now to haplodiploid organisms in which the coefficients of relatedness shown in Table 9.3 are asymmetric and sex-dependent. In applying inclusive fitness to kin selection in haplodiploids when the donor and the recipient are of different sexes we must also take into account that a male is only worth half as much as a female in transmitting genes into future generations, because he only transmits genes to daughters while she transmits genes to both sexes. To formalize this argument following Taylor (1988a) let a_{ij} denote the proportion of genes a sex-i offspring gets from its sex-j parent ($i, j = 1$ for female and 2 for male). The matrix $\mathbf{A} = \{a_{ij}\}$ is

$$\mathbf{A} = \begin{bmatrix} \frac{1}{2} & \frac{1}{2} \\ 1 & 0 \end{bmatrix} \tag{17}$$

Now $\mathbf{A}^2$ represents the proportion of genes a sex-i grandoffspring gets from his or her sex-j grandparents, and $\mathbf{A}^n$ for large n represents the proportion of genes a sex-i individual receives from his or her sex-j ancestors in the distant past. But

$$\mathbf{A}^n \rightarrow \begin{bmatrix} \frac{2}{3} & \frac{1}{3} \\ \frac{2}{3} & \frac{1}{3} \end{bmatrix} \tag{18}$$

(This result follows from the fact that (2, 1) is the dominant left eigenvector of the matrix **A** giving the relative reproductive values of females and males.) Thus a female is worth twice as much as a male in projecting genes into the distant future. Hence Hamilton's rule states that J will be selected to sacrifice fitness c if the benefit b to the relative K satisfies

$$r_{\mathrm{JK}} b v_{\mathrm{K}} > c v_{\mathrm{J}} \tag{19}$$

where v_{J} and v_{K} are the reproductive values of J and K in transmitting genes into distant generations, which are proportional to their ploidies (1 for a male, 2 for a female).

Thus in haplodiploids the benefit to the recipient must be devalued by the quantity $r_{\mathrm{JK}} v_{\mathrm{K}}/v_{\mathrm{J}}$ compared with the cost to the donor. Hamilton (1972) called this quantity the *life-for-life* coefficient of relatedness. It is tabulated in Table 9.5.

As an example, consider colony founding by single or multiple foundresses in social insects. Metcalf and Whitt (1977a,b) have studied this question in the paper wasp *Polistes metricus*. Nests are founded in the spring, either by a single overwintering female or by a pair of females who came from the same nest the previous year. In the latter case, one of them, the α-female, is dominant and lays most of the eggs while the other, the

Table 9.5. Life-for-life coefficients of relatedness in outbred haplodiploids.

Relationship	$r_{\mathrm{JK}} \frac{v_{\mathrm{K}}}{v_{\mathrm{J}}}$	**Relationship**	$r_{\mathrm{JK}} \frac{v_{\mathrm{K}}}{v_{\mathrm{J}}}$
Mother–daughter	$\frac{1}{2}$	Daughter–mother	$\frac{1}{2}$
Mother–son	$\frac{1}{2}$	Son–mother	1
Father–daughter	1	Daughter–father	$\frac{1}{2}$
"Father"–son	0	Son–"father"	0
Sister–sister	$\frac{3}{4}$	Brother–brother	$\frac{1}{2}$
Sister–brother	$\frac{1}{4}$	Brother–sister	$\frac{1}{2}$

β-female, does most of the foraging. Nests with two foundresses produce about 3.1 times as many reproductives at the end of the season as those with a single foundress. Why does the β-female remain in the nest to help her sister rather than leaving to found her own nest? Taking the unit of fitness as the number of reproductives produced by a colony with a single foundress, the β-female confers on her sister a benefit $b = 2.1$ at a cost to herself of $c = 1$. If they are sisters, their life-for-life coefficient of relatedness is $r = 0.75$ and $rb - c = 0.575$; since this is positive, it pays the β-female to stay. Why she should be content to adopt the subordinate role is a problem in game theory. (It may be noted that if the same costs and benefits held in a diploid species with a coefficient of relatednes $r = 0.5$ between sisters, then $rb - c = 0.05$; the advantage to the β-female would still be positive, but it would be very slight. The increased relatedness between sisters in haplodiploids may provide a genetic predisposition toward the evolution of eusociality (Hamilton, 1972). I shall return to this possibility in the next chapter.)

In fact, things are not quite so simple. The β-female in a joint colony lays about 20% of the eggs, and partly for this reason and partly because of multiple insemination, the two females have a relatedness of 0.63 rather than 0.75, as estimated from electrophoretic studies. Taking these facts into account, $b = 3.1 \times 0.8 - 1 = 1.48$, $c = 1 - 0.2 \times 3.1 = 0.38$ and $rb - c = 0.5524$, which does not alter the conclusion.

The result that females have twice the reproductive value of males in haplodiploids is based on the assumption that in social insects the queen lays all the eggs. In some cases the workers lay some of the eggs, that are always haploid and therefore male since the workers' eggs are unfertilized. Suppose that a proportion p of the males develop from queen-laid eggs, the remaining $1 - p$ from worker-laid eggs. All the genes in males from queen-laid eggs are derived from the queen, but only half the genes in males from worker-laid eggs come from the queen, the other half coming from her mate. Hence

$$\mathbf{A} = \begin{bmatrix} \frac{1}{2} & \frac{1}{2} \\ \frac{1}{2}(1+p) & \frac{1}{2}(1-p) \end{bmatrix} \tag{20}$$

The dominant left eigenvector of this matrix is $(1 + p, 1)$ so that the reproductive value of females compared with males is $1 + p$ (Benford, 1978). Thus females are twice as valuable as males if all the eggs are queen-laid, but this advantage declines as the proportion of worker-laid eggs increases.

When considering the evolution of altruism in haplodiploid organisms it is important to distinguish carefully between the basic coefficient of relatedness r_{JK} and the life-for-life coefficient that also takes into account the relative reproductive value of females compared with males. The life-for-life coefficient is the key quantity in the evolution of altruism, but the two factors incorporated into it may vary independently of each other. In particular, worker-laying changes the relative value of females compared with males without changing the basic relatedness, while inbreeding changes the basic relatedness without changing the relative values of females and males.

Limitations of inclusive-fitness models

Inclusive-fitness models and Hamilton's rule are popular because they can be used without knowing the underlying genetics of the trait and are easier to apply than a population genetic model. It is therefore important to know when they will predict the same answer. In deriving Hamilton's rule we made two assumptions that may not be satisfied.

First, the inclusive-fitness model assumes that the regression in Equation 14, derived for a neutral locus, can be applied to the altruism locus that is under selection. This may not be exactly true, since selection may cause changes in gene frequencies between generations as well as departures from Hardy-Weinberg equilibrium. Thus the inclusive-fitness approach can only be guaranteed to be accurate under weak selection when the departures from the assumption of neutrality will be minimal.

A more serious constraint on the validity of the inclusive-fitness model is that the costs and benefits must be additive constants if individuals interact with several other individuals or if they may interact with the same individual as both donor and recipient. To illustrate this point consider the evolution of siblicide when the clutch size is two, with juveniles reared on their own or with their sibling having chances of S and s, respectively, of surviving to reproduce. If siblicide is restricted to the older sib killing the younger, as in the model for the evolution of siblicide in birds, the cost of altruism to the older sib is $S - s$, and the benefit to the younger is s. By Hamilton's rule, siblicide is expected to evolve when $S > 1.5s$, which agrees exactly with the result from the population genetic model. Thus Hamilton's rule is exact in this case even under strong selection.

Now suppose that the sibs are of the same size and that each may try to kill the other, as in the model for the evolution of siblicide in parasitic wasps but ignoring the complication of haplodiploidy. In *aa* sibships, with two siblicidal young, each has fitness $\frac{1}{2}S$ since they each have a probability of $\frac{1}{2}$ of surviving; in *Aa* sibships with one offspring altruist and one siblicidal, the altruist has fitness 0 and the other has fitness S; in *AA* sibships each has fitness *s*. Comparison of *Aa* with *aa* sibships shows that the cost of altruism to a siblicidal sib is $\frac{1}{2}S$ and the benefit to the sib is also $\frac{1}{2}S$; altruism could not evolve unless the sib were in fact an identical twin, and even in that case it would be selectively neutral. On the other hand, comparison of *Aa* with *AA* sibships shows that the cost of altruism toward an altruistic sib is $S - s$ and the benefit to the sib is *s*, as in the previous paragraph. Thus the costs and benefits are frequency-dependent; it is much more difficult for altruism to evolve when its recipient is likely to be siblicidal than when the recipient is likely to be altruistic. There is no straightforward way in which Hamilton's rule can be applied in this type of situation, and a population genetic analysis is required. However, the analysis of costs and benefits in the two situations is valuable because it explains why the condition for the evolution of altruism in a siblicidal population is much more stringent than the condition for resistance to the invasion of siblicide in an altruistic population, so that the evolution of siblicide is irreversible.

The evolution of siblicide provides an extreme example of a situation in which inclusive-fitness arguments cannot be used, because the costs and benefits of altruism are not constant. In many situations, particularly when selfishness takes a less extreme form than killing the relative, the costs and benefits are likely to be approximately constant in different backgrounds and approximately additive over interactions with different individuals. In this case Hamilton's rule can be used in the confidence that it will give approximately the same answer as would be obtained from analyzing a population genetic model; the approximation is likely to be better under weak selection than under strong selection.

Under these conditions suppose that a particular individual (the actor) performs an act that affects a number of related individuals (the recipients), the benefit to the *i*th individual being b_i (a cost being counted as a negative benefit). We define the inclusive fitness of the actor as

$$\text{Inclusive fitness} = \sum_i r_i v_i b_i \qquad (21)$$

where r_i is the relatedness of the actor and the *i*th individual and v_i is the value of the *i*th individual in transmitting genes into future generations. (In an outbred haplodiploid population $r_i v_i$ is the life-for-life coefficient of relatedness shown in Table 9.5.) Hamilton's rule states that the altruistic act will be favored by selection if its inclusive fitness is positive. This generalization covers the original formulation of the rule in Equations 1 and 19 if the cost to the actor is counted as a negative benefit, since the relatedness of the actor to himself is unity.

AN INCLUSIVE-FITNESS MODEL FOR THE DISPERSAL OF OFFSPRING

Most plants and animals have mechanisms for the dispersal of young from their natal patch, such as wings in insects and devices to assist wind dispersal in seeds. Three possible advantages of dispersal are as follows: (i) risk-spreading in an environment that varies in both space and time (see the discussion of bet-hedging in Chapter 5); (ii) avoidance of mating with close relatives, that is disadvantageous if there is inbreeding depression; and (iii) avoidance of competition for resources with close relatives. There may also be a substantial cost of dispersal from not finding a suitable habitat; the very fact that an individual has been born and has survived in a particular patch is evidence that this patch is suitable for its needs.

We shall here consider the third advantage of dispersal, the avoidance of competition with relatives, in the absence of inbreeding depression or environmental instability, but taking into account its potential cost. This problem was first discussed by Hamilton and May (1977) by a game-theoretic argument, that was formalized by Motro (1982, 1983) using a population genetic model. However, the postulated advantage of dispersal is essentially due to kin selection (you benefit your relatives by dispersing since you are not competing with them in the natal patch), and it is both simpler and more informative to formulate it in this context. I shall here describe the inclusive-fitness model of dispersal developed by Taylor (1988b).

Offspring control of dispersal

Suppose that the world consists of a large number of patches, each capable of supporting n mated pairs. Each pair produces a large, fixed number of offspring. A fraction d_1 of the female offspring disperse; dispersing females incur a cost c_1 that we may interpret as an additional mortality such that the survival rate of nondispersing females is s_1, while that of dispersing females is $s_1(1 - c_1)$. After dispersal there are N_1 females within each patch who compete with each other for the n available breeding sites. The problem is to find the dispersal probability d_1 that maximizes the inclusive fitness of a female offspring. Likewise, we suppose that a fraction d_2 of the male offspring disperse, each incurring a cost c_2. After dispersal, the N_2 males in each patch compete with each other to mate with the females in the n available breeding sites, and the problem is to find the dispersal probability d_2 that maximizes their inclusive fitness.

Consider female dispersal. We first calculate the probability k_1 that a young female, after dispersal, is in her native patch. Since a proportion d_1 of all females disperse, and are replaced by $1 - c_1$ as many immigrants, we find that

$$k_1 = \frac{1-d_1}{1-d_1+(1-c_1)d_1} = \frac{1-d_1}{1-d_1c_1} \tag{22}$$

The chance that a female present on a patch after dispersal will breed is

$$p_1 = n/N_1 \tag{23}$$

We now calculate the change in the inclusive fitness of a nondispersing female if she had chosen instead to disperse. The cost to her is c_1p_1. (If $N_1 >> n$ we can ignore the fact that by dispersing she would increase the pool of competing females by 1 in her new patch.) The benefit to each of the other females in the patch resulting from the decrease by 1 in the number of competitors is

$$\frac{n}{N_1-1} - \frac{n}{N_1} \cong \frac{p_1}{N_1} \tag{24}$$

This benefit accrues to all the k_1N_1 females native to the patch, but it must be devalued by R, the average coefficient of relatedness between two individuals native to the same patch. (Under diploidy, the coefficient of relatedness will not depend on the sex of the individuals.) Hence the change in inclusive fitness is

$$\Delta w = p_1(Rk_1 - c_1) \tag{25}$$

If d_1 is the optimal dispersal rate, the change in inclusive fitness must be zero, so that $Rk_1 = c_1$, which leads to the result

$$d_1 = \frac{R-c_1}{R-c_1^2} \tag{26}$$

The same argument goes through for male dispersal, with the subscript changed, since, under this model, the dispersal behavior of one sex will not affect the fitness of members of the other. As expected, the optimum dispersal rate is smaller for the sex with the higher cost of dispersal.

To complete the analysis we must calculate the relatedness R; it will depend on the rates of dispersal and their costs. Assume for simplicity that male and female costs of dispersal are the same, $c_1 = c_2 = c$, so that $d_1 = d_2 = c$ and $k_1 = k_2 = k$. It is shown in Appendix 9.2 that at equilibrium

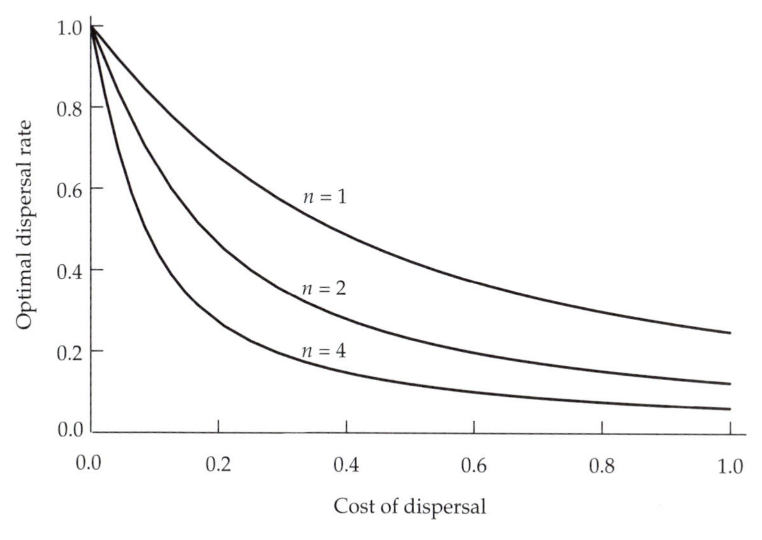

Figure 9.3. Optimal dispersal rate as a function of the cost of dispersal for different patch sizes. After Taylor, 1988b.

$$R = \frac{1}{2n - (2n-1)k^2} \tag{27}$$

This result, together with the result that $Rk = c$, can be used to find the optimal dispersal rate as a function of the cost of dispersal (see Equation 49 in Appendix 9.2). Some numerical values are shown in Figure 9.3. Significant dispersal is predicted by this model even when the cost of dispersal is high.

Maternal control of dispersal

We have assumed in this account that the dispersal behavior of the offspring is determined by the offspring itself, but it may sometimes be determined by the mother. In this case the costs and benefits of dispersal should be weighted by the relatedness of the individual incurring the cost or receiving the benefit to the mother. Thus the change in inclusive fitness analogous to Equation 25 is

$$\Delta w = p(R_1 k - R_2 c) \tag{28}$$

where R_1 is the relatedness of the mother and a random patchmate of one of her offspring and R_2 is the relatedness of the mother and one of her own offspring. Thus the previous analysis goes through if we replace R by

R_1/R_2. It is shown in Appendix 9.2 that

$$\frac{R_1}{R_2} = \frac{1}{n-(n-1)k^2} \tag{29}$$

This is equivalent to R in Equation 27 if $2n$ is replaced by n, so that the optimal dispersal rate under maternal control is the optimal rate under offspring control with $2n$ replaced by n. A mother wants a higher dispersal rate than her offspring; the optimal rate from the offspring's point of view coincides with the optimal rate from the mother's point of view for a patch twice the size.

Mating before dispersal

Finally we consider what happens if mating occurs in the patch before dispersal of mated females, assuming that dispersal is under the control of the dispersing females. Equation 26 follows by the same argument as before, but the calculation of relatedness differs. It turns out that the optimal dispersal rate is the same as that which mothers would want for their daughters if mating followed dispersal.

KIN SELECTION IN VISCOUS POPULATIONS

Hamilton (1964) defined a viscous population as one in which individuals disperse only a short distance from their place of birth, so that they tend to interact with their relatives. One might expect this population structure to be favorable to the evolution of altruism. Hamilton wrote: "Over a range of different species we would expect to find giving-traits commonest and most highly developed in the species with the most viscous populations whereas uninhibited competition should characterize species with the most freely mixing populations." However, this is not necessarily the case, since in a viscous population the offspring of close relatives are in competition with those of a potential altruist. Thus an increase in the number of offspring of a relative will benefit the altruist because they carry genes identical by descent to the altruist's genes, but they also harm him by competing for resources with his own offspring. Several recent models (Taylor, 1992a,b; Wilson et al., 1992) show that these two factors exactly cancel, so that population viscosity is not per se favorable to the evolution of altruism.

I describe here the model of Taylor (1992a) that is similar to his model for dispersal discussed above except that mating occurs before rather than after dispersal: (i) Mated females breed on patches with n females per patch. (ii) The offspring mate at random on each patch, and then (iii) a fraction d of the mated female offspring disperse to distant patches. Finally, (iv) on each patch, the mated female offspring, native and immigrant, compete for the n breeding sites, and the cycle begins again. There is no cost of dispersal in this model. In the absence of altruism, each breeding mated female produces f daughters and m sons. Breeding females interact at random in the patch. Each altruistic act incurs a small cost c, such that the altruist produces $(1 - c)f$ daughters and $(1 - c)m$ sons, and provides a small benefit b to one of the other breeding females (chosen at random) such that she produces $(1 + b)f$ daughters and $(1 + b)m$ sons.

Write r for the relatedness of a mother and her offspring and R for the relatedness of a mother and the offspring of another female in the same patch. The inclusive fitness of a breeding female can be calculated in terms of the number of breeding females in the next generation weighted by their relatedness. In the absence of altruism,

$$w_{\text{wild type}} = [fr + (n-1)fR]/f = r + (n-1)R \tag{30}$$

(There are f daughters of a particular mother and $(n - 1)f$ daughters of other mothers. Each of them has a chance $1/f$ of breeding since there are nf females competing for n breeding sites). Now consider a rare mutant female who is altruistic. Her inclusive fitness is affected both directly by the change in the numbers of offspring and indirectly by the consequent change in their breeding success. It is

$$w_{\text{mutant}} = f[r + (n-1)R - rc + Rb] \times \left[\frac{d}{f} + \frac{(1-d)n}{dnf + (1-d)(n+b-c)f}\right] \tag{31}$$

(If a mated female disperses, her chance of breeding is $1/f$, as before, but if she does not she must compete for the n breeding sites with the dnf immigrant and the $(1 - d)(n + b - c)f$ resident females.) Linearizing this expression with b and c small, we find that

$$\begin{aligned} \Delta w \equiv w_{\text{mutant}} - w_{\text{wild type}} &\cong (R - \alpha)b - (r - \alpha)c \\ \alpha &= [r + (n-1)R](1-d)^2/n \end{aligned} \tag{32}$$

Evaluation of the coefficients of relatedness (see Appendix 9.2) shows that

$$\begin{aligned} R &= \alpha \\ r &> \alpha \end{aligned} \tag{33}$$

Thus the coefficient of b in Equation 32 is zero, and the coefficient of c is negative. Thus "altruism" cannot spread unless $c < 0$, which is by definition a nonaltruistic act. (In calculating the inclusive fitness in Equations 30 and 31 I have ignored contributions through male offspring. They are the same as the contributions through female offspring and would therefore double the inclusive fitness without changing the conclusion.)

Altruism does not evolve in this model because its benefits are exactly balanced by the increased competition for resources that occur between related individuals. The same result has been obtained by Taylor (1992b) for a population in a homogeneous inelastic environment rather than a patch-structured environment and seems likely to be of general validity. Two things should be noted. First, there is still room in a viscous population for the evolution of altruism between individuals who are more closely related to each other than the average within a local population. Second, the result is confined to viscous populations characterized by limited dispersal with no global mixing phase. Wilson et al. (1992) contrast this type of viscosity with what they call alternating viscosity. Consider as an example of the latter an insect population in which single females lay eggs in clusters. Siblings hatch and interact among themselves as larvae before dispersing as adults. If density-dependent factors operate not at the local level of the larval cluster, but at the population-wide level after dispersal, then one would expect altruism to evolve between the siblings in a larval cluster because it does not generate any competitive side-effect on the donors or the recipients of altruism. Social insects provide another example of the operation of kin selection in alternating viscosity, since there is population-wide random mating between reproductives released at the end of the season followed by dispersal of mated females.

FURTHER READING

Wade (1985) and Breden (1990) discuss how the inclusive-fitness argument can be developed through group selection; Hamilton (1970) and Queller (1992) describe an alternative derivation through the use of Price's covariance mathematics. Brown (1987), Stacey and Koenig (1990), and Emlen (1991) review cooperative breeding.

EXERCISES

1. Consider the model of Maynard Smith (1989) with *aa* individuals being altruistic and *A* being a rare gene causing selfishness. Find the analogue to Equation 2, and hence show that the condition for altru-

ism to resist invasion when it is common is the same as the condition for it to invade when rare under this model of altruism between sibs.

2. (a) Verify the calculations leading to Equation 4.
(b) Show that the dominant eigenvalue of the transition matrix is 2C.
(c) Show that, under the further assumptions made by Godfray, siblicide will evolve when the clutch size is 2 or 3, but not when it is greater than 4.

3. Derive the results shown in Table 9.2.

4. Derive the results shown in Table 9.3.

5. Find the coefficient of relatedness r_{JK} in an outbred haplodiploid population where J is a female and K is (a) her sister's son, (b) her sister's daughter, (c) her mother's sister's son, and (d) her mother's sister's daughter.

6. (a) Justify the interpretation of $\mathbf{A}$, $\mathbf{A}^2$, $\mathbf{A}^3$, and so on, where $\mathbf{A}$ is defined in Equation 17. Verify numerically that $\mathbf{A}^n$ for large n tends to the matrix in Equation 18. Confirm this result by finding the dominant left eigenvector of $\mathbf{A}$.
(b) Find the dominant left eigenvector for the matrix in Equation 20, appropriate where there is some laying by workers in social haplodiploids.

7. Derive in detail the inclusive-fitness model for the optimal dispersal of offspring that is sketched in the text.

8. Derive in detail the the model of kin selection in a viscous population that is sketched in the text.

APPENDIX 9.1. REGRESSIONS BETWEEN RELATIVES

In an outbred diploid population a pair of relatives may share zero, one, or two pairs of identical genes. These three cases may be characterized as {(1, 2), (3, 4)}, {(1, 2), (1, 3)}, and {(1, 2), (1, 2)}, respectively, the use of distinct numbers indicating genes that are not identical by descent, and they provide a complete classification in the absence of inbreeding (so that two homologous genes in the same individual are never identical). The probabilities of these three possibilities can be calculated from the pedigree for any type of relative by elementary arguments, and are denoted as P_0, P_1, and P_2. For example, for a pair of sibs, $P_0 = \frac{1}{4}$, $P_1 = \frac{1}{2}$, and $P_2 = \frac{1}{4}$. The coefficient of consanguinity between a pair of outbred diploid relatives is

$$F_{JK} = \tfrac{1}{4} P_1 + \tfrac{1}{2} P_2 \tag{34}$$

Consider a neutral locus with two alleles, *A* and *a*, with frequencies p and $q = 1 - p$ in an outbred diploid population. Under random mating, the frequencies of the three genotypes, *AA*, *Aa*, and *aa*, are p^2, $2pq$, and q^2, by the Hardy-Weinberg law. Define X as a variable taking the values 1, 0.5, and 0 for the genotypes *AA*, *Aa*, and *aa* respectively; it is the proportion of *A* genes in an individual's genotype. X is a random variable with the Hardy-Weinberg probabilities (i.e., taking the values 1, 0.5, and 0 with probabilities p^2, $2pq$, and q^2), and has mean and variance

$$\begin{aligned} E(X) &= p \\ \mathrm{Var}(X) &= \tfrac{1}{2} pq \end{aligned} \tag{35}$$

(These parameters can be calculated from first principles or from the properties of the binomial distribution.)

The values of X in two relatives, J and K, are correlated. We shall now find from first principles the regression of X_K on X_J, that is to say, the expected value of X_K given X_J. We first write down the conditional probabilities of X_K given X_J (see Table 9.6); for example, the probability that K is *AA* given that J is *AA* is $p^2P_0 + pP_1 + P_2$ since K has probability p^2 of being *AA* if it has no genes identical with J, while this probability becomes p if they share one identical gene (that must be *A*), the other gene being chosen at random from the population, and it becomes unity if both genes are identical. The expectation of X_K in each row is then calculated and simplified by elementary algebra (see the last column in Table 9.6), whence the regression of X_K on X_J is

Table 9.6. The conditional distribution of X_K given X_J; $r = r_{JK}$.

	X_K			
X_J	1	0.5	0	$E(X_K \mid X_J)$
1	$p^2P_0+pP_1+P_2$	$2pqP_0+qP_1$	q^2P_0	$r+p(1-r)$
0.5	$p^2P_0+0.5pP_1$	$2pqP_0+0.5P_1+P_2$	$q^2P_0+0.5qP_1$	$0.5r+p(1-r)$
0	p^2P_0	$2pqP_0+pP_1$	$q^2P_0+qP_1+P_2$	$p(1-r)$

$$E(X_K | X_J) = p(1 - r) + rX_J = p + r(X_J - p) \quad (36)$$

where $r \equiv r_{JK} = \frac{1}{2}P_1 + P_2$ is the coefficient of relatedness. Thus the regression is linear with slope r.

A similar argument goes through for an outbred haplodiploid population, with X taking values 1 and 0 for the genotypes A and a in males and with the coefficient of relatedness defined in Equations 12a and 12c.

We consider finally the extension of this result to inbred populations. Define X_J and X_K as the proportions of A genes in J and K, as before. We now wish to find the regression of X_K on X_J. There is no guarantee under inbreeding that this regression will be linear, but a well-known result in regression theory states that the best linear approximation to the true regression will have the form of Equation 14 if

$$r_{JK} = \mathrm{Cov}(X_J, X_K)/\mathrm{Var}(X_J) \quad (37)$$

We shall now show that the coefficient of relatedness defined in Equation 12 is equal to the right-hand side of Equation 37.

Suppose that J and K are diploid. Define $Y = 1$ for A and 0 for a. Order the genes in J randomly and write Y_{1J} and Y_{2J} for their Y values, and likewise for K. Then,

$$\mathrm{Cov}(Y_{1J}, Y_{1K}) = F_{JK}\,\mathrm{Var}(Y) = pq\,F_{JK} \quad (38)$$

since the two variables are independent with probability $1 - F_{JK}$ and identical with probability F_{JK}, and similarly if we replace either or both of the 1's by 2's. Hence,

$$\begin{aligned}\mathrm{Cov}(X_J, X_K) &= \mathrm{Cov}[\tfrac{1}{2}(Y_{1J} + Y_{2J}), \tfrac{1}{2}(Y_{1K} + Y_{2K})] \\ &= \frac{1}{4}\sum_{i,j=1,2} \mathrm{Cov}(Y_{iJ}, Y_{jK}) \\ &= pqF_{JK}\end{aligned} \quad (39)$$

Also,

$$\begin{aligned}\mathrm{Var}(X_J) &= \mathrm{Cov}(X_J, X_J) \\ &= pqF_{JJ}\end{aligned} \quad (40)$$

Hence,

$$\mathrm{Cov}(X_J, X_K)/\mathrm{Var}(X_J) = F_{JK}/F_{JJ} = r_{JK} \quad (41)$$

Suppose now that J or K is haploid (or that both are haploid), with X taking values 1 or 0 for the genotypes A and a in a haploid. It is straightforward to show that

$$\mathrm{Cov}(X_J, X_K) = pq\, F_{JK} \tag{42}$$

and that

$$\mathrm{Var}(X_J) = pqF_{JJ} \tag{43}$$

regardless of the ploidies of J and K. Hence,

$$\mathrm{Cov}(X_J, X_K)/\mathrm{Var}(X_J) = r_{JK} \tag{44}$$

for all ploidies if r_{JK} is defined as in Equation 11, leading to the results in Equation 12.

APPENDIX 9.2. CALCULATION OF THE RELATEDNESS WITH LIMITED DISPERSAL

Offspring control of dispersal

Write $F(t)$ for the coefficient of inbreeding of an individual in generation t, and $G(t)$ for the coefficient of consanguinity between two individuals born on the same patch in generation t. Then,

$$F(t+1) = k^2 G(t) \tag{45}$$

since the inbreeding coefficient of an individual in generation $t + 1$ is the coefficient of consanguinity between his parents [$G(t)$ if they were born in the same patch and zero otherwise]. Also,

$$G(t+1) = \frac{1}{n}\left[\frac{1+F(t)+2k^2G(t)}{4}\right] + \frac{n-1}{n}k^2G(t) \tag{46}$$

This recurrence relationship follows from the following argument. The probability that two individuals in the same patch are sibs is $1/n$. In this case, two genes picked at random from them have probability $\frac{1}{2}$ of coming from the same parent and probability $\frac{1}{2}$ of coming one from the father and the other from the mother; if they come from the same parent they are equally likely to be copies of the same gene in that parent [being identical with certainty] or to be copies of different genes [having probability $F(t)$ of being identical]; if they come from different parents they have probability $G(t)$ of being identical if the parents were born in the same patch. This gives the first term on the right-hand side in Equation 46. If the two individuals are not full sibs, the probability that two genes drawn randomly (one from each of them) are identical is the coefficient of consanguinity between the two individuals in the previous generation who transmitted those genes. This gives the second term in Equation 46.

Setting $F(t+1) = F(t)$ and $G(t+1) = G(t)$ in Equations 45 and 46, we find that at equilibrium

$$F = k^2 G$$
$$G = \frac{1}{4n - (4n-1)k^2} \tag{47}$$

We now calculate R as

$$R = \frac{2G}{1+F} = \frac{1}{2n - (2n-1)k^2} \tag{48}$$

Finally, the solution of the equation $Rk = c$ in terms of d is

$$d = \frac{(4n-2)c + 1 - H}{(4n-1-H)c} \tag{49}$$

where

$$H = \sqrt{1 + 8n(2n-1)c^2} \tag{50}$$

Maternal control of dispersal

The ratio R_1/R_2 is the same as the ratio of the corresponding consanguinities F_1/F_2. The coefficient of consanguinity between a mother and her offspring is

$$F_2 = \tfrac{1}{4}(1 + F + 2k^2 G) \tag{51}$$

[If the randomly chosen offspring gene is of maternal origin (with probability $\frac{1}{2}$), it is equally likely to be a copy of the randomly chosen maternal gene (identical with certainty) or of the other maternal gene (identical with probability F); if it is of paternal origin (with probability $\frac{1}{2}$) it has probability G of being identical with a randomly chosen maternal gene if both parents are native to the patch.] The coefficient of consanguinity between the mother and a random patch-mate of one of her offspring is

$$F_1 = G \tag{52}$$

[If the patch-mate is a sib of the offspring (with probability $1/n$), then $F_1 = F_2$. Otherwise $F_1 = k^2 G$. Write J for the mother and K for the parent from which the random patchmate is derived. Then $F_1 = F_{JK}$, which is G if both J and K are native to the patch. Equation 52 now follows from Equation 46.] Hence,

$$\frac{R_1}{R_2} = \frac{F_1}{F_2} = \frac{4G}{1+F+2k^2G} = \frac{4G}{1+3F} = \frac{1}{n-(n-1)k^2} \tag{53}$$

Mating before dispersal

The analogues of Equations 45 and 46 are

$$F(t+1) = G(t) \tag{54}$$

and

$$G(t+1) = \frac{1}{n}\left[\frac{1+F(t)+2G(t)}{4}\right] + \frac{n-1}{n}k^2G(t) \tag{55}$$

since a male and his mate must have been born in the same patch. At equilibrium,

$$F = G = \frac{1}{4n-3-4(n-1)k^2} \tag{56}$$

We now follow the argument for dispersal before mating under maternal control and find that

$$\frac{R_1}{R_2} = \frac{F_1}{F_2} = \frac{4G}{1+F+2G} = \frac{1}{n-(n-1)k^2} \tag{57}$$

Kin selection in viscous populations

The consanguinity between the mother and one of her offspring is $\frac{1}{4}$ $(1 + F + 2G)$, where F and G are defined in Equation 56 with $k = 1 - d$. (An offspring gene is equally likely to be of maternal or paternal origin. In the second case it has probability G of being identical to a maternal gene. In the first case it is equally likely to be a direct copy of the randomly chosen maternal gene or of the other maternal gene, which has probability F of being identical to it.) The consanguinity between a mother and one of the offspring of another parent is G. Relatedness is twice the consanguinity divided by $(1 + F)$. Hence we find that

$$\begin{aligned} R - \alpha &= 0 \\ r - \alpha &> 0 \end{aligned} \tag{58}$$

CHAPTER 10

The Sex Ratio

"Let us now take the case of a species producing from the unknown causes just alluded to, an excess of one sex—we will say of males—these being superfluous and useless, or nearly useless. Could the sexes be equalised through natural selection? We may feel sure, from all characters being variable, that certain pairs would produce a somewhat less excess of males over females than other pairs. The former, supposing the actual number of the offspring to remain constant, would necessarily produce more females, and would therefore be more productive. On the doctrine of chances a greater number of the offspring of the more productive pairs would survive; and these would inherit a tendency to procreate fewer males and more females. Thus a tendency towards the equalisation of the sexes would be brought about."

—Charles Darwin, *The Descent of Man, and Selection in Relation to Sex* (1871)

"In no case, as far as we can see, would an inherited tendency to produce both sexes in equal numbers or to produce one sex in excess, be a direct advantage or disadvantage to certain individuals more than to others; for instance, an individual with a tendency to produce more males than females would not succeed better in the battle for life than an individual with an opposite tendency; and therefore a tendency of this kind could not be gained through natural selection. ... I formerly thought that when a tendency to produce the two sexes in equal numbers was advantageous to the species, it would follow from natural selection, but I now see that the whole problem is so intricate that it is safer to leave its solution for the future."

—Charles Darwin, *The Descent of Man, and Selection in Relation to Sex* (1874, 2nd Ed.)

One would expect natural selection to equalize the sex ratio for the following reason. If there is an excess of one sex in the population, then an individual of the rare sex will on average have more offspring than one of the common sex since every offspring has a father and a mother. Thus a parent who specializes in producing offspring of the rare sex will on average have more grandchildren than other parents and genes to equalize the sex ratio will increase in frequency. This is an example of frequency-dependent selection, but the effect of selection is delayed a generation.

This explanation of how natural selection acts to equalize the sex ratio is usually attributed to Fisher (1930). It is not generally appreciated that Charles Darwin came close to it in the first edition of *The Descent of Man* (see the first quotation above). Taken in isolation this passage cannot be faulted; unfortunately it is followed by a passage that obscures the issue. In the second edition Darwin retracted this explanation (see the second quotation above). The second passage is better known, and was quoted by Fisher (1930) in his discussion of the problem. Second thoughts are not always for the better!

In this chapter I shall derive this result more formally, using both a population genetic approach and an inclusive-fitness approach, and I shall then extend the theory to situations where departures from an equal sex ratio are predicted. Sex-ratio theory is one of the most successful applications of modeling in evolutionary ecology, largely because the fitnesses of different strategies can be calculated from first principles and do not have to be estimated empirically. There are many examples in which animals modify the sex ratio adaptively in accordance with theoretical predictions, but most of them occur in haplodiploid organisms in which the method of sex determination permits the mother to determine the sex of her offspring when she lays the egg. (Fertilized eggs are female; unfertilized eggs are male.) The chromosomal mechanism of sex determination found in most organisms exerts a constraint on the facultative manipulation of the sex of the offspring, though some species have found a way around this (Figure 10.5).

EVOLUTION OF AN UNBIASED SEX RATIO UNDER STANDARD CONDITIONS

A population genetic model

To formalize the intuitive argument presented above, consider a gene acting in females to control the sex ratio of their offspring (defined as the relative frequency of males). The locus has two alleles, *a* and *A*, with *A* being rare so that *AA* individuals have negligible frequency under random mating; *aa* and *Aa* females have sex ratios r and s, respectively, among their offspring. We shall now determine whether the *A* gene will increase in frequency, using the fact that it is rare to obtain an approximate linear recurrence relationship.

Write p_1 and p_2 for the relative frequencies of *Aa* among females and males, respectively, in the current generation. (It is necessary to keep track of these frequencies separately because the gene has an effect in females but not in males.) Under random mating the frequencies of the mating types and the numbers of offspring produced (if each mating produces on average n offspring) are shown in Table 10.1. The relative frequencies of *Aa* in females and males in the next generation (again keeping only linear terms in p_1 and p_2) are

$$\begin{aligned} p_1' &= \frac{1}{2}\left(\frac{1-s}{1-r}\right)p_1 + \frac{1}{2}p_2 \\ p_2' &= \frac{1}{2}\left(\frac{s}{r}\right)p_1 + \frac{1}{2}p_2 \end{aligned} \quad (1)$$

Table 10.1. Mating table under a model of sex ratio evolution.

Mating type		Offspring per mating			
		Females		Males	
(F × M)	Frequency	*Aa*	*aa*	*Aa*	*aa*
$Aa \times aa$	p_1	$\frac{1}{2}(1-s)n$	$\frac{1}{2}(1-s)n$	$\frac{1}{2}sn$	$\frac{1}{2}sn$
$aa \times Aa$	p_2	$\frac{1}{2}(1-r)n$	$\frac{1}{2}(1-r)n$	$\frac{1}{2}rn$	$\frac{1}{2}rn$
$aa \times aa$	$1-p_1-p_2$	0	$(1-r)n$	0	rn

Note: The frequencies are given for the mating types (in the second column of the table) keeping only linear terms in p_1 and p_2.

In matrix terminology

$$\begin{bmatrix} p_1' \\ p_2' \end{bmatrix} = \begin{bmatrix} \frac{1}{2}\left(\frac{1-s}{1-r}\right) & \frac{1}{2} \\ \frac{1}{2}\left(\frac{s}{r}\right) & \frac{1}{2} \end{bmatrix} \begin{bmatrix} p_1 \\ p_2 \end{bmatrix} \tag{2}$$

The eigenvalues of the transition matrix in Equation 2 are real, and the mutant will increase in frequency if the larger of them is greater than 1. It can be shown that this is the case either if $r < 0.5$ and $s > r$ or if $r > 0.5$ and $s < r$. In other words, the population is open to invasion by any rare mutant that increases the sex ratio when it is below 0.5 and by any rare mutant that decreases the sex ratio when it is above 0.5. When $r = 0.5$ the dominant latent root is unity, which implies that any rare mutant with a different sex ratio is selectively neutral; however, such a mutant will become selectively disadvantageous as it increases in frequency enough to change the population sex ratio. The equilibrium at 0.5 is an evolutionarily stable strategy that satisfies the condition of continuous stability defined in Chapter 8.

The mathematics becomes rather easier if we consider only mutants that change the sex ratio slightly. Linearizing the dominant latent root from Equation 2 about $s = r$, we find

$$\lambda \cong 1 + \frac{(s-r)(1-2r)}{4r(1-r)} \tag{3}$$

from which the conclusions already reached follow immediately. However, by this analysis they are only guaranteed for mutants of small effect.

There is a useful general technique for simplifying this calculation. The condition that r be at equilibrium is that the dominant eigenvalue of the transition matrix be less than or equal to 1 for all s. Write $\mathbf{A}(r, s)$ for the transition matrix and $\lambda(r, s)$ for its dominant eigenvalue, and note that $\lambda(r, r) = 1$, since in in this case the mutant has no effect and must therefore be selectively neutral. Thus a necessary condition that $\lambda(r, s) \le 1$ for all s is that $\partial\lambda/\partial s = 0$ when $s = r$. Write

$$f(r,s,\lambda) = \det[\mathbf{A}(r,s) - \lambda\mathbf{I}] \tag{4}$$

(where $\det[\cdot]$ denotes the determinant of a matrix and $\mathbf{I}$ is the identity matrix; see Appendix E). Note that λ is defined by $f(r, s, \lambda) = 0$. Differentiating f with respect to s yields

$$\frac{df}{ds} = \frac{\partial f}{\partial s} + \frac{\partial f}{\partial \lambda}\frac{\partial \lambda}{\partial s} = 0 \tag{5}$$

Thus the condition $\partial\lambda/\partial s = 0$ is equivalent to $\partial f/\partial s = 0$. From the definition of the function f, the condition becomes

$$\frac{\partial}{\partial s}\det[\mathbf{A}(r,s) - \mathbf{I}] = 0 \qquad \text{at } s = r \tag{6}$$

Applying this technique to the transition matrix in Equation 2 yields

$$\frac{\partial}{\partial s}\det[\mathbf{A}(r,s) - \mathbf{I}] = \frac{1}{4(1-r)} - \frac{1}{4r} \tag{7}$$

Setting $s = r$ (which is redundant in this case since the expression does not involve s) and equating to zero, we obtain immediately the unbeatable sex ratio $r = 0.5$. Having found this equilibrium sex ratio, it is now necessary to check whether $\partial^2\lambda/\partial s^2 \leq 0$, which means that any mutant would be either selected against or neutral. In fact $\partial^2\lambda/\partial s^2 = 0$, which means that mutants are neutral, as we have already seen. The secondary condition $\partial^2\lambda/\partial s\,\partial r < 0$ establishes continuous stability.

It has been assumed that males and females cost the same to produce. This is implicit in the assumption that the total number of offspring per mating is n, which may be distributed between sons and daughters in any way; a female may produce m sons and f daughters subject to the constraint that $m + f = n$. It may be that one sex is more costly than the other, so that producing an extra daughter, for example, can only be done at the expense of producing two fewer sons. In this case we may reinterpret the sex ratio r as the proportion of resources allocated to producing males. If a daughter is k times as costly as a son to produce ($k < 1$ meaning that daughters are cheaper than sons and $k > 1$ that they are more expensive), we may suppose that a mother can produce m sons and f daughters subject to the constraint that $m + kf = R$, where R is the total resource allocated to reproduction, measured in units of sons. The sex ratio defined as relative investment in sons is $r = m/R$, and the numbers of sons and daughters produced by a mother with this sex ratio are $m = rR$ and $f = (1 - r)R/k$. This can be taken into account in the mating tables by substituting R for n in the male-offspring columns and R/k for n in the female-offspring columns. But note that this makes no difference to the recurrence relations since p_1 and p_2 are defined as relative frequencies among females and males, respectively, so that k cancels out in the numerator and denominator of p_1. It can be concluded that it is the investment sex ratio, not the numerical sex ratio that should evolve to 0.5.

It is usual to measure the investment sex ratio by the biomass of the offspring, on the assumption that this represents cost to the parent. For example, Metcalf (1980) studied two species of paper wasp, *Polistes metri-*

cus and *P. variatus.* The sex ratio was measured as the proportion of males among reproductives (excluding workers). Female reproductives are smaller than males in the former but not in the latter species. As predicted, the numerical sex ratio is biased in favor of females in the former but not significantly so in the latter (the observed sex ratios being 0.45 and 0.48, respectively). The investment sex ratio, measured in terms of biomass, is about equal in both species (being 0.50 and 0.49). (I ignore for the time being the complication that paper wasps are haplodiploid.)

The prediction that the investment sex ratio should be 0.5 refers to the primary sex ratio and is unaffected by differential mortality after parental investment is completed. The numbers of offspring in the mating tables could be changed to refer to numbers of offspring surviving to reproductive age by multiplying them by the probability of survival to this age. Even if the survival rate differs in males and females, it will not affect the difference equations because it will cancel in the numerator and the denominator. Intuitively, if males have only half the survival rate of females, so that the sex ratio among breeding individuals is biased two to one in favor of females, it would not pay to produce more sons because the fact that each surviving male has twice the expected reproductive success as a surviving female is exactly counterbalanced by the fact that males have only half the chance of surviving to reproduce. It is expected reproductive success of an offspring at the time that investment is committed to it that matters to a parent.

This account only considers the conditions for the invasion of a rare allele. The more complex question of what happens when the mutant is no longer rare can be investigated in a simple way by simulation and has been studied in great detail from a theoretical viewpoint by Karlin and Lessard (1986). The general conclusion is that the sex ratio will evolve to 0.5 under the simple conditions of population-wide random mating, unless it is prevented from doing so by some bizarre feature of the genetic system. For example, suppose that the sex of a zygote is determined by a single locus with two alleles, the three genotypes having different probabilities of being male. If the two homozygote probabilities straddle both the heterozygote probability and 0.5, so that there is no overdominance and a sex ratio of 0.5 is possible, the population sex ratio at equilibrium will indeed be 0.5, but otherwise this may not be true. For example, if both homozygote probabilities of being male are 0.1 and the heterozygote probability is 0.75, the population sex ratio at equilibrium is not 0.5, but 0.425. This might seem to contradict Fisher's principle, but it does not really do so. With so much overdominance, additive genetic variability is not available to produce the Fisherian sex ratio; 0.425 is the highest sex ratio attainable in Hardy-Weinberg equilibrium. However, a third allele leading to an even sex ratio can always invade the system; on the other hand, a system with an even sex ratio at equilibrium cannot be invaded by a new allele leading to a departure from the Fisherian value. This property has been called evolutionary genetic stability (Eshel and Feldman, 1982). The conclusion is that one would expect a sex ratio of 0.5 to evolve under these ideal assumptions, provided that adequate genetic variability is available.

An inclusive-fitness model

These results can also be obtained by an inclusive-fitness argument that is more heuristic but that offers more insight into what is happening. In a diploid population we may represent the fitness of a female as

$$\begin{aligned}\text{fitness} &= \text{number of daughters}\\ &\quad + \text{number of females inseminated by sons}\end{aligned} \tag{8}$$

Suppose also that there are two types of female: wild-type females who produce F daughters and M sons, and rare mutant females who produce f daughters and m sons. Their fitnesses are

$$\text{fitness}_{\text{wild type}} = F + \frac{F}{M}M = 2F \tag{9}$$

and

$$\text{fitness}_{\text{mutant}} = f + \frac{F}{M}m \tag{10}$$

If the total number of offspring is constant, so that $f + m = F + M = n$, and if we write $s = m/(f + m)$ and $r = M/(F + M)$ for the mutant and wild-type sex ratios, then the fitness of the mutant relative to that of the wild type is

$$w(s,r) = \frac{\text{fitness}_{\text{mutant}}}{\text{fitness}_{\text{wild type}}} = \frac{f}{2F} + \frac{m}{2M} = \frac{r + (1-2r)s}{2r(1-r)} \tag{11}$$

We now seek the sex ratio r^* that is the best reply to itself in the sense that if the population sex ratio is $r = r^*$, then a mutant cannot do better than by playing $s = r^*$. This can be done either graphically or analytically. To find the solution graphically, note that Equation 11 is linear in s with a coefficient that is positive or negative depending on whether r is less than or greater than 0.5. The value of s that maximizes mutant fitness is therefore $s = 1$ if $r < 0.5$ and $s = 0$ if $r > 0.5$, while the fitness is independent of s if $r = 0.5$. The best mutant response is shown graphically in Figure 10.1 (compare Figure 8.2); the unique sex ratio that is the best

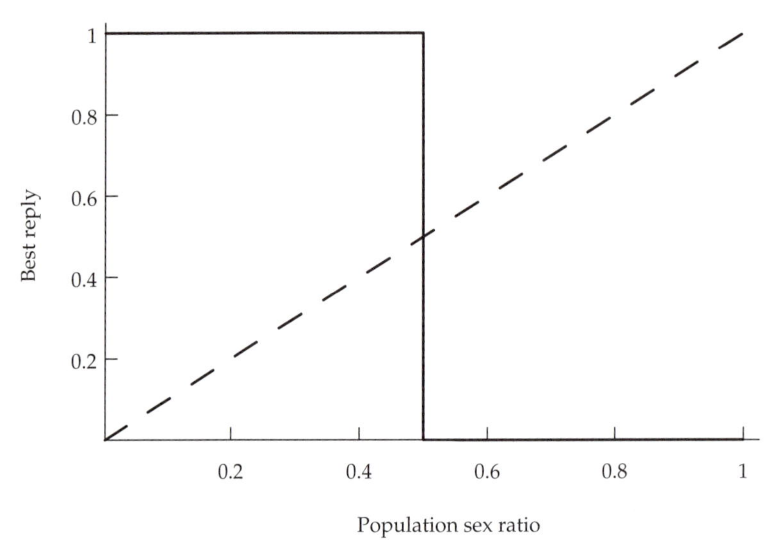

Figure 10.1. Best-reply diagram for the sex ratio under standard conditions.

reply to itself is $r^* = 0.5$.

To find the solution analytically we solve the equation

$$\partial w/\partial s\big|_{s=r=r^*} = 0 \tag{12}$$

which again leads to the solution $r^* = 0.5$. The second derivative $\partial^2 w/\partial s^2$ evaluated at $s = r = 0.5$ is zero, which reflects the fact that when the population sex ratio is 0.5 any reply has equal fitness. It is therefore important to check that the criterion for continuous stability in this situation, $\partial^2 w/\partial s\,\partial r < 0$, is satisfied. (See the discussion at the end of Chapter 8.)

EVOLUTION OF THE SEX RATIO IN HAPLODIPLOIDS

Much of the empirical work on sex ratios concerns haplodiploid organisms because their method of sex determination gives them more opportunity than organisms with chromosomal sex determination to bias the sex ratio adaptively when occasion arises, and many of them have life histories in which such adaptive bias might be expected. (Males are haploid and females diploid in haplodiploids; the female determines the sex of an offspring at the time she lays the egg by either fertilizing it or not, unfertilized eggs becoming males and fertilized eggs females.) It is therefore

Table 10.2. Mating table for a sex ratio model in haplo-diploids.

Mating type		Offspring per mating			
		Females		Males	
(F × M)	Frequency	Aa	aa	A	a
$Aa \times a$	p_1	$\frac{1}{2}(1-s)n$	$\frac{1}{2}(1-s)n$	$\frac{1}{2}sn$	$\frac{1}{2}sn$
$aa \times A$	p_2	$(1-r)n$	0	0	rn
$aa \times a$	$1-p_1-p_2$	0	$(1-r)n$	0	rn

Note: The frequencies are given for the mating types (in the second column of the table) keeping only linear terms in p_1 and p_2.

important to see whether haplodiploidy per se is expected to lead to a biased sex ratio in the absence of other disturbing factors such as nonrandom mating.

We can set up the same population genetic model as before, with the exception that p_1 is the relative frequency of Aa among females but p_2 is the relative frequency of A among males. The mating table, allowing for the asymmetry of haplodiploid inheritance, is shown in Table 10.2. The linearized recurrence relations in the next generation are

$$\begin{bmatrix} p_1' \\ p_2' \end{bmatrix} = \begin{bmatrix} \frac{1}{2}\left(\frac{1-s}{1-r}\right) & 1 \\ \frac{1}{2}\left(\frac{s}{r}\right) & 0 \end{bmatrix} \begin{bmatrix} p_1 \\ p_2 \end{bmatrix} \tag{13}$$

The dominant latent root, linearized about $s = r$, is

$$\lambda \cong 1 + \frac{(s-r)(1-2r)}{3r(1-r)} \tag{14}$$

leading to the same conclusion as before, that selection will lead to an unbiased sex ratio.

This result can also be obtained by an inclusive-fitness argument. The representation of the inclusive fitness of a female in Equation 8 is justified for diploids by their symmetrical genetics, but in haplodiploids the genetic asymmetry between males and females must be taken into account. In this case we may write the reproductive fitness of a mutant female producing f daughters and m sons in a population of individuals producing F daughters and M sons as

$$\text{fitness}_{\text{mutant}} = v_f r_{MD} f + v_m r_{MS} \frac{F}{M} m \quad (15)$$

where v_f (v_m) is the value of a mated daughter (son) in projecting genes into the distant future and r_{MD} (r_{MS}) is the relatedness between mother and daughter (son). The terms $v_f r_{MD}$ and $v_m r_{MS}$ are the life-for-life coefficients of relatedness, which in an outbred population are the same (see Table 9.5). The unbeatable sex ratio is 0.5 by the same argument as before.

In the rest of this section I shall consider the sex ratio in a colonial eusocial hymenopteran, such as a bumblebee, a paper wasp, or an ant, in which sterile workers do most of the work in building up the size and strength of the colony. In this context, sex ratio means the sex ratio among the winged reproductives (alates) that the colony produces at the end of the season. If the queen controls this sex ratio it should be unbiased, on a biomass basis, by the above argument, because she is their mother, but several complications may arise. If the queen lays all the eggs, the workers may nonetheless control the sex ratio by allocating resources preferentially to one sex or the other; this will give rise to a conflict of interest between the queen and the workers over the sex ratio, since the workers are rearing sibs and will prefer a female-biased sex ratio because of their higher relatedness to their sisters than to their brothers. The situation is further complicated if the queen has mated twice or more, since the workers may then be rearing half sibs rather than full sibs (Page, 1986). Another way in which workers may try to affect the outcome is by laying their own eggs (Bourke, 1988); because workers are unmated their eggs will always develop into males, and the queen may then try to balance the sex ratio by laying an excess of diploid eggs that will develop into female reproductives. I shall now examine these questions in more detail, and I shall conclude the section by considering whether the asymmetries in relatedness coefficients in haplodiploids may have predisposed them toward the evolution of eusociality.

Worker control in social Hymenoptera

Consider a colony of a eusocial hymenopteran in which workers feed the larvae and may thus control the sex ratio (of the alates produced at the end of the season) by allocating resources preferentially to one sex or the other. If the queen has mated only once then the workers are full sibs of the larvae with life-for-life coefficients of relatedness to female and male larvae of $\frac{3}{4}ns$ and $\frac{1}{4}$, respectively (Table 9.5). Hence,

$$\text{fitness}_{\text{mutant}} = \frac{3}{4} f + \frac{1}{4}\left(\frac{F}{M}\right) m \quad (16)$$

By the same argument as before, the relative fitness of the mutant is

$$w(s,r) = \frac{3r + (1-4r)s}{4r(1-r)} \tag{17}$$

At equilibrium the coefficient of s must vanish, giving $r^* = 0.25$, or a 3:1 female-biased sex ratio. This result can be confirmed by a population genetic model if one assumes that in a colony with two types of workers (which will happen when the queen is heterozygous) the colony sex ratio is intermediate between their individual desires (Charnov, 1978; Bulmer, 1983a).

Thus there is a conflict of interest between the queen, who would like an unbiased sex ratio, and the workers, who would like a female-biased sex ratio. Empirical evidence, following the work of Trivers and Hare (1976), suggests that the sex ratio among alates in social insects is often female-biased on a biomass basis, particularly in ants, indicating that the workers may be in control (Nonacs, 1986; Boomsma, 1989). On the other hand, we have already seen that there is an unbiased sex ratio, suggesting maternal control, in the paper wasp *Polistes*. The situation is also complicated by the possibilities of worker laying, multiple mating of the foundress, and multiple foundresses.

Worker laying

Taylor (1981) discussed the effect of worker laying, using a population genetic model. Forsyth (1981) had studied the sex ratio in colonies of a species of ant. Most colonies (35 out of 53 studied) had a functional egg-laying queen (i.e., they were "queenright"), and these colonies had produced 269 female and 164 male reproductives. In the remaining 18 colonies the queen had died, and eggs were being laid by workers; these colonies produced 121 males. Male and female reproductives had the same dry weight, so that no correction is needed for differential investment. Thus the sex ratio is 164/433 = 0.38 in queenright colonies and 285/554 = 0.51 in all colonies. Forsyth suggested that the female-biased sex ratio in queenright colonies was a compensation by the queen to counter the all-male production of queenless colonies and so restore the population sex ratio to 0.5.

Taylor (1981) constructed a population genetic model to determine the appropriate strategy of the queen in this situation. Boomsma and Grafen (1991) developed a general inclusive-fitness approach that reaches the same conclusion. The assumption is that the queen lays all the eggs and is in control of the sex ratio in queenright colonies, but that workers lay eggs in colonies that lose the queen. The application of the inclusive fitness approach to find the optimal strategy of the queen in queenright colonies goes as follows. Let β denote the ratio of worker reproductive output to queen reproductive output in the population. Forsyth's estimate of β would be 121/433 = 0.279. If the sex ratio in queenright colonies is r, the population ratio of females to males, which gives the average number of matings each male can expect, is $(1 - r)/(r + \beta)$, and the proportion of queen-produced males is $p = r/(r + \beta)$. From Chapter 9 we recall that the ratio of the reproductive values of females and males is $v_f/v_m = 1 + p$. The fitness of a rare mutant queen with sex ratio s is

$$w(s,r) \propto v_f\, r_{MD}(1-s) + v_m\, r_{MS}(N_f/N_m)s$$
$$= \frac{(2r+\beta)-(4r+\beta-2)s}{2(r+\beta)} \tag{18}$$

where N_f and N_m are the numbers of females and males. The optimal sex ratio in queenright colonies is

$$\begin{aligned} r^* &= \tfrac{1}{2} - \tfrac{1}{4}\beta \quad &&\text{if } \beta < 2 \\ r^* &= 0 &&\text{if } \beta > 2 \end{aligned} \tag{19}$$

In order to compensate completely for worker laying, the queen would have to use a sex ratio of $\frac{1}{2}(1 - \beta)$ to produce a population sex ratio of 0.5. Thus the extent of sex ratio compensation is to go halfway toward a population sex ratio of 0.5; the reason that the queen only goes halfway is the increase in the relative reproductive value of a male due to worker laying. For Forsyth's data in the previous paragraph the predicted sex ratio in queenright colonies should be 0.43, whereas it is in fact 0.39.

The excess female bias could reflect some degree of worker control over the sex ratio in queenright colonies. (We are supposing that the workers are not laying eggs in these colonies but can control the number of queen-laid male and female eggs that become reproductive adults.) The optimal sex ratio in queenright colonies under worker control can be calculated by substituting the sister–sister and sister–brother relatedness instead of the mother–daughter and mother–son relatedness in Equation 18. (See the coefficients of relatedness in Table 9.3.) The optimal sex ratio under worker control in queenright colonies is

$$\begin{aligned} r^* &= \tfrac{1}{4} - \tfrac{3}{8}\beta \quad &&\text{if } \beta < \tfrac{2}{3} \\ r^* &= 0 &&\text{if } \beta > \tfrac{2}{3} \end{aligned} \tag{20}$$

This gives a predicted sex ratio for Forsyth's data of 0.15. The observed sex ratio in queenright colonies is much closer to the prediction under queen control than under worker control.

Multiple mating

Another complication is that the foundress of the colony may have been inseminated by more than one male. Suppose that she has been inseminated by many males so that her offspring are half sibs, with the same mother but (in the case of female offspring) with different fathers. The relatedness of a female to her half sister is $\frac{1}{4}$, while it remains unchanged at $\frac{1}{2}$ to her "half brother"; in the absence of worker laying, females are twice as valuable as males, so that the life-for-life coefficients of relatedness of a worker to male and female larvae are the same and the optimum sex ratio is 0.5. Thus the conflict of interest between the queen and the workers is abolished under multiple mating if the queen has mated with many males.

Suppose now that the foundress always mates with two males so that a worker is equally likely to be rearing a full or a half sib. If she can distinguish between them, she should rear only her full sibs, to whom she is more closely related, and would prefer a 3:1 female-biased sex ratio among them. It is more likely that she cannot distinguish between them, so that the average relatedness of a worker to female and male larvae is $\frac{1}{2}$ for both sexes. Since females are twice as valuable as males, this leads to a preferred sex ratio biased 2:1 in favor of females under worker control.

The most interesting situation arises if there is variability in the number of mates who inseminate a foundress (Boomsma and Grafen, 1991). Suppose that a proportion p_i of the foundresses have been inseminated by i males who contribute equally to their progeny. Suppose also that the sex ratio is under the control of the workers, who cannot distinguish whether the individual larvae they are rearing are full or half sibs, but who can determine, perhaps through the variability of genetic odor cues, the proportion of full sibs among them (that is to say, the type of colony they are in). The life-for-life coefficients of relatedness of a worker to an average male or female larva in the ith type of colony are

$$\begin{aligned} l_{im} &= 1 \\ l_{if} &= 1 + 2/i \end{aligned} \tag{21}$$

Suppose that most colonies produce F_i females and M_i males in the ith type of colony, while a rare colony type produces f_i females and m_i males (with $F_i + M_i = f_i + m_i = n$ for all colonies). The inclusive fitness of the workers in the mutant colonies is

$$\text{fitness}_{\text{mutant}} = \sum_i p_i l_{if} f_i + \sum_i p_i l_{im} m_i \frac{\sum_i p_i F_i}{\sum_i p_i M_i} \tag{22}$$

Write $s_i = m_i/n$ and $r_i = M_i/n$ for the mutant and wild-type sex ratios and consider a mutant that changes only the sex ratio in the jth type of colony, so that $s_i = r_i$ ($i \neq j$). The fitness of this mutant relative to the wild type is

$$w = 1 + k(s_j - r_j)\left[\frac{l_{jm}}{\sum_i p_i r_i} - \frac{l_{jf}}{\sum_i p_i (1 - r_i)}\right] \tag{23}$$

where k does not depend on s_j and is strictly positive. The equilibrium sex ratios must satisfy one of the following conditions:

$$\begin{aligned} r_j^* &= 1 \quad \text{and} \quad \lambda_j > \alpha \\ 0 < r_j^* &< 1 \quad \text{and} \quad \lambda_j = \alpha \\ r_j^* &= 0 \quad \text{and} \quad \lambda_j < \alpha \end{aligned} \tag{24}$$

where

$$\lambda_j = l_{jm}/l_{jf} = j/(j+2) \text{ and}$$

$$\alpha = \frac{\sum_i p_i r_i^*}{\sum_i p_i (1 - r_i^*)} \tag{25}$$

Thus the optimum strategy is to produce only males in colonies with a high relatedness of workers to males compared with females, λ, and only females in colonies with a low relative relatedness of workers to males. There can be at most one colony type (the balancing type) with an intermediate sex ratio. A numerical solution can be found by trial and error.

Intuitively, the workers in the balancing type of colony can arrange that the population sex ratio is optimal for them; in this case workers in colonies with a more promiscuous queen want a more male-biased sex ratio, while workers in colonies with a less promiscuous queen want a more female-biased sex ratio, and in both cases they go as far as they can to obtain what they want. The significance of these results is that they predict great variability in sex ratio in different types of colony, correlated with the multiplicity of inseminations of the foundress. This prediction has been confirmed by Sundström (1994) in a Finnish population of a wood ant whose colonies have a single queen, mated to one or several males. Colonies show a bimodal distribution of sex ratios, with most colonies headed by a multiply mated queen producing a preponderance of males and most colonies headed by a singly mated queen producing a preponderance of females among the reproductives.

The predisposition to eusociality in haplodiploids

Eusociality is defined by the presence of three characteristics: (i) cooperative brood care, (ii) a caste of sterile workers or helpers, and (iii) overlapping generations with offspring helping their parents. It has evolved independently many times in Hymenoptera (at least 11 according to one estimate), once in termites, and once in mammals (the naked mole rat). Colony members are closely related in all eusocial animals, and it is clear that kin selection has been the driving force in its evolution. But why should it be so frequent in Hymenoptera? Two reasons have been suggested, the first genetic and the second ecological. Members of this order are haplodiploid and Hamilton (1964, 1972) argued that the asymmetry of their relatedness coefficients facilitates kin selection and the evolution of eusociality. This view won rapid acceptance, but opinion today places greater emphasis on ecological preconditions that lower the costs and increase the benefits of reproductive altruism in Hymenoptera (Evans, 1977; Andersson, 1984). I shall discuss the genetic argument first.

The life-for-life coefficient of relatedness between sisters is $\frac{3}{4}$ in haplodiploids, whereas it is only $\frac{1}{2}$ in diploids. As we saw in the last chapter in discussing colony foundation by a pair of sisters rather than by a single foundress in a paper wasp, this may provide a genetic predisposition toward the evolution of cooperation between sisters in colony formation. Suppose that a nest founded by a single female produces n_1 offspring while a nest founded by two females can produce n_2 offspring. From the point of view of a β-female who has the choice of founding a nest by herself or of helping her sister but not reproducing herself, the cost of helping is $c = n_1$, while the benefit to her sister is $b = n_2 - n_1$. Helping will evolve through kin selection in a diploid species if $n_2 > 3n_1$, while it will evolve in haplodiploids if $n_2 > 2.33n_1$; there is thus a genetic predisposition for this type of behavior, that is called *semi-social*, to evolve more easily in haplodiploids.

The evolution of full-blown eusocial behavior requires that daughters stay behind in the nest to help their mother rather than leaving to found their own nests. Consider a daughter who has the choice between founding her own nest and raising c offspring and staying in the maternal nest to help her mother raise an additional b offspring who are her full sibs. (I assume single insemination of the mother, which is likely to be the case in a solitary or primitively social insect.) In diploids a female is equally closely related to her offspring and her sibs, so that helping will be favored if $b > c$. In haplodiploids a female has a life-for-life relatedness of $\frac{1}{2}$ to her offspring, $\frac{3}{4}$ to her sisters, and $\frac{1}{4}$ to her brothers (Table 9.5). If she rears equal numbers of brothers and sisters in her mother's nest, her average relatedness to them is $\frac{1}{2}$, the same as her relatedness to her own offspring, so that helping will be favored when $b > c$, as in diploids. (I assume a 1:1 sex ratio under the control of the mother.) But if she has the option of helping to rear only sisters in her mother's nest, helping will be favored when $b > 0.67c$.

However, this advantage to helping is only temporary. If the population sex ratio is r, the inclusive fitness of a female who helps her mother to produce f more sisters and m more brothers is

$$\text{fitness}_{\text{helper}} = \frac{3}{4}f + \frac{1}{4}\left(\frac{1-r}{r}\right)m \tag{26}$$

while the inclusive fitness of a female who produces f daughters and m sons is

$$\text{fitness}_{\text{nonhelper}} = \frac{1}{2}f + \frac{1}{2}\left(\frac{1-r}{r}\right)m \tag{27}$$

When $r = 0.5$, helping will be favored when $b > 0.67c$, if the helper rears b extra sisters and the nonhelper produces c offspring of either sex. As this type of helper increases in frequency the population sex ratio will become increasingly female-biased until it reaches the 3:1 ratio favored under worker control. At this point the helpers will have an inclusive fitness of $0.75b$ whether they help rear sisters or brothers, while the nonhelpers will have inclusive fitness of $0.75c$ if they also have a 3:1 sex ratio, and an inclusive fitness of $1.5c$ if they produce only sons. Thus helping mother to rear one's sisters loses its advantage as it becomes common, owing to the female bias in the sex ratio that this produces.

It can be concluded that there is a genetic predisposition toward the evolution of semi-social behavior in haplodiploids but that the genetic predisposition toward the evolution of full eusocial behavior is more limited. As Seger (1991) remarks: "To exploit their greater relatedness to sisters than to offspring, workers must invest at [sex] ratios that are more female biased than the population as a whole." Seger (1991) considers a number of scenarios of how this could happen. The simplest is that, in a solitary bee, for example, some females fail to mate and therefore produce all-male broods (Godfray and Grafen, 1988). Mated females will produce a brood that is sufficiently female-biased to give an even population sex ratio. (This differs from the result with all-male broods produced by worker laying considered above because there is no change in the relative reproductive values of males and females.) A daughter who stays behind to help rear her siblings will therefore have an average life-for-life relatedness to them greater than $\frac{1}{2}$, which would be her relatedness to her own offspring. This might be the first step toward the evolution of eusociality. Seger (1983) discusses another scenario giving rise to alternating sex-ratio biases that may favor eusociality.

The ecological theory argues that there are a number of ecological preconditions favoring the evolution of eusociality in Hymenoptera by lowering the costs or increasing the benefits of helping mother to rear one's siblings. The most important of them is nest-building; nearly all eusocial animals build nests, as did their solitary ancestors. This acts as a preadaptation for offspring to stay behind in the nest and help their mother to rear their siblings in several ways. (1) The nest is a place where individuals who meet each other are almost certainly related; no other cues to kin recognition are required. (2) The nest is a place where one adult can increase the reproductive success of another by foraging and bring food back to it. (3) The nest allows adults to cooperate in guarding the brood from external enemies. (4) Nests are likely to be expensive to build, and an offspring who stays behind to help her mother rather than breeding herself avoids this cost. The first factor makes it possible to recognize and thus to help one's kin; the other three factors facilitate the evolution of reproductive altruism through kin selection by lowering the cost to a daughter who stays behind in the nest to help her mother and increasing the benefit to her siblings.

Under this interpretation the frequency of eusocial Hymenoptera is attributed to the frequency of nest building in their solitary ancestors, rather than to their mode of sex determination. It is also possible that haplodiploidy favors the evolution of eusociality because it enables mothers to produce worker daughters who are adapted to care for the young; the fact that workers are always female in Hymenoptera may have more to do with their preadaptation to the maternal role than to genetic asymmetries.

LOCAL MATE COMPETITION

Fisher's principle leading to an even sex ratio depends, among other things, on the assumption of population-wide competition for mates, whereas many natural populations have a geographical population structure in which limited dispersal imposes constraints on mating patterns. Hamilton (1967) in an influential paper on "extraordinary sex ratios" considered the following model to investigate this question. The world consists of a large number of patches, each of which is colonized by n mated females; their offspring mate at random within each patch, and the newly mated females disperse in population-wide competition for the n breeding sites available in each patch in the next generation. The essential feature of this model is that mating occurs *before* dispersal.

This model is intended to capture the life history of organisms such as fig wasps. A small number of mated foundress wasps enter a fig syconium (an enclosed inflorescence) nearly simultaneously, pollinate the flowers, lay eggs and die. Their offspring mature and mate within the fig, the wingless males then dying while the winged females disperse to restart the cycle. In this situation, brothers are competing with each other for mating opportunities (hence the name of local mate competition). In consequence there are diminishing returns for the mother in producing sons; the more sons a mother produces, the fewer matings each of them will on average expect, because they are in competition with each other. Thus one would expect selection for a female-biased sex ratio. In the extreme case in which there is only one foundress in a patch, she will maximize her reproductive output by producing as many daughters as she can and only enough sons to ensure that all the daughters are mated.

Hamilton derived what he called the "unbeatable" sex ratio in this situation by an inclusive-fitness argument. Assume that the fitness of a female can be represented as in Equation 8:

$$\text{fitness} = \text{number of daughters} + \text{number of females inseminated by sons}$$

Suppose as before that there are two types of female, wild-type females who produce F daughters and M sons, and rare mutant females who produce f daughters and m sons. Because mutants are rare, they will occur in a patch with $n - 1$ wild-type females, and their fitness will be

$$\text{fitness}_{\text{mutant}} = f + \left[\frac{(n-1)F + f}{(n-1)M + m}\right] m \tag{28}$$

where the term in square brackets is the average number of matings by each male in that patch. The fitness of wild-type females, who will almost always occur in a patch without mutant females, is $2F$ so that the relative fitness of the mutant is

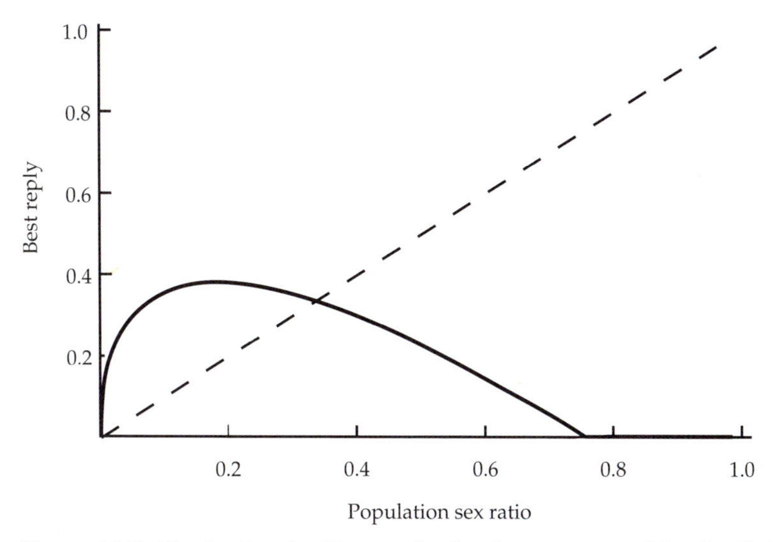

Figure 10.2. The best-reply diagram for local mate competition in diploids with $n = 3$.

$$w(s,r) = \left\{(1-s) + \frac{[(n-1)(1-r)+(1-s)]s}{(n-1)r+s}\right\} \Big/ 2(1-r) \tag{29}$$

The unbeatable sex ratio is found from Equation 12 as

$$r^* = \frac{n-1}{2n} \tag{30}$$

The best-reply diagram is shown in Figure 10.2 with $n = 3$. In this case the equilibrium is a strict Nash equilibrium, whereas it was only a weak Nash equilibrium in Figure 10.1.

Hamilton (1967) simulated his model with $n = 2$ under haplodiploid sex determination (males haploid, females diploid) and found the unbeatable sex ratio to be between 0.205 and 0.215, rather than 0.25 as predicted by his formula. The reason for the discrepancy is that Equation 8 does not take into account the asymmetry of haplodiploidy. The fitness should instead be represented as

$$\text{fitness} = v_f r_{MD} f + v_m r_{MS}\left[\frac{(n-1)F+f}{(n-1)M+m}\right]m \tag{31}$$

(Compare Equation 15.) By the same argument as before the unbeatable sex ratio is found to be

$$r^* = \frac{n-1}{2n}\rho \tag{32}$$

where

$$\rho = 2v_m r_{MS} / (v_m r_{MS} + v_f r_{MD}) \tag{33}$$

For diploids, $v_m = v_f$ and $r_{MS} = r_{MD}$ by symmetry so that $\rho = 1$, and the analysis given previously that ignores these complications is valid. For haplodiploidy $v_f = 2v_m$ while the ratio r_{MS}/r_{MD} varies with the level of inbreeding between 2 (no inbreeding) and 1 (complete inbreeding). It is shown in Appendix 10.1 that

$$\rho = \frac{4n-2}{4n-1} \tag{34}$$

so that

$$r^* = \frac{(n-1)(2n-1)}{n(4n-1)} \tag{35}$$

With $n = 2$ this gives $r^* = 0.214$, in agreement with Hamilton's simulations. Taylor and Bulmer (1980) re-examined the model using a population genetic argument and confirmed the result in Equation 30 for diploids and in Equation 35 for haplodiploids.[1]

Hamilton (1967) compiled a list of 26 haplodiploid insects and mites that usually mate with their sibs and have extreme female-biased sex ratios. The most bizarre is a mite that has one male and about 14 females in a batch of progeny; the male mates with his sisters inside his mother and dies before he is born!

Much experimental work has been done on the parasitic wasp *Nasonia vitripennis* which lays batches of eggs in the pupae of flies such as blowflies. Males are flightless, and mating occurs on or near the host. The hosts can be either dispersed (leading to sib-mating) or highly aggregated (leading to mating among the offspring of many foundresses). Theory predicts that the wasp should vary its sex ratio with the degree of host aggregation. Both experiments and field observations show this effect. Werren (1983) placed a variable number (1, 2, 3, 4, 6, 8, or 12) of recently emerged, mated female wasps in a petri dish with four blowfly pupae. As seen in

[1] This argument is reproduced in the *Mathematica* solutions.

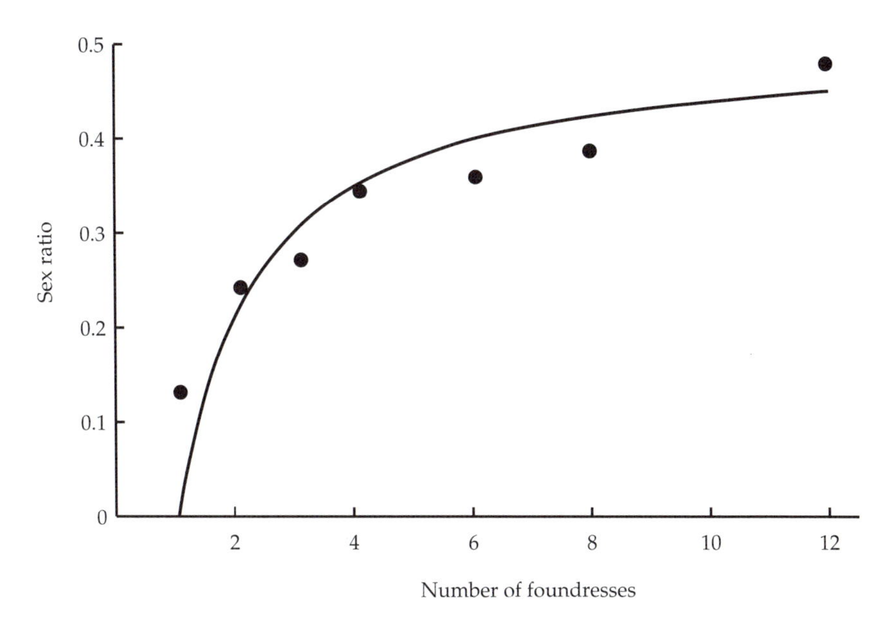

Figure 10.3. Sex ratios of a parasitic wasp from patches containing different numbers of foundresses in a laboratory experiment. The curve shows the theoretical prediction. After Werren, 1988.

Figure 10.3, he found that the sex ratio of the progeny varied with the number of foundresses according to the prediction in Equation 35, except that for obvious reasons the sex ratio with one foundress was not zero but about 0.13; this can be interpreted as the minimum number of males necessary to inseminate all the sisters. Thus *N. vitripennis* females have a sophisticated ability to monitor the number of co-foundresses and adapt the sex ratio accordingly.

The assumptions of Hamilton's model are followed closely in the natural history of fig wasps, discussed above. In a natural population the number of foundresses will vary between figs. Herre (1985) shows that this can be taken into account to a good approximation by using the harmonic mean of the number of foundresses in calculating ρ; denoting this harmonic mean by $\tilde{n}$, the optimal predicted sex ratio is

$$r^* = \frac{n-1}{n}\left(\frac{2\tilde{n}-1}{4\tilde{n}-1}\right) \tag{36}$$

The sex ratio should vary with the number of foundresses, n, and the bias

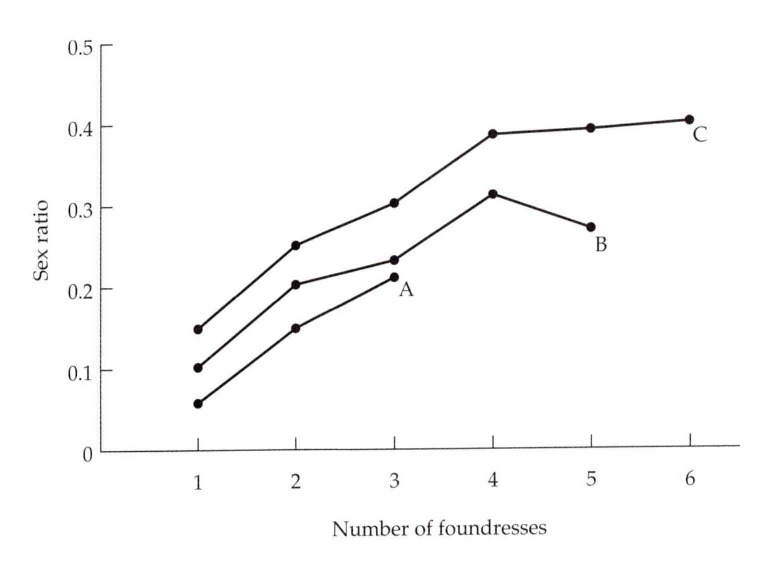

Figure 10.4. The relationship between sex ratio and foundress number in three species of fig wasp with harmonic mean foundress number 1.13 (species A), 1.40 (species B), and 1.99 (species C). After Herre, 1985.

should be more extreme the greater the degree of inbreeding, as evidenced by a low value of $\tilde{n}$. Herre (1985) shows the relationship between sex ratio and foundress number in three species of fig wasp with harmonic mean foundress numbers of 1.13, 1.40, and 1.99. The observed relationship shown in Figure 10.4 agrees well with prediction.

The force driving selection for a female-biased sex ratio under local mate competition is that there are diminishing returns on producing sons; the more sons a mother produces, the fewer matings each of them will on average expect, because they are competing with each other for mating opportunities. It has been suggested by Colwell (1981) that group selection is implicated in the evolution of of female-biased sex ratios for the following reason. Suppose that there are two kinds of females in the population: "Fisher" females who produce offspring with a sex ratio of 0.5 and "Hamilton" females with female-biased sex ratios. In a mixed patch containing both types, Fisher females are fitter than Hamilton females in the sense that their genes are disproportionately represented among the offspring of that patch. However, patches with a large number of Hamilton females are fitter than those with a small number, because they produce more mated females to compete for breeding sites in the next generation. In other words, individual selection within a patch favors Fisher females, while group selection between patches favors Hamilton females. The outcome depends on the balance between these forces. This is an alternative, and equally valid, way of looking at the problem.

LOCAL RESOURCE COMPETITION

Clark (1978) reported a male-biased sex ratio in bush babies that she explained as follows. Young males disperse from the area in which they are born, but young females stay in their natal area where they compete both with each other when they reach maturity and with their mother (if she is still alive) for food resources necessary for reproductive success. Thus there will be a linearly increasing return on investment in sons (a mother with four sons is likely to have twice as many grandchildren as one with two sons), but there will be diminishing returns on investment in daughters because they compete with each other for resources required to reproduce. This will select for a male-biased sex ratio. Clark (1978) called this situation local resource competition since it is analogous to local mate competition, with competition between males for mates being replaced by competition between females for resources.

Johnson (1988) provides further evidence of the effect of local resource competition on the sex ratio in primates. The reproductive success of females in several species of primates is limited by the availability of food, and the observation that reproductive success generally declines with group size suggests that females in groups compete with one another for resources. Johnson (1988) used the slope of the regression of reproductive success on group size as an index of the intensity of within-group competition and showed that the sex ratio in seven genera of primates increases linearly with intensity of competition, from about 0.50 in genera with little or no competition to about 0.67 in the genus in which competition is most intense. Her results are shown in Figure 10.5.

An open question is how primates manage to bias their sex ratio in the face of the even numbers of X and Y male gametes produced by meiosis. It may be that females can favor fertilization by X rather than Y gametes by some unidentified mechanism, or it may be that male embryos are selectively aborted. However, it is likely that there would be some reproductive cost of abortion, that should be taken into account in modeling the situation.

Charnov (1982) discusses the following simple model of local resource competition (based on unpublished work by Hamilton). The world consists of a large number of patches in each of which there are n available breeding sites. Each mated pair produces a large number of offspring. The female offspring remain in their natal patch and compete with each other

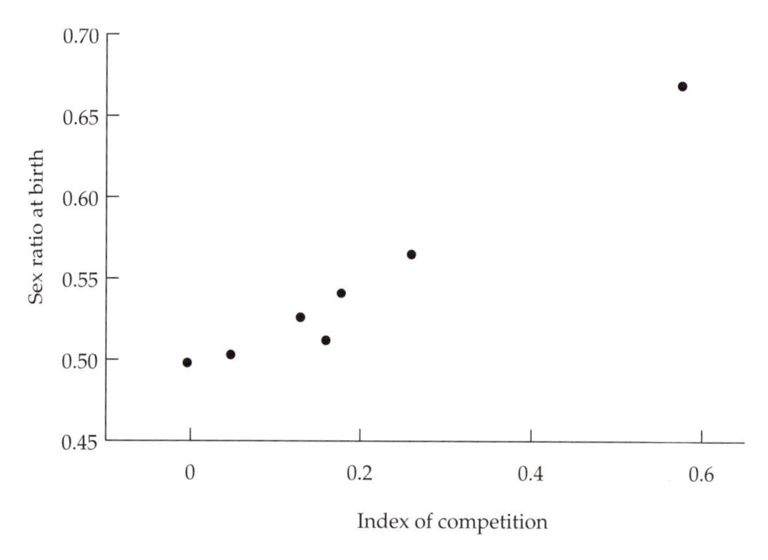

Figure 10.5. Relation between the intensity of competition among females and the sex ratio at birth in genera of primates with female-based philopatry. After Johnson, 1988.

to occupy the n available sites in the next generation; the male offspring disperse to mate with females in other patches. Suppose that most pairs produce M sons and F daughters, while a rare mutant form produces m sons and f daughters. The inclusive fitness of the mutant form is

$$\text{fitness}_{\text{mutant}} = f\frac{n}{(n-1)F+f} + m\frac{n}{nM} \tag{37}$$

since a mutant mother will nearly always find herself in a patch with $n-1$ wild-type mothers whose daughters compete with each other for the available breeding sites, while a male offspring (regardless of whether he is the son of a wild-type or a mutant mother) nearly always finds himself in a patch with nM males competing for the available females holding breeding sites. If $f + m = F + M$, the relative fitness of the mutant, apart from a factor of 2, is

$$w(s,r) = \frac{n(1-s)}{(n-1)(1-r)+(1-s)} + \frac{s}{r} \tag{38}$$

The unbeatable sex ratio is found by solving Equation 12 to be

$$r^* = n/(2n-1) \tag{39}$$

With only one site per patch the optimal sex ratio is $r^* = 1$, an all-male ratio. In practice this means that the single pair in each patch should have nearly all sons and just enough daughters to ensure that at least one of them survives to inherit the patch. This is the strategy adopted by social insects which reproduce by swarming or colony division (honey bees, army ants, and many polybiine wasps). There is only one queen in each new swarm. The old colony only produces a few queens, because any further investment in queens would be wasted since a queen without a swarm is valueless, but it produces a large number of males to mate with virgin queens produced by other colonies (Craig, 1980; Charnov, 1982; Bulmer, 1983b).

SEX RATIO WHEN FITNESS VARIES

Offspring fitness (in terms either of survival or future reproductive success) will often depend on the environment in which they develop, and it may be that this effect will be greater for one sex than for the other. Suppose that some juveniles encounter good conditions and grow into large adults, while others are less fortunate and grow into small adults. Suppose also that adult size matters more to a female than to a male because female fecundity is directly proportional to size whereas male mating ability is little affected by size. (The last assumption presupposes the absence of intrasexual competition mediated through size.) Then it would be advantageous (both to the mother and to her offspring) for juveniles to be male or female depending on whether they develop in a poor or a good environment, because the reward for being large is greater for a female whereas the penalty for being small is less for a male. On the other hand, if there is substantial intrasexual selection such that large males have much greater reproductive success than small ones, the situation is reversed, and it would be advantageous for a juvenile to be male in a good environment and female in a poor environment.

To model this situation, suppose that the environment contains k patch types, the fitness of a female reared in the ith patch being w_{if} and that of a male w_{im}. Write p_i for the relative frequency of the ith patch, and suppose that wild-type mothers produce F_i females and M_i males in the ith patch, while rare mutant mothers produce f_i females and m_i males (with $F_i + M_i = f_i + m_i = n$ for all patches, so that the clutch size is independent of patch type). The inclusive fitness of the mutant mother is

$$\text{fitness}_{\text{mutant}} = \sum_i p_i w_{if} f_i + \sum_i p_i w_{im} m_i \frac{\sum_i p_i w_{if} F_i}{\sum_i p_i w_{im} M_i} \tag{40}$$

Write $s_i = m_i/n$ and $r_i = M_i/n$ for the mutant and wild-type sex ratios, and consider a mutant that changes only the sex ratio in the jth patch, so that $s_i = r_i$ $(i \neq j)$. (The argument is similar to that developed above for the effect of foundresses with a variable number of inseminations on colony sex ratios for social insects with worker control of the sex ratio.) The fitness of this mutant relative to the wild type is

$$w = 1 + k(s_j - r_j)\left[\frac{w_{j\mathrm{m}}}{\sum_i p_i w_{i\mathrm{m}} r_i} - \frac{w_{j\mathrm{f}}}{\sum_i p_i w_{i\mathrm{f}}(1 - r_i)}\right] \tag{41}$$

where k does not depend on s_j and is strictly positive. The equilibrium sex ratios must satisfy one of the following conditions:

$$\begin{aligned} r_j^* &= 1 \quad \text{and} \quad \rho_j > \alpha \\ 0 < r_j^* &< 1 \quad \text{and} \quad \rho_j = \alpha \\ r_j^* &= 0 \quad \text{and} \quad \rho_j < \alpha \end{aligned} \tag{42}$$

where

$$\begin{aligned} \rho_j &= w_{j\mathrm{m}} / w_{j\mathrm{f}} \\ \alpha &= \frac{\sum_i p_i w_{i\mathrm{m}} r_i^*}{\sum_i p_i w_{i\mathrm{f}}(1 - r_i^*)} \end{aligned} \tag{43}$$

Thus the optimum strategy is to produce only males in patches with a high relative fitness of males compared with females, ρ, and only females in patches with a low relative fitness of males. There can be at most one patch type with an intermediate sex ratio. A numerical solution can be found by trial and error. If there is continuous variability in patch quality, this theory predicts an abrupt transition from an all-male to an all-female sex ratio as the relative fitness of males falls below a threshold value.

Trivers and Willard (1973) argued for mammals that a son would gain more from being in good condition than a daughter because of male–male competition for females and that a female in good condition should therefore have sons, while a female in poor condition should have daughters. Clutton-Brock et al. (1984) provide evidence that this happens in red deer: dominant females with access to good feeding sites tend to have sons, while subordinate females tend to have daughters.

It is easier to understand how haplodiploid organisms manipulate the sex of their offspring to suit the circumstances. Charnov (1982) reviews a number of cases in which hymenopteran parasitoids lay a male egg in a small host and a female egg in a large host; large size confers greater benefit to females than to males in insects because it is highly correlated with female fecundity, while there is little male–male competition.

The theory is also applicable to organisms with environmental sex determination, the sex of the offspring being determined by an environmental cue. Some parasitic nematode worms develop as males in crowded conditions, in which they are constrained to be small, but as females in uncrowded conditions, in which they can grow larger (Charnov, 1982); again size is more important to females because of its correlation with female fecundity. As another example of environmental sex determination, consider the Atlantic silverside, a marine fish inhabiting the Atlantic seaboard of North America. Sex determination is temperature-dependent, the young developing into females in colder water and into males in warmer water (Conover, 1984). Fish spawn in their second year between May and July, and the consequence of temperature-dependent sex determination is that young spawned early in the season when the water is colder are female, while those spawned later are male. Thus females have a longer growth period and grow larger than males. This is adaptive since size is more important to females in these fish, because of its association with female fecundity. Indeed, it is likely that this is the selective force leading to temperature-dependent sex determination in silversides: at the northern end of their range off Nova Scotia, where the spawning season is much shorter so that there would be little advantage in temperature-dependent sex determination, they have abandoned it in favor of a more usual genotypic mechanism (Conover and Heins, 1987).

Sequential hermaphroditism

Consider now a perennial organism that continues to grow after maturity. If size makes more difference to female fecundity than to male reproductive success, it would be of advantage to an individual to be a male when it is young and small and to become a female when it is old and large. On the other hand, if size makes more difference to male than to female fecundity because of size-dependent competition between males for mates, it would be advantageous for an individual to start reproductive life as a female and to become a male when it is old and large. These types of sex change, known as sequential hermaphroditism, are not uncommon in invertebrates and fish.

The theory developed above can be applied to this situation if we interpret being in the ith patch as being an adult of age i and p_i as the proportion of adults of this age. Charnov (1979) has applied this theory to sex change in Pandalid shrimp, which are protandrous hermaphrodites, changing sex from male to female. In some species, some individuals always breed as females, even in their first breeding year, while the remainder start breeding as males and change to females in their second breeding year. Data are available on the proportion of early breeding females, P, in a number of species. If $P > 0$, we may substitute $r_1^* = 1 - P$, $r_i^* = 0$ $(i > 1)$ in Equations 42 and 43 to obtain

$$\frac{w_{1m}}{w_{1f}} = \frac{p_1 w_{1m}(1-P)}{p_1 w_{1f} P + \sum_{i=2}^{\infty} p_i w_{if}} \tag{44}$$

from which follows Charnov's equation,

$$P/(1-P) = \begin{cases} 1 - 2\beta & \text{if } \beta < \frac{1}{2} \\ 0 & \text{if } \beta > \frac{1}{2} \end{cases} \tag{45}$$

where

$$\beta = \frac{\sum_{i=2}^{\infty} p_i w_{if}}{\sum_{i=1}^{\infty} p_i w_{if}} \tag{46}$$

The parameter β is the relative contribution to female function of individuals who change sex to that of early-breeding females. To estimate this quantity Charnov (1979) assumed constant mortality and fecundity proportional to the cube of length, with length increasing according to the von Bertalanffy equation (see Chapter 5), the life-history parameters being estimated separately for 27 shrimp populations belonging to six species. Figure 10.6 shows good agreement between the observed proportion of early-breeding females and the prediction from Equation 45.

FURTHER READING

Charnov (1982) is the best account of the subject. Godfray (1994) reviews sex ratios in parasitoid wasps. Bull (1983) reviews the evolution of sex-determining mechanisms.

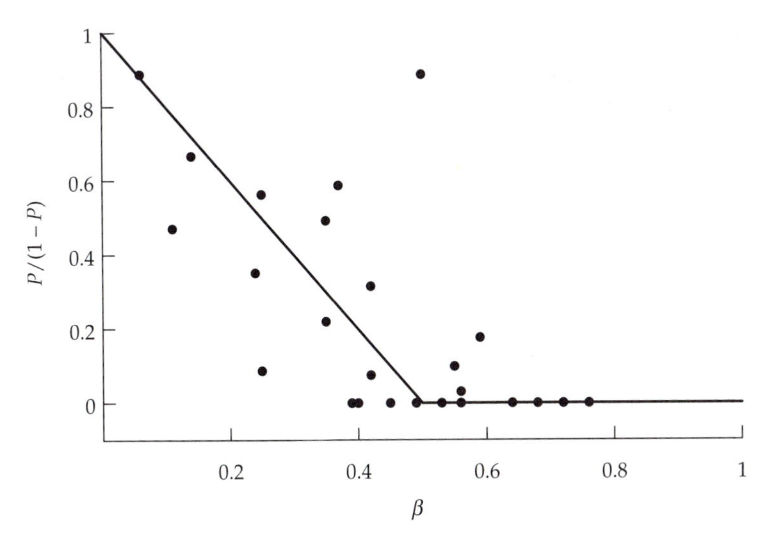

Figure 10.6. Occurrence of early-breeding females in Pandalid shrimp compared with predicted relationship. After Charnov, 1979.

EXERCISES

1. Verify Equation 2. Plot the dominant eigenvalue of the transition matrix against s for $r = 0.4$, 0.5, and 0.6. Comment.

2. (a) Verify Equation 3 when s is close to r.
   (b) Verify Equation 7 and hence find the optimal sex ratio.

3. Verify Equation 11, find the optimal sex ratio from Equation 12, and check that the criterion for continuous stability is satisfied.

4. Verify Equation 13 for haplodiploids and repeat the analogues of Exercises 1 and 2 for this transition matrix.

5. The above results only hold for the invasion of a rare allele; the theory becomes more complex when this assumption is relaxed but can be investigated by simulation. Write a computer program to simulate the evolution of the sex ratio in a haplodiploid organism with sex ratios produced by *aa*, *Aa*, and *AA* females being r, s, and t, respectively. Use this program to find the equilibrium genotype frequencies and hence the equilibrium sex ratio in the following cases: (a) $r = 0.3$, $s = 0.4$, $t = 0.4$; (b) $r = 0.4$, $s = 0.6$, $t = 0.8$; (c) $r = 0.1$, $s = 0.75$, $t = 0.1$.

6. Consider a eusocial hymenopteran colony with two foundresses who lay equal numbers of eggs with no worker contribution to egg-laying. The foundresses may be (a) unrelated or (b) sisters. (The latter is biologically more likely.) Find the average coefficient of relatedness of a foundress and a worker to male and female larvae in these two cases (see Table 9.3 and Exercise 9.5), and hence find the predicted sex ratio under queen control and worker control.

7. Verify the arguments leading to Equations 19 and 20.

8. (a) Verify the argument leading to Equation 29 giving the relative fitness of a rare sex-ratio mutant under local mate competition in a diploid organism.
(b) Find the unbeatable sex ratio from this equation and check that it is a strict Nash equilibrium and that it is continuously stable.

9. (a) Verify the argument leading to Equation 35, giving the unbeatable sex ratio under local mate competition in a haplodiploid organism.
(b) If you have the *Mathematica* solutions, use the program there to verify that Equation 35 follows from a population genetic argument.

10. Verify that Equation 39, giving the optimal sex ratio under a model of local resource competition, follows from Equation 38.

11. The optimal strategy in a patchy environment can be found from Equations 42 and 43 as follows. Order the patches so that $\rho_1 > \rho_2 > \ldots > \rho_k$. Suppose that the mth patch is the switch patch with an intermediate sex ratio, so that

$$r_j^* = 1 \qquad j < m$$

$$r_j^* = 0 \qquad j > m$$

From Equation 43, r_m^* can now be evaluated, and the problem is solved if this value is between 0 and 1. The solution can be found by repeating this procedure for $m = 1, 2, \ldots$ until the switch patch is identified.

Find the optimal sex-ratio strategy for three equally frequent patches with the following fitnesses:

	Patch		
	1	2	3
Male fitness:	1	1	1
Female fitness:	1	2	3

12. (a) Verify the argument leading to Equation 45.
    (b) Consult Charnov (1979) to find out how he estimated β.

APPENDIX 10.1. RELATEDNESS BETWEEN MOTHER AND OFFSPRING IN A HAPLODIPLOID UNDER INBREEDING DUE TO LOCAL MATE COMPETITION

First consider the relatedness between mother and son. It is clear from Figure 10.7a that

$$F_{\mathrm{MS}}(t) = \tfrac{1}{2}[1 + F_{\mathrm{M}}(t)] \tag{47}$$

where $F_{\mathrm{MS}}(t)$ is the coefficient of consanguinity between mother and son and $F_{\mathrm{M}}(t)$ is the coefficient of inbreeding in the mother in generation t. (See Chapter 9 for the definition of these coefficients.) It follows from Equation 9.12b that

$$r_{\mathrm{MS}}(t) = 1 \qquad \text{for all } t \tag{48}$$

Now consider the relatedness between mother and daughter. It can be seen from Figure 10.7b that

$$\begin{aligned} F_{\mathrm{MD}}(t) &= \tfrac{1}{4}[1 + F_{\mathrm{M}}(t)] + \tfrac{1}{2}F_{\mathrm{MF}}(t) \\ F_{\mathrm{M}}(t) &= F_{\mathrm{MF}}(t-1) \end{aligned} \tag{49}$$

At equilibrium,

$$\begin{aligned} F_{\mathrm{MD}} &= \tfrac{1}{4}(1 + F_{\mathrm{M}}) + \tfrac{1}{2}F_{\mathrm{M}} \\ r_{\mathrm{MD}} &= \frac{1}{2} + \frac{F_{\mathrm{M}}}{1 + F_{\mathrm{M}}} \end{aligned} \tag{50}$$

To find the coefficient of inbreeding, note that the probability that father and mother are sibs is $1/n$. It can be seen from Figure 10.7c that

$$F_{\mathrm{MF}}(t) = \tfrac{1}{4n}\left[1 + 2F_{\mathrm{MF}}(t-1) + F_{\mathrm{M}}(t-1)\right] \tag{51}$$

Using the second part of Equation 49, it is easily shown that at equilibrium

$$F_{\mathrm{M}} = \frac{1}{4n - 3} \tag{52}$$

Hence, from the second part of Equation 50,

$$r_{\mathrm{MD}} = \frac{n}{2n - 1} \tag{53}$$

It follows from Equations 48 and 53 that the value of ρ defined in Equation 33 is

$$\rho = \frac{4n-2}{4n-1}$$

as stated in Equation 34.

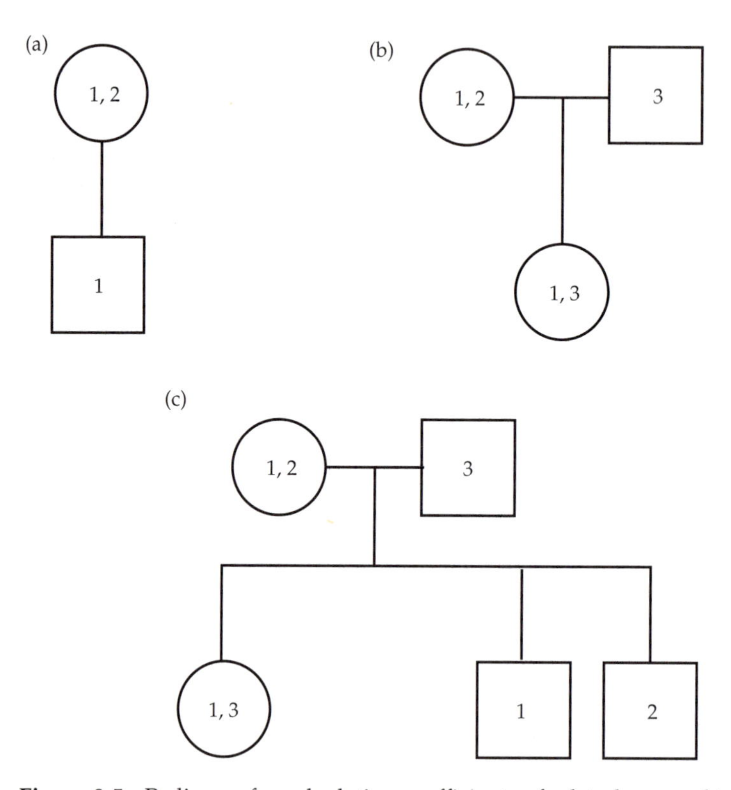

Figure 9.5. Pedigrees for calculating coefficients of relatedness and inbreeding.

CHAPTER 11

Sexual Selection

"Sexual selection depends on the success of certain individuals over others of the same sex in relation to the propagation of the species. … The sexual struggle is of two kinds; in the one it is between the individuals of the same sex, generally the male sex, in order to drive away or kill their rivals, the females remaining passive; whilst in the other, the struggle is likewise between the individuals of the same sex, in order to excite or charm those of the opposite sex, generally the females, which no longer remain passive, but select the more agreeable partners."

—Charles Darwin, *The Descent of Man, and Selection in Relation to Sex* (1871)

In some species there is competition between males for opportunities to mate with females. Males who are successful in this competition will mate with many females and have many descendants, while unsuccessful males will have few or no descendants. This will give rise to selection for characters leading to success in competition for mates. Darwin (1859, 1871) called this *sexual selection*, and he pointed out in the passage above that it can take two forms. The first form, *intrasexual selection*, involves selection for characters (such as antlers in deer) useful in direct competition between males; the female passively accepts whichever male wins the contest, and there is no active female choice between males. The second form, *intersexual selection*, involves selection for characters (such as the peacock's tail) which are valuable indirectly because they are preferred by females; the competition between males is mediated through female choice. In this chapter I shall discuss these two types of sexual selection in turn.

INTRASEXUAL SELECTION

Intrasexual selection occurs through direct competition between males for mates, the female passively accepting the winner. This leads to the evolution of male characters, such as large size in elephant seals or large antlers in deer, which will help them to win contests with other males. An increase in size x will carry a cost, which I represent by supposing that the probability of a male of eventual size x surviving to reproductive maturity is $s(x)$, a decreasing function of x. To counterbalance this cost, a large male will have greater mating success, which will depend not on his absolute size, but on his size relative to that of other males. As Darwin (1871) observed: "Unarmed ... males would succeed equally well ... in leaving behind numerous progeny, but for the presence of better endowed males." Male mating success may be represented as $m(x, p)$, where p is the distribution of size in the population; m will be an increasing function of x but a decreasing function of the mean value of p. Male fitness is the product of survival and mating success, $s(x) \times m(x, p)$. This is frequency-dependent because of its dependence on the population size distribution, and the evolution of size should therefore be studied through evolutionary game theory.

Size is really a continuous variable, but suppose for simplicity that there are only three size classes: small, medium, and large, with sizes x_1, x_2, and x_3, respectively. I shall consider a simplified version of a game-theoretic model analyzed by Maynard Smith and Brown (1986) with the payoff matrix in Table 11.1. The idea is that males compete in pairs for females with the larger one winning; if they are of the same size they each have the same chance of winning. The behavior of the system depends on the rela-

Table 11.1. Payoff matrix for the size game with three size classes.

	Small	**Medium**	**Large**
Small	$\frac{1}{2}s(x_1)$	0	0
Medium	$s(x_2)$	$\frac{1}{2}s(x_2)$	0
Large	$s(x_3)$	$s(x_3)$	$\frac{1}{2}s(x_3)$

tive survival rates $\rho_1 = s(x_2)/s(x_1)$ and $\rho_2 = s(x_3)/s(x_2)$. There is always an ESS with all males large, and if both relative survival rates exceed 0.5 this is the unique ESS. If $\rho_1 < 0.5$ there is also an ESS with all males small, and if $\rho_2 < 0.5$ there is an ESS with all males of medium size. If the population starts with small males, at the optimum size under natural selection, male size will evolve to become medium if $\rho_1 > 0.5$, and it will continue to evolve to become large if $\rho_2 > 0.5$.

Let us now increase the number of size classes to 50, with smaller differences between each class. Suppose we use the same model, so that the ijth element in the payoff matrix is

$$a_{ij} = \begin{cases} 0 & j > i \\ \frac{1}{2}s(x_i) & j = i \\ s(x_i) & j < i \end{cases} \tag{1}$$

The behavior of the system is determined, as before, by the relative survival rates of successive sizes, $\rho_i = s(x_{i+1})/s(x_i)$. These ratios will be nearly 1 since the size differences are small, so that the system will evolve to the largest size possible. This conclusion is an artifact of the assumption in Equation 1 that very small differences in size determine the winner with certainty. In fact there will be considerable uncertainty in the interaction, so that if one player is slightly larger than the other his chance of winning will be only slightly greater than 0.5, rather than 1. A more realistic model would be

$$a_{ij} = \begin{cases} s(x_i)(1 - P_{ji}) & j > i \\ \frac{1}{2}s(x_i) & j = i \\ s(x_i)P_{ij} & j < i \end{cases} \tag{2}$$

where P_{ij} is the probability that a male of size x_i wins against a male of size x_j ($x_i \geq x_j$), which will increase from 0.5 to 1 as $x_i - x_j$ increases from 0 to infinity. Analysis of this model will show the expected outcome, that male size will increase under sexual selection until it is balanced by its increasingly high cost under natural selection (Parker, 1983; Maynard Smith and Brown, 1986).

INTERSEXUAL SELECTION

Darwin correctly proposed that if females prefer certain male characteristics, there will be selection on males to develop them, but he did not have a convincing explanation for the existence of female choice. In some cases females choose a character that is of direct benefit to them. For example, female bullfrogs choose males holding good territories in which their eggs have a high chance of survival; there is also intrasexual selection between males for possession of good territories that give them access to females as well as high offspring survival rates (Howard, 1978a,b [quoted in Krebs and Davies, 1993]). In species with paternal care of the young, females are selected to choose males who are likely to make good fathers (the good-parent model of Hoelzer [1989]). In the House Finch, males have red breasts, but there is much variation in plumage coloration ranging from pale yellow to bright red, and females prefer the reddest male available (Hill, 1990, 1991); males feed their mates as well as their offspring during incubation. The amount of plumage coloration depends on the type and quantity of carotenoids ingested by males at their annual molt and is a reliable indicator of their foraging ability. It seems likely that red breasts in male House Finches have evolved through sexual selection as an honest and reliable signal of their foraging ability. (Signaling theory will be discussed later in this chapter.)

The evolution of female choice is more difficult to understand in other situations in which the male does not provide care for the offspring or provide any direct benefit to the female. In the Long-tailed Widowbird, an African polygynous weaverbird in which there is no paternal care for the young, the male develops an elongated tail during the breeding season. Andersson (1982) experimentally increased the tail length of males by gluing an extra piece in the middle and showed that females preferred to mate with such super-birds rather than with normal birds. There was also evidence, though not statistically significant, that males with increased tails flew less well and that males with shortened tails flew better than normal males. Indirect evidence that elongated tails are a disadvantage for flight is that they only occur in males, and only in the breeding season. How can we account for the evolution of exaggerated male sexual characters that are disadvantageous to the male and seem to provide no obvious benefit to the female?

There are two theories for the evolution of this type of male character, the good-taste model and the good-genes model. In the good-taste model, originally proposed by Fisher (1930), the male character is arbitrary; it is selected in males because it is the object of female preference so that males with the character will have more offspring, and this in turn reinforces the female preference, since females mating with males that other females prefer will tend to have attractive sons. In the good-genes model, which has been developed more recently, the male character is a marker for the possession of good genes that he will pass on to his offspring; this is similar to the good-parent model, except that females are selecting males not for their ability as fathers to help look after the young, but for their possession of good genes that they will pass on to their offspring. The main difficulty with this idea is in understanding how genetic variability for fitness is maintained. We shall consider these models in turn.

Good-taste models

Fisher (1930) proposed the theory that, once some arbitrary female preference for, say, long tails has begun, the species can become involved in a runaway process, females being selected to evolve ever stronger preference for long tails because the sons of choosy females will have long tails and will therefore have more mates, and males being selected to follow suit to attract females. However, he did not give an explicit model for his idea, and the argument is not easy to follow. The idea was not properly understood until it was modeled explicitly in a population genetic setting by Lande (1981) using a polygenic model and by Kirkpatrick (1982) using a two-locus haploid model. I shall consider the latter, which is easier to explain.

The two loci are a P locus, expressed only in females, that determines the female mating preference and a T locus that determines a trait expressed only in males. There are two alleles at each locus. T_1 males lack some secondary sexual characteristic (such as a long tail) and T_2 males have it; T_2 males have viability $1 - s$ (with $s > 0$) compared with T_1 males. P_1 females mate at random, while P_2 females favor mating with T_2 males. Write t_1 and t_2 for the frequencies of the two types of male before viability selection, and t_1^* and t_2^* for the corresponding frequencies after selection. Suppose that mating occurs after viability selection on males. P_1 females give a proportion t_2^* of their matings to T_2 males, whereas P_2 females give a proportion

$$at_2^*/(t_1^* + at_2^*) \tag{3}$$

of their matings to such males. This expresses the fact that T_2 males are more conspicuous or more attractive to P_2 females than T_1 males by a factor a ($a > 1$), which measures the degree of female preference.

The behavior of this model can be investigated by evaluating the recurrence relationships for the frequencies of the four genotypes T_1P_1, T_1P_2, T_2P_1, and T_2P_2, which we denote $x_1, \ldots, x_4$. However, it is more informative to do the simulations in terms of the two gene frequencies, $t_2 = x_3 + x_4$ and $p_2 = x_2 + x_4$, and the coefficient of linkage disequilibrium, $D = x_1x_4 - x_2x_3$, which measures the extent to which the two genes are associated; if there is no correlation between the presence of T_1 and P_1 (as we would expect in the absence of selection), then $D = 0$, while $D > 0$ reveals a positive correlation and $D < 0$ a negative correlation between them.

One may then simulate the evolution of the process starting with different values of t_2, p_2, and D. Some insight into what is happening is obtained by looking at the algebraic formulas for the recurrence relations:

$$\begin{aligned} t_2' &= t_2 + t_2(1-t_2)A \\ p_2' &= p_2 + DA \\ D' &= \text{something complicated} \end{aligned} \tag{4}$$

where A is a rational function of t_2 and p_2 whose numerator is linear in these gene frequencies and whose denominator is quadratic in t_2. For example, when $s = 0.4$ and $a = 2$,

$$A = (2t_2 - 15p_2 + 10)/2(t_2 + 5)(2t_2 - 5) \tag{5}$$

Two important consequences follow from Equation 4. First, the recurrence for p_2 shows that there is no direct selection operating on female choice but that any response at the P locus is a correlated response to selection on the T locus induced by the small positive correlation (D) that builds up between them. Secondly, the gene frequencies are at equilibrium when $A = 0$, which is equivalent to the numerator of A being zero. Since this numerator is linear in the gene frequencies, there is a line of internal equilibria defined by $A = 0$. When $s = 0.4$ and $a = 2$, for example, equating (5) to zero implies $2t_2 - 15p_2 = -10$, provided that both gene frequencies lie between 0 and 1. This is illustrated in Figure 11.1, which shows the line of internal equilibria between $\{t_2, p_2\} = \{0, 0.67\}$ and $\{t_2, p_2\} = \{1, 0.8\}$. When $p_2 < 0.67$, the equilibrium value of t_2 is zero, and when $p_2 > 0.8$, it is unity. If the system starts at some point not on the line, it will move under selection to the "nearest" point on it and will then stay there, except that it may drift up and down the line with no selective force acting on it. Thus the line as a whole is stable in the sense that selection prevents movements away from it, but individual points on the line are only neutrally stable.

The behavior of this model follows from the following facts. There are two selective forces acting on T_2 males: viability selection against them and reproductive success selecting for them, the strength of the latter depending on the frequency of females preferring them. Thus whether males with the trait are favored or disfavored depends on the frequency of females with the P_2 gene. There is no direct selection on the P_2 gene, but a positive association (linkage disequilibrium D) will build up between the P_2 gene and the T_2 gene because P_2 females tend to mate with T_2 males; this means that P_2 females tend to have more T_2 sons than P_1 females, so that there will be indirect selection for or against P_2 females according to whether there is net selection for or against T_2 males.

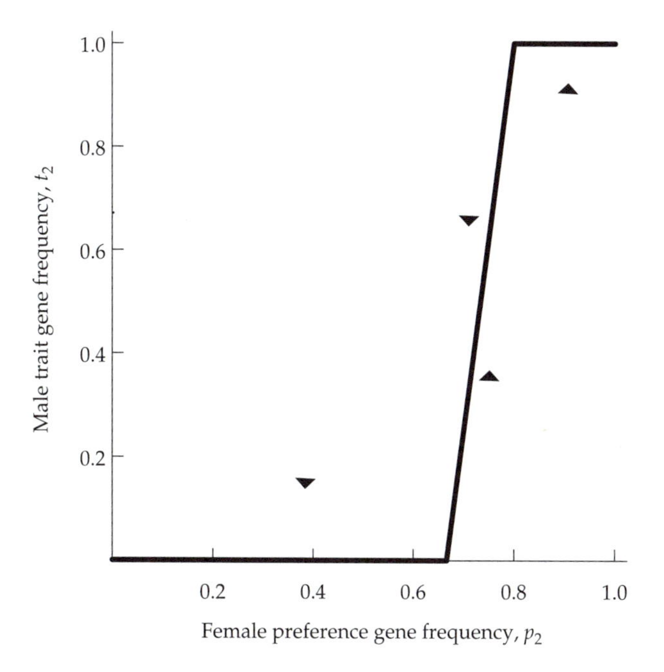

Figure 11.1. Kirkpatrick's model of sexual selection with $s = 0.4$ and $a = 2$. The velocity vectors are schematic.

Kirkpatrick's model of female choice is a *fixed-relative-preference model*. Seger (1985) considered an alternative model, the *better-of-two model*, in which a P_2 female chooses a mate from a randomly chosen pair of males; if they are both of the same trait type she has no choice, but if they are of opposite types she choses the T_2 male with probability $\frac{1}{2}(1 + c)$. Thus $c = 0$ represents random choice, and a positive value of c quantifies the extent of her preference, with $c = 1$ denoting complete preference when she has a choice. A P_2 female gives on average a proportion

$$(t_2^*)^2 + 2t_1^*t_2^* \times \tfrac{1}{2}(1+c) = t_2^*(1+ct_1^*) \tag{6a}$$

of her matings to T_2 males and a proportion

$$t_1^*(1-ct_2^*) \tag{6b}$$

to T_1 males so that her relative preference for T_2 males, corresponding to a in Kirkpatrick's model, is

$$\frac{1+ct_1^*}{1-ct_2^*} \tag{7}$$

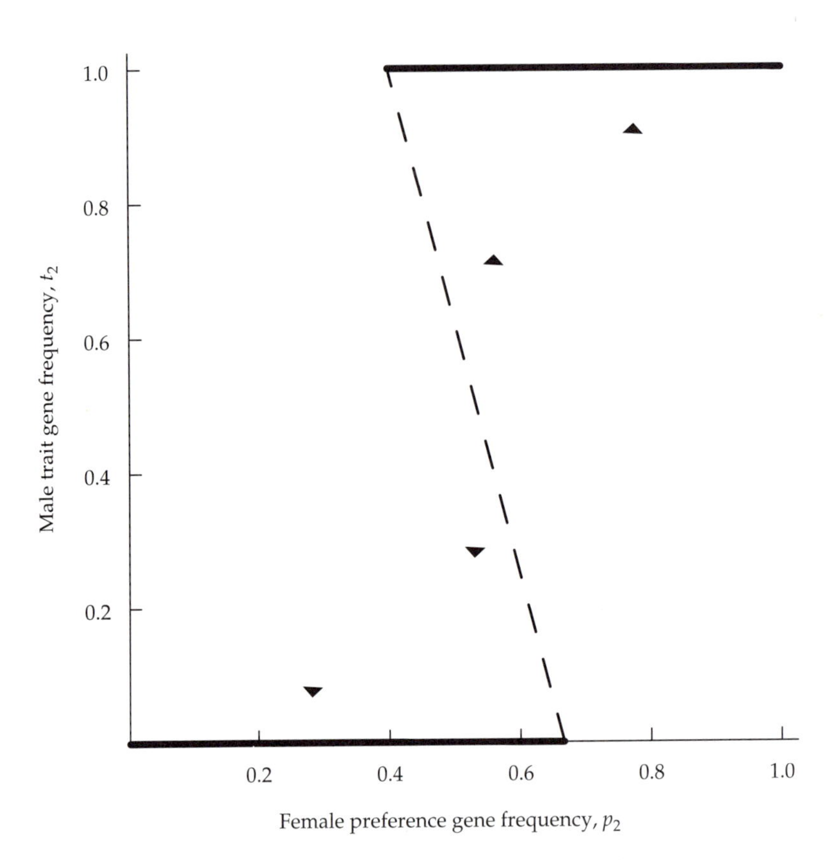

Figure 11.2. Seger's better-of-two model with $s = 0.4$ and $c = 1$.

The calculation of the recurrence relations is similar, but the internal line of equilibria turns out to have negative slope and to be unstable, behaving more like Fisher's runaway process.

These models have been used to explicate the coevolution of female choice and an arbitrary male trait, leading to sexual selection for the trait once the frequency of the female preference gene has risen above a critical threshold value. But why should this have happened? One suggestion is that there may be a preexisting female preference due to a bias in females' sensory system, which may have evolved for other reasons. A female bird may choose a male "at random," but may nevertheless "prefer" a bright to a dark male because she is more likely to notice him; likewise a female frog may "prefer" a loud to a quiet potential partner. In either case this will lead to selection for the male trait (bright birds or loud-calling frogs). In the terminology of the models above there is a preexisting high frequency of the female preference gene that leads to an increase in the frequency of the male trait gene from zero as soon as the appropriate mutation occurs, as shown in the right-hand parts of Figures 11.1 and 11.2. Ryan et al. (1990) call this *sexual selection for sensory exploitation*. A good example of a preexisting female preference for an arbitrary male character is provided by the work of Burley et al. (1982) and Burley (1986) on Zebra Finches. She found that the colored bands placed on the birds' legs for individual identification affected their attractiveness to the opposite sex. Females preferred red-banded males over unbanded ones, and they avoided blue- and green-banded males. Although the Zebra Finch is monogamous, red-banded males had greater reproductive success than unbanded males. If a mutation arose causing males to develop a red band on their legs it would therefore spread through sexual selection, provided that its viability cost did not exceed its reproductive advantage.

The results in Figures 11.1 and 11.2, and the existence of a neutrally stable line of equilibria, are crucially dependent on the assumption that female choice has no direct benefit or cost. The cost of female choice can be introduced into the model by supposing that P_2 females have fitness $1 - k$ relative to P_1 females, so that there is a cost of choice if $k > 0$ and a direct benefit of being choosy if $k < 0$. This situation has been examined by Pomiankowski (1987) and Bulmer (1989). The line of equilibria no longer exists, since the recurrence relations no longer have the form of Equation 4. There is a unique stable equilibrium with both loci fixed for one or the other allele. The outcome is dominated by direct selection on the preference locus, P_1 or P_2 being fixed according to whether $k > 0$ or $k < 0$. The trait locus is then carried along with the successful preference allele, whichever trait allele is fitter under the joint effects of viability selection and reproductive success being fixed. Under the fixed-relative-preference model, if $k > 0$ so that P_1 is fixed, then T_1 is fixed since only viability selection acts on the trait locus in the absence of female choice; if $k < 0$ so that P_2 is fixed then T_1 is fixed if $a(1 - s) < 1$, since viability selection against T_2 outweighs its reproductive success whereas T_2 is fixed if $a(1 - s) > 1$. When there is direct selection on female choice, one should imagine female preference finding its optimal level under this selection pressure, dragging the male trait with it along the line of equilibria.

There is thus a problem in understanding the evolution of exaggerated male ornaments through female choice under the good-taste model, since the process will not work if there is any direct cost of choice. It seems plausible that there will be a cost of choice when it entails the consumption of time or energy by the female, and two recent studies suggest that this is the case (Slagsvold et al., 1988; Gibson and Bachman, 1992). The good-taste model is likely to be most relevant when there is a preexisting female preference that has evolved for other reasons and which males are selected to exploit.

Good-genes and good-parent models

When there is a cost to female choice, females will not be choosy unless they get some benefit in return from the chosen father. The benefit may be a good parent for their offspring in species with paternal care or good genes for their offspring in species without any physical contribution from the father except his genes. The latter idea presupposes that some males have better genes than others. A problem with this idea is how genetic variability for fitness is maintained, since selection is always eliminating inferior genes. However, mutation is always introducing deleterious mutations, so that at equilibrium substantial variability for fitness may be maintained through selection–mutation balance. An alternative suggestion is that there may be substantial genetic variability for parasite resistance because host and parasite are engaged in a continual arms race, with hosts developing resistance to prevailing parasite genotypes and parasites evolving new ways to outwit the host.

Grafen (1990a,b) has developed a general model of sexual selection in this context from the ideas of Zahavi (1975, 1977). The model treats sexual selection as a special case of a biological signaling problem. Its full treatment is mathematically complex, but the underlying idea is simple enough that a verbal description together with an illustrative example will be adequate. The conclusions are supported by the results of more conventional population genetic models (Pomiankowski, 1988; Iwasa et al., 1991).

Males vary continuously in quality q. Females cannot observe quality directly, but males advertise their quality by some continuous ornamental trait, such as plumage coloration or tail length, which is a function of quality, $a = a(q)$. Females prefer to mate with highly advertising males because this signals high quality that will confer some benefit on them, but there is an associated cost determined by the degree of female choice. This is an evolutionary game in which a male decides his advertisement level, $a(q)$, conditional on his quality, and a female decides her preference $p(a)$ for a male with advertisement level a. What can be said about the properties of an evolutionarily stable strategy, if it exists?

At equilibrium $p(a)$ must be the best response to $a(q)$. It follows that the advertisement must be an honest signal of quality, so that a is an increasing function of q; otherwise, females would not use it to determine their mating behavior. (Remember the boy who cried "Wolf!") But at the same time $a(q)$ must be the best response to $p(a)$. What prevents males from cheating by developing a more exaggerated trait than their quality warrants and so persuading more females to mate with them? There are two possible answers. The first is that low-quality males *cannot* develop a higher trait value. Maynard Smith (1976) called this the *revealing-handicap model*.

Hamilton and Zuk (1982) showed that there is a correlation between brightness of male plumage and parasite load in birds, and they suggested that bright colors were being used as a revealing handicap indicating male freedom from parasites, sick males being unable to have bright colors. They also suggested that this resistance to parasites was heritable, so that females were choosing for good genes as revealed by bright color. In sticklebacks the male has a red belly in breeding condition, which is used in courtship. Milinski and Bakker (1990) show that females prefer brighter red males in paired-choice experiments, but that this preference disappears in green light; therefore the females are using color directly as a cue. They also showed that brighter males tend to be in better physical condition. Finally, they infected the brighter male of the pair and found that his coloration and condition dropped to that of the duller male; this abolished the preference of the female for the original male in white light while she remained indifferent between them in green light. This experiment provides strong evidence that females are using red color as an indicator of condition. The idea is that males in poor condition cannot develop as bright a color as those in good condition, and that females can therefore use color as a reliable guide to condition. Under the parasite-resistance hypothesis, the advantage to a female in choosing a male in good condition is that he is likely to have genes for parasite resistance that he will transmit to his offspring. It is possible in this case that the female may be choosing a male in good condition for direct benefit to the brood, since there is male parental care in sticklebacks. Red breasts in House Finches discussed above may be another example of a revealing handicap, revealing male foraging ability that will make him a good parent. As a final example, there is evidence that females may prefer males who are bilaterally symmetrical because symmetry reveals developmental stability, which may be a reflection of genetic quality (Swaddle and Cuthill, 1994; Watson and Thornhill, 1994).

In the revealing-handicap model the male trait has evolved through sexual selection as an honest signal of his quality because he *cannot* cheat. It is likely that the trait, such as bright coloration, will be somewhat deleterious under natural selection, which is why it is called a handicap, but this is coincidental to its function of revealing the true quality of the male. The other possible reason why males do not cheat is that they *could* do so but that it would not *pay* them to do so. This can happen if advertising is costly to the male and is more costly to low-quality males than to high-quality males. In this case a male of quality q will adjust his advertising level a until the reproductive advantage he could gain by increasing it is just balanced by the increased cost. Since high-quality males have a lower cost of advertising, they can afford a higher level of advertisement than low-quality males. This leads to an honest signaling system that the females can use as a basis for mate choice. Grafen (1990b) called this the *strategic-choice handicap model*. In this model the cost of advertising is essential to the evolutionary stability of the system since it ensures the honesty of the male signal, which in turn ensures that it will be used by females in their choice of mates. This formalizes and validates the *handicap principle* of Zahavi (1975, 1977) according to which one can conclude from the evolutionary stability of signals that they are honest (otherwise they would not be respected by their recipient) and that they will be costly, and costly in a way that relates to the true quality revealed (otherwise they would not resist cheating by the sender of the signal and so would not be honest). This principle can be applied in many areas of signaling between animals; for example, Godfray (1991) applies it to the signaling of need by offspring to their parents, and Fitzgibbon and Fanshaw (1988) interpret "stotting" by Thompson's gazelles as an honest signal of their condition directed to their predators.

Grafen (1990a) gives the following example of the strategic-choice handicap model of sexual selection. Male quality, q, is equally likely to take any value within the continuous range from 0.3 to 1.3. Males choose an advertising level a depending on their quality, and their chance of surviving until the breeding season starts is

$$\exp[q - 1.3 - a(1.5 - q)] \tag{8}$$

There is a cost of advertising that increases with decreasing quality as well as a direct cost of low quality. The breeding season lasts from time $t = 0$ to $t = 1$, and during this time a female meets males at random at a rate of $20p$ males per unit time, where p is the proportion of surviving males. If a female meets a male with advertising level a at time t, she must have a decision rule telling her whether or not to mate with him. If she mates with him she leaves the pool of unmated females, and her fecundity is

$$q \times m(t) \tag{9}$$

where $m(t)$ is the seasonal function

$$m(t) = [4(t - 0.5)^4 - 2(t - 0.5)^2 + 0.25] \tag{10}$$

The seasonal function increases from zero at $t = 0$ to a maximum at $t = 0.5$ and falls back to zero at $t = 1$. This represents the idea that conditions for breeding are best in the middle of the season. The female's fecundity is the product of the male's quality and the seasonal function; male quality is environmentally determined, so that this is a model of good-parent rather than good-genes sexual selection, though there is unlikely to be a qualitative difference between them. The cost of female choice is reflected in the

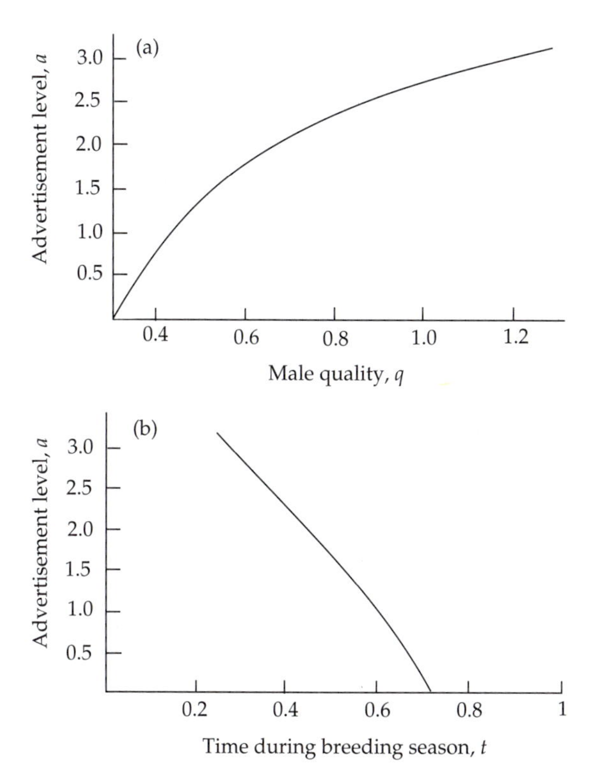

Figure 11.3. Evolutionarily stable levels of male advertisement and female choice at the advertising equilibrium for the model described in the text. (a) The optimal level of male advertisement as a function of his quality given the level of female choice. (b) The optimal threshold level of female choice during the season in response to the above level of advertising. Before $t = 0.25$, no male is accepted; between $t = 0.25$ and $t = 0.715$, a male is accepted if he is advertising at more than the threshold level shown in the figure; after $t = 0.715$, all males are accepted. From Grafen, 1990b.

fact that a choosy female who accepts only the best males must either be prepared to mate early in the season before conditions are optimal or accept the risk of not mating by the end of the season. (In human terms, a woman might decide to settle for second best rather than wait indefinitely for Mr. Right.)

There are two equilibrium solutions (ESS's) to this model; this is a general feature of Grafen's signaling models. The first is the nonadvertising solution, with no males advertising and hence with no female choice. The second is the advertising solution shown in Figure 11.3. Male level of advertisement increases with male quality (Figure 11.3a); females become less choosy as the season progresses (Figure 11.3b). This is an ESS because the level of advertisement by a male of a particular quality is optimal given the strategies of females and of other males; and the female strategy is the optimal response to the male level of advertisement. This equilibrium has two counterintuitive features. First, males of intermediate quality advertise so much that they have lower survival than males of either low or high quality. (Male fitness, the product of survivorship and mating success, is nevertheless a strictly increasing function of quality.) Secondly, the mean fitness of females is lower at the advertising equilibrium (0.184) than at the nonadvertising equilibrium (0.211), so that sexual selection is not good for the species. The reason is that the overall survival of males is reduced by advertising, which reduces the ability of females to mate at the best time of the season; this reduction in their fitness more than compensates for the increase in fitness obtained through mating with higher-quality males.

It was suggested above that the revealing-handicap model is more appropriate than the strategic-choice handicap model in several cases of sexual selection for bright coloration in males. However, the distinction may be rather artificial, since it is often difficult to decide whether a male *cannot* develop a more extreme value of the trait or whether he could but it would be *very* expensive. It seems more plausible to associate sexual selection for tail length in birds with the strategic-choice handicap model since it is likely that a male *could* always grow a slightly longer tail by sacrificing something else.

Møller (1988, 1989, 1990) has made a thorough study of sexual selection on tail length in Barn Swallows. The outermost tail feathers are about 16% longer in male than in female swallows. Møller (1988) experimentally shortened the tails of some males and elongated those of others by sticking on an extra piece with superglue. Females mated preferentially with males with elongated tails, which suggests that this character has evolved through sexual selection. Swallows are monogamous, but males with long tails have greater reproductive success for two reasons: they mate earlier, which gives them a greater chance of having two broods during the season, and they enjoy more matings in extra-pair copulations.

Do females gain anything from mating with long-tailed males? Swallows suffer from parasitism by a blood-sucking mite. By experimental manipulation of mite loads and by cross-fostering experiments, Møller (1990) showed that these mites affect the fitness of their hosts, that there is heritable variability for resistance to mites, and that long tail length in males is a marker for the possession of genes for resistance that leads to a lower mite burden in their offspring. These results are in accord with the good-genes model of sexual selection.

In another experiment Møller (1989) showed that males with artificially elongated tails had abnormally short tails in the following year after the annual molt, and that as predicted from his first experiment they had lower reproductive success (due to a lower frequency of second broods and fewer extra-pair copulations) than unmanipulated males. Thus possession of an elongated tail exacts a physiological cost that is paid in the following year by the inability to grow a tail of normal length.

These elegant experiments with Barn Swallows are in general agreement with a good-genes model of sexual selection. There is genetic variability for resistance to blood-sucking mites. Males with good genes for resistance are in better condition than males without them, which allows them to grow longer tails; this character is used by females as an honest marker of the possession of good genes that will be transmitted to their offspring. The experiments do not show whether males in poor condition because of a large mite burden *cannot* grow long tails (the revealing-handicap model) or whether it would not be in their interest to do so because the gain in reproductive success would be more than offset by the differential physiological cost (the strategic-choice handicap model).

FURTHER READING

Bradbury and Andersson (1987), Pomiankowski (1988), Clayton (1991), Kirkpatrick and Ryan (1991), Maynard Smith (1991), and Andersson (1994) review different aspects of sexual selection. Møller (1994) reviews work on Barn Swallows.

EXERCISES

1. Find the ESS solutions to the game in Table 11.1 when (a) $\rho_1 = \rho_2 = 0.8$; (b) $\rho_1 = 0.8$, $\rho_2 = 0.2$; (c) $\rho_1 = 0.2$, $\rho_2 = 0.8$; (d) $\rho_1 = \rho_2 = 0.2$. (A procedure for finding ESS solutions is given in the *Mathematica* model solutions.)

2. Find the algebraic recurrence relations for the frequencies of the four genotypes in Kirkpatrick's model of sexual selection, and hence find the recurrence relations for the gene frequencies and linkage disequilibrium in Equation 4.

3. Evaluate A, defined implicitly in Equation 4, for $s = 0.4$, $a = 2$ and

hence verify the line of neutral equilibria shown in Figure 11.1. Investigate the behavior of the recurrence relations with these parameter values and with $r = 0.5$ in the light of the expectation that this is a neutrally stable line of equilibria.

4. Repeat the calculations in Exercises 2 and 3 for Seger's better-of-two model, using $s = 0.4$ and $c = 1$ for the numerical parameter values.

5. Consider Kirkpatrick's model with the modification that P_2 females have fitness $1 - k$ relative to P_1 females. Find recurrence relations for the genotype frequencies. Show by iterating these recurrence relations with $s = 0.4$, $a = 2$, $k = 0.2$, and $r = 0.5$ from different initial values that the only stable equilibrium is with P_1 and T_1 fixed.

(Note: Exercises 2–5 are demanding. They would be made easier by using a computer language with an algebraic capability. The *Mathematica* solutions would be useful as models even to those using another language.)

CHAPTER 12

The Evolution of Sex

Most eukaryotes reproduce sexually; that is to say, each new individual contains genetic material from two parents. However, a simple argument suggests that in most organisms a mutant that reproduced parthenogenetically rather than sexually would have a twofold selective advantage over its sexual conspecifics. This paradox has stimulated evolutionists to look for the possible advantages of sexual reproduction that might offset its apparent twofold cost. I shall begin by discussing how the cost of sex arose from the evolution of anisogamy, in which the zygote results from the fusion of a small microgamete with a large macrogamete; I shall then consider possible benefits of sex to counter its twofold cost; and I shall finally discuss the evolution of gender, asking why some species (in particular, most flowering plants) are hermaphrodite (the same individual producing both micro- and macrogametes), while others (for example, most higher animals) have separate male and female sexes specializing in the production of the two types of gamete.

THE EVOLUTION OF ANISOGAMY AND THE COST OF SEX

The essential features of sex in eukaryotes are brought about by a two-step process, in which the number of chromosomes in the nucleus is first halved from the diploid to the haploid number in meiosis and subsequently restored to the diploid number by the fusion of two haploid gametes during fertilization. Some mechanism for the avoidance of selfing usually ensures that two fusing gametes originate from different individuals, so that the zygote contains genetic material from two parents.

In unicellular eukaryotes, such as the green alga *Chlamydomonas*, uniting gametes are usually of the same size (isogamy), but in multicellular organisms fertilization involves the fusion of a small motile microgamete with a large immobile macrogamete (anisogamy). The evolution of anisogamy brings with it the twofold cost of sex, since the "male" microgametes are essentially parasitic on the resources of the "female" macrogametes. In this section I shall first discuss the evolution of anisogamy, and I shall then explain in more detail how this leads to the twofold cost of sex.

The evolution of anisogamy: Disruptive selection on gamete size

There are two main theories to account for the evolution of anisogamy. The first, originally proposed by Parker et al. (1972), is that there is disruptive selection acting on gamete size. I shall here follow the approach of Maynard Smith (1978b), but I shall modify his model by assuming that anisogamy evolved from an isogamous population with two mating types; that is to say, there are two types of gamete, + and –, which are morphologically indistinguishable, but mating can only occur between + and – gametes. Hoekstra (1987) argues that this is the correct starting point for models of the evolution of anisogamy.

Suppose that an individual produces n gametes each of size m; the number and size of gametes are variable, subject to the constraint that their product, the total investment in gamete production, is a constant, $nm = M$. All + individuals produce n_1 gametes of size m_1, and – individuals produce n_2 gametes of size m_2, with $n_i = M/m_i$. Zygotes are of size $m_1 + m_2$, and the survival probability of a zygote is an increasing function of its size, $s(m_1 + m_2)$. For purposes of illustration I shall use two survival functions (see Figure 12.1):

$$\begin{aligned} s_1(x) &= 1 - \exp(-x) \\ s_2(x) &= 1 - \exp(-x^2) \end{aligned} \qquad (1)$$

The first function is concave (with diminishing returns on investment), the second function is convex (with increasing returns) for $x < 0.71$, only becoming concave after that.

The situation can be treated as an asymmetric game. If – gametes are of

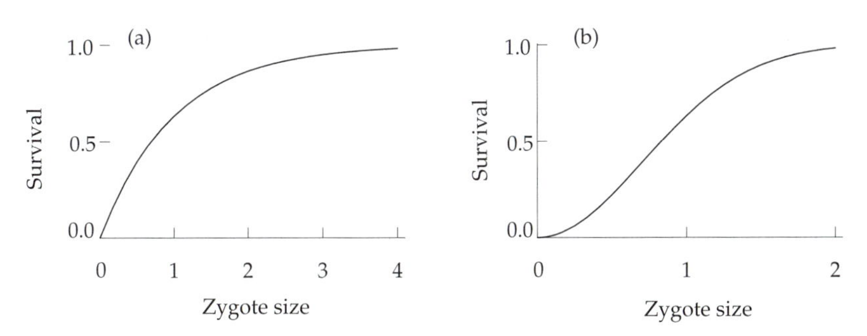

Figure 12.1. Survival as a function of zygote size: (a) $s_1(x) = 1 - \exp(-x)$; (b) $s_2(x) = 1 - \exp(-x^2)$.

size m_2, the fitness of + individuals with gametes of size m_1 is

$$w_1(m_1) = \frac{M}{m_1} s(m_1 + m_2) \tag{2}$$

The size of + gametes will be selected to maximize $w_1(m_1)$ for fixed m_2. Similarly, if + gametes are of size m_1, the fitness of – individuals with gametes of size m_2 is

$$w_2(m_2) = \frac{M}{m_2} s(m_1 + m_2) \tag{3}$$

The size of – gametes will be selected to maximize $w_2(m_2)$ for fixed m_1. The evolutionary outcome will be the resultant of these two processes.

We first note that if $m_1 << m_2$

$$w_1(m_1) \cong \frac{M}{m_1} s(m_2) \tag{4}$$

which for fixed m_2 can be made as large as we like by making m_1 small enough, provided that $s(m_2) > 0$. Thus + gametes will be selected to become infinitely small, and so will – gametes, whatever the form of the zygote survival function. In reality there is likely to be a lower limit on the size of a viable gamete. Suppose that gametes cannot be less than 0.01 in size. Incorporation of this constraint completely changes the nature of the game and ensures that there will be isogamy under the first survival function in Equation 1, with both types of gamete having their minimum size of 0.01, but that anisogamy will evolve under the second survival function, with microgametes having size 0.01 and macrogametes 1.10; it is of course arbitrary whether + gametes become large and – gametes remain small or vice versa.

For the first survival function, it is not difficult to show that $w_1(m_1)$ is a decreasing function of m_1 for any value of m_2 (Figure 12.2a). Thus + gametes will be selected to be of size 0.01 whatever the size of – gametes, and – gametes will be selected to be of size 0.01 whatever the size of + gametes, if this is the smallest size they can be. The reason is that, with diminishing returns on investment, it is always more profitable to produce twice as many gametes of half the size, since the survival rate of the zygote will be cut by less than half. The process will only be halted by the fact that there must be a lower limit to the size of a viable gamete.

For the second survival function, $w_1(m_1)$ has a maximum at $m_1 = 1.10$ when $m_2 = 0.01$, if we ignore values of $m_1 < 0.01$ (Figure 12.2b). On the other hand $w_2(m_2)$ has a maximum at $m_2 = 0.01$ when $m_1 = 1.10$, if we ignore values of $m_2 < 0.01$. Thus if you know that your mate will be a microgamete, your best strategy is to be a macrogamete, and vice versa. There is no stable solution with both types of some intermediate size, and so anisogamy will evolve. The microgametes are exploiting the macrogametes who contribute far more to the zygote, but there is nothing that the macrogametes can do about it. It seems plausible that the survival function should change from one having diminishing returns on investment in zygote size as in Figure 12.1a in unicellular organisms to one having increasing returns on investment at small zygote sizes as in Figure 12.1b in multicellular organisms that grow to a larger size.

The evolution of anisogamy: Uniparental cytoplasmic inheritance

An alternative theory proposes that anisogamy is an adaptation to ensure that cytoplasmic particles are inherited only through female gametes and are excluded from male gametes because they have very little cytoplasm (Cosmides and Tooby, 1981). This has the advantage of preventing both the spread of harmful cytoplasmic parasites and the evolution of selfish behavior in cytoplasmic organelles (mitochondria and chloroplasts). Under this view male gametes are selected for small size because this is associated with preventing cytoplasmic inheritance through them. This is one example of the possible importance of intragenomic conflict as an evolutionary force, an idea powerfully championed by Hurst (1992).

Law and Hutson (1992) consider a model of the dynamics of cytoplasmic parasites (viruses and bacteria, for example). Adults are either infected or not, and infection can be transmitted to offspring through the cyto-

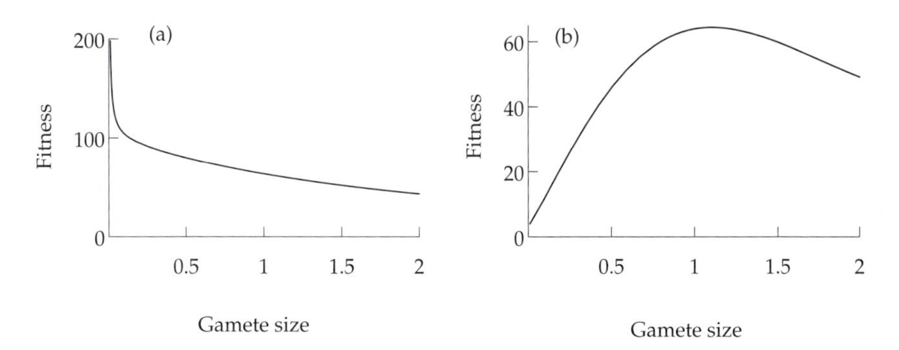

Figure 12.2. Fitness of + gametes as a function of size when – gametes have size 0.01 for (a) the first and (b) the second survival function in Equation 1 and Figure 12.1 with $M = 100$.

plasm of the adults' gametes. Write p_t for the proportion of infected adults censused just before gamete production at the end of generation t. Suppose first that infection is transmitted through the cytoplasm of both male and female gametes. The proportion of uninfected offspring at birth in the next generation under random mating will be $(1 - p_t)^2$ because an offspring is infected unless both its parents are uninfected, and the proportion of infected offspring is $1 - (1 - p_t)^2$. Suppose also that the relative probability that an infected offspring will survive to reproduce is $1 - s$ compared to an uninfected offspring. The proportion of infected adults from these offspring is

$$p_{t+1} = \frac{(1-s)[1-(1-p_t)^2]}{(1-s)[1-(1-p_t)^2]+(1-p_t)^2} = \frac{(1-s)p_t(2-p_t)}{1-sp_t(2-p_t)} \tag{5}$$

When infection is rare this is approximately

$$p_{t+1} \cong 2(1-s)p_t \tag{6}$$

so that it can spread if $s < 0.5$. Thus under biparental inheritance cytoplasmic parasites can spread in the host population even though they have a strong deleterious effect on the host. The reason is that the parasite has an inbuilt advantage; even if present in only one of the gametes that unite to produce a zygote, it is present in all of the gametes produced by that zygote, irrespective of whether they are male or female.

Suppose on the other hand that infection can be transmitted through the cytoplasm of the female but not of the male gametes. If p_t is the proportion of affected adults in generation t, the proportion of affected offspring is also p_t, and the recurrence relation is

$$p_{t+1} = \frac{(1-s)p_t}{(1-s)p_t + (1-p_t)} = \frac{(1-s)p_t}{1-sp_t} \tag{7}$$

so that the infection cannot spread unless $s < 0$. Thus uniparental cytoplasmic inheritance prevents the spread of parasites that are harmful to the host, whereas biparental inheritance allows the spread of parasites with quite strong deleterious effects. Under uniparental inheritance the parasite is transmitted vertically from parent to offspring, so that the interest of the parasite is completely identified with that of the host and no conflict of interest can arise between them. Under biparental inheritance there is not strict vertical transmission, so that a conflict of interest can arise between genes in the nucleus of the host and genes in the parasite.

One would therefore expect that selection will favor a nuclear gene that resolves the conflict of interest by suppressing cytoplasmic inheritance in males, but a population genetic model is needed to demonstrate that this will actually happen. Consider a nuclear gene with two alleles *U* and *B* (uniparental and biparental); *U* acts in male gametes to suppress transmission of any cytoplasmic parasites. There are four gametic types depending on their infection status, I (infected) or N (not infected), and their nuclear transmission gene, *U* or *B*. Denote the four types I*U*, I*B*, N*U*, and N*B* as 1, 2, 3, and 4, with frequencies $x_1, \ldots, x_4$. The mating table for the 16 types of gamete union can be written down, from which the following set of recurrence relations can be derived:

$$\begin{aligned}
x_1' &= (1-s)(x_1 - \tfrac{1}{2}x_1x_4 + x_2x_3)/(1-k) \\
x_2' &= (1-s)(x_2 + \tfrac{1}{2}x_1x_4 + x_2x_4)/(1-k) \\
x_3' &= (x_3 + \tfrac{1}{2}x_1x_4 - x_2x_3)/(1-k) \\
x_4' &= (x_4 - \tfrac{1}{2}x_1x_4 - x_2x_4)/(1-k) \\
k &= s(x_1 + x_2 + x_2x_3 + x_2x_4)
\end{aligned} \tag{8}$$

The behavior of this system is rather subtle. If an infection with $0 < s < 0.5$ is introduced into a population with a preponderance of uninfected biparental individuals, it will spread to fixation because of its advantage in a biparental population, but in so doing it will increase the relative frequency of uniparental individuals who fare better while the infection is spreading. Suppose, for example, that the population starts with no infection and with 1% of the individuals uniparental, so that the frequencies of the four types are {0, 0, 0.01, 0.99}, and that 1% of the population becomes infected with a disease with $s = 0.25$, so that the frequencies of the four types become {0.0001, 0.0099, 0.0099, 0.9801} if uniparental and biparental individuals are equally likely to be infected. After 20 generations under the recurrence relations in Equation 8 the frequencies would become {0.013, 0.987, 0, 0}; the infection has spread to fixation while the percentage of uniparental individuals has increased from 1% to 1.3%. Successive waves of infection by different pathogens will gradually increase the uniparental frequency until it becomes the dominant type and can resist invasion by harmful pathogens.

A similar argument applies to the cytoplasmic organelles, the mitochondria that are responsible for energy conversion in all eukaryotes and the chloroplasts that are the organs of photosynthesis in plants. These organelles have their own genes, and under biparental inheritance there would be a conflict of interest between their genes and genes in the nucleus; this potential conflict is avoided under uniparental inheritance. Hastings (1992) has explored a simulation model of mitochondrial evolution. He finds that under biparental inheritance a selfish mitochondrion can evolve which increases its own rate of replication at the cost of reduced metabolic activity, which is deleterious to the host cell, but that this cannot occur under uniparental vertical transmission. This factor creates a selection pressure favoring the evolution of a nuclear gene for uniparental inheritance. Hurst and Hamilton (1992) discuss a variant of this scenario in which, under biparental inheritance, there is an active conflict between the organelles inherited from the two parents as to which set will survive.

Uniparental inheritance of mitochondria and chloroplasts is almost universal in eukaryotes. In anisogamous organisms these organelles are usually maternally inherited, the main exception being the paternal inheritance of chloroplasts in gymnosperms. Uniparental inheritance of organelles is also found in isogamous eukaryotes. In *Chlamydomonas*, chloroplasts are inherited from the + parent and mitochondria from the − parent; this is brought about by selective elimination of one type of organelle in the zygote. Thus the evolution of uniparental inheritance of organelles predates the evolution of anisogamy, which makes it unlikely that small microgametes evolved as an adaptation to prevent paternal inheritance of cytoplasmic organelles. It seems more plausible that they evolved to prevent biparental transmission of cytoplasmic parasites.

In conclusion, the first theory of the evolution of anisogamy stresses disruptive selection on gamete size, while the second views the evolution of small male gametes as a by-product of selection for uniparental vertical transmission of cytoplasmic parasites and (possibly) organelles. These two theories are not exclusive, and it is likely that both factors contributed to the evolution of anisogamy. Once this has happened sexual reproduction incurs a twofold cost over parthenogenesis for reasons which will now be discussed.

The twofold cost of sex

Suppose that, in a sexual species, with separate sexes and with primary sex ratio r, a mutation A occurs causing females to produce only parthenogenetic females like themselves (normally by apomixis, in which meiosis is suppressed and the ovum is diploid). Then asexual females will produce k offspring of which a fraction S survive to reproduce asexually in the next generation. Suppose that fecundity and survival are the same in sexual females, so that they will produce k offspring of which a fraction r are male and $1 - r$ female. Thus each asexual female will produce kS asexual females in the next generation, whereas each sexual female will produce $(1 - r)kS$ sexual females in the next generation. If n and N are the numbers of asexual and sexual females in generation t, then in the next generation

$$\begin{aligned} n' &= kSn \\ N' &= (1-r)kSN \end{aligned} \tag{9}$$

so that

$$\frac{n'}{N'} = \frac{1}{1-r}\left(\frac{n}{N}\right) \tag{10}$$

Thus the frequency of asexuals relative to sexuals increases by a fraction $1/(1 - r)$ each generation. This is the cost of sex; it represents a selective force of this magnitude acting against the sexual form each generation.

At the equilibrium sex ratio of 0.5, the ratio of asexuals doubles each generation, giving a twofold cost of sex. The cost disappears as $r \to 0$ (complete female bias in the sex ratio) and becomes very high for a highly male-biased sex ratio ($r \to 1$). The sex ratio will not be at equilibrium unless $r = 0.5$ since an equal cost of producing males and females has been assumed in postulating that sexuals and asexuals produce the same total number of offspring. However, this argument shows that the cost of sex under this model is due to the cost of producing males, which contribute nothing to population growth under the assumption of female reproductive dominance.

This argument has most force in species with no paternal care of the offspring. If there is paternal care of the young then it is likely that an asexual female will have substantially lower survival rate for her offspring unless she can trick a male into believing that he is the father. Similarly, the argument will not operate under isogamy, since an asexual female will produce ova of only half the size of the zygote from a sexual female; alternatively the asexual female could produce half as many ova that are double the size of those of a sexual female and hence the same size as a zygote. The argument works under anisogamy because the contribution of the male gamete to the size of the zygote is negligible.

The above argument presupposes a species with separate sexes, but a similar argument holds for an outcrossing hermaphrodite species, with the cost of allocating resources to the production of male gametes replacing the cost of producing males. Consider a hermaphrodite plant species with obligate outcrossing; each individual produces both pollen and ovules, but selfing is prevented by some mechanism such as self-incompatibility. Suppose that sexual plants produce k seeds of which a fraction S survive to reproduce in the next generation, and that a mutant arises which produces K seeds parthenogenetically but no pollen; because of the reallocation of resources from male to female function, we may suppose that $K > k$. If n and N are the numbers of asexual and sexual females in generation t, then we find as before that

$$\frac{n'}{N'} = \frac{K}{k}\left(\frac{n}{N}\right) \tag{11}$$

so that the asexual type is at a relative advantage of K/k. It may be that the reallocation of resources from male to female function would not produce a twofold increase in seed production, as we shall see in the last section of this chapter, but it is nevertheless likely to produce a substantial advantage.

It is of interest to consider two other asexual strategies open to a hermaphrodite plant (Charlesworth, 1980). First, it might produce seeds parthenogenetically but continue to produce pollen to fertilize sexual plants in the population. Because there is no reallocation of resources, we suppose that both sexual and asexual plants produce k seeds of which a fraction S survive to reproduce. Write *aa* and *Aa* for the sexual and asexual genotypes, with the latter having frequency x ; *AA* plants cannot arise. *Aa* plants produce only *Aa* offspring, while a proportion is $\frac{1}{2}x$ of the offspring of *aa* plants will be *Aa* since a proportion $\frac{1}{2}x$ of pollen is *A*. Hence the recurrence relation for the *Aa* genotype frequency is

$$x' = x + \tfrac{1}{2}x(1 - x) \tag{12}$$

When the asexual genotype is rare it has a selective advantage of 1.5 over the sexual genotype, but this advantage declines as its frequency increases and vanishes when the sexual genotype becomes rare. The advantage of asexuality in this model does not derive from the cost of male function, since there has been no reallocation of resources in this model. It derives from a different factor, which has been called the *cost of meiosis* (Williams, 1975) or the *cost of genome dilution*, which is explained as follows by Charlesworth (1980) in the present context: "The value of 1.5 for the advantage of a dominant gene for asexuality in an outbreeding hermaphrodite can be understood as follows. An *Aa* individual with the same male fertility as the *aa* sexual population will transmit an average of half a copy of *A* via its male gametes, whereas each offspring produced via its ameiotic eggs will contain one copy of *A*. Hence 1.5 copies of *A* are transmitted, compared with 1 if reproduction were entirely sexual." This advantage declines as the frequency of sexual individuals declines because *A* is only transmitted via male gametes to sexually produced offspring.

Second, our hermaphrodite plant might permit self-fertilization. Consider a dominant gene *A* that permits selfing; *aa* plants require pollination by pollen from other plants, while *AA* and *Aa* plants are self-fertilized and also pollinate *aa* plants. All plants produce k seeds, but because of inbreeding depression a fraction s of the selfed seeds survive, while a fraction S of the outcrossed seeds survive, with $s < S$. The *coefficient of inbreeding depression* is defined as

$$\delta = 1 - \frac{s}{S} \tag{13}$$

Suppose that the *A* gene is rare and write x and y for the frequencies of the *AA* and *Aa* genotypes, respectively, which are both assumed to be small. The frequencies of the three genotypes in the next generation will, to the first order, be

$$\begin{aligned} AA:&\quad ksx + \tfrac{1}{4}ksy \\ Aa:&\quad kSx + \tfrac{1}{2}ksy + \tfrac{1}{2}kSy \\ aa:&\quad kS \end{aligned} \tag{14}$$

The linearized recurrence relationship is

$$\begin{bmatrix} x' \\ y' \end{bmatrix} = \begin{bmatrix} 1-\delta & \tfrac{1}{4}(1-\delta) \\ 1 & 1-\tfrac{1}{2}\delta \end{bmatrix} \begin{bmatrix} x \\ y \end{bmatrix} \tag{15}$$

The dominant eigenvalue of the transition matrix is $1.5 - 0.5\delta$, so that selfing will increase in frequency when it is rare if $\delta > 0.5$. In fact, it will go to fixation under this condition (Charlesworth, 1980).

In anisogamous sexual species there are thus powerful selective forces favoring the evolution of asexuality, either through parthenogenesis or through self-fertilization in hermaphrodite species. Selfing carries with it the cost of inbreeding depression that may be large enough to prevent its spread, but there is no obvious immediate cost preventing the spread of parthenogenesis.

MODELS FOR THE ADVANTAGE OF SEX

The twofold cost of sex in anisogamous species makes it difficult to understand the widespread persistence of sex. A bewildering number of theories have been suggested to resolve this paradox. They can be classified into short-term and long-term models. Short-term models seek to identify a counterbalancing twofold advantage of sex acting each generation that will negate its twofold cost and prevent parthenogenetic mutants from spreading when they arise. Long-term models accept that a parthenogenetic mutant will spread to fixation, because of its twofold advantage once it has arisen in a sexual population, but argue that an asexual lineage will go extinct much more rapidly than a sexual lineage, so that sex is maintained by a balance between the short-term advantage and the long-term disadvantage of asexual forms. Another classification is into genetic and ecological models. Genetic models view the disadvantage of asexual forms arising through the accumulation and expression of deleterious genetic mutations, while ecological models view it as arising through their inability to cope with a variable environment.

In this section I shall first describe two genetic models, Muller's ratchet and Kondrashov's model, the first being a long-term and the latter a short-term mechanism. I shall then discuss in more detail a group-selection model of long-term models. I shall finally review ecological models rather briefly.

Muller's ratchet

> "An asexual population incorporates a kind of ratchet mechanism, such that it can never get to contain, in any of its lines, a load of mutations smaller than that already existing in its at present least-loaded lines. However, the latter lines can ... become more heavily loaded by mutation."
>
> —H. J. Muller (1964)

Consider a haploid asexual population of size N. Each individual has l loci at each of which there is a probability u per generation that a deleterious mutation will occur; back mutation is sufficiently rare that it can be ignored. An individual carrying i deleterious mutations has fitness $(1 - s)^i$, so that fitness is multiplicative.

Let n_i be the number of individuals with i harmful mutations. Suppose that the smallest number of mutations in any individual is j, so that $n_i = 0$ for $i < j$. In the absence of back mutation, no individual with fewer than j mutations can arise. However, if n_j is small, then all these individuals may die out. There is now a new optimal class with $j + 1$ mutations; the ratchet has moved one notch. In a sexual population, individuals with fewer than j mutations can arise through recombination, since the individuals in the optimal class will be mutant at different loci; hence the ratchet does not operate.

A mathematical treatment has been given by Haigh (1978). Consider first a very large population in which changes in the n_i's can be treated deterministically. The "deterministic equilibrium" that reproduces itself under selection and mutation is a Poisson distribution with mean U/s where $U = lu$ is the mutation rate per genome:

$$\frac{n_i}{N} = P_i = e^{-U/s}(U/s)^i/i! \qquad i = 0, 1, 2, \cdots \tag{16}$$

(See Appendix K for an account of the Poisson distribution.)

To prove this, suppose that the probabilities before selection are Poisson with mean U/s as above. After selection the probability P_i will be multiplied by $(1 - s)^i$ and must then be divided by a constant to make the probabilities sum to unity; this gives a Poisson distribution with mean $U(1 - s)/s$. Write X for the random variable representing the number of mutations in a random individual after selection, which is a Poisson variate with this mean, and Y for the number of new mutations in this individual, which will be Poisson with mean U. Then $X + Y$ is the number of mutations in a random individual in the next generation, which is the sum of two independent Poisson variates, with means $U(1 - s)/s$ and U, and is therefore a Poisson variate with mean equal to the sum of these means, U/s. This reproduces the postulated equilibrium distribution. It is of interest to calculate the mean fitness in the population relative to that of the zero class:

$$\overline{w} = e^{-U/s}\sum_{i=0}^{\infty}(U/s)^i \frac{(1-s)^i}{i!} = e^{-U/s}e^{U(1-s)/s} = e^{-U} \tag{17}$$

It has been suggested by Bell (1988), based on the work of Mukai in

Table 12.1. Deterministic distribution of number of individuals (n_x) with x mutations in a population of size 10^7 when $U/s = 15$.

x	n_x	x	n_x
0	3	10	486,108
1	46	15	1,024,359
2	344	20	418,103
3	1721	25	49,799
4	6452	30	2211
5	19,358	35	43

Drosophila (Mukai, 1964; Mukai et al., 1972), that

$$\begin{aligned} U &\cong 0.39 \\ s &\cong 0.025 \end{aligned} \tag{18}$$

so that

$$\begin{aligned} &\text{mean number of mutants per individual } = U/s \cong 15 \\ &\text{probability of no mutants } = e^{-U/s} \cong 3\times 10^{-7} \\ &\text{mean fitness } = e^{-U} \cong 0.68 \end{aligned} \tag{19}$$

Part of the deterministic equilibrium distribution is shown in Table 12.1, assuming a Poisson distribution with mean 15 and a population size of $N = 10^7$.

If the population size N is small enough that the expected number in the zero class n_0 is small, as in Table 12.1, then X_0 (the actual number in the zero class) will fluctuate stochastically and will eventually drift to zero. When this happens, the class with one mutant becomes optimal, and the distribution will shift to reflect this: X_1 will fall from about 46 to about 3, X_2 from about 344 to about 46, and so on. This is called the establishment phase. Once this new distribution is established, X_1 will fluctuate stochastically until it goes extinct during the extinction phase, and the process repeats itself with the two mutant class being optimal, and so on.

The important question in determining how fast the ratchet moves is the length of the extinction phase. If the total population size is kept approximately constant by density-dependent factors, then each member of the optimal class must have on average e^U offspring, since the fitness of the population relative to this class is e^{-U}; however only a fraction e^{-U} of these offspring will be free of new mutations. Thus each member of the optimal class will have, on average, one offspring in that class. The actual number of such offspring will be a random variable, and we assume that its distribution is Poisson with mean 1. This is called a Poisson branching process, and its properties are thoroughly studied in the theory of stochastic processes.

Write $q(t,a)$ as the probability that the optimal class will be extinct by generation t if it starts with a members at time zero. If $a = 1$, a standard result is that this extinction probability can be calculated from the recurrence relation:

$$q(t,1) = \exp[q(t-1,1)-1]$$
$$q(0,1) = 0 \tag{20}$$

(see Appendix K). This result can be used to calculate the extinction probabilities for any initial number from the relation

$$q(t,a) = q^a(t,1) \tag{21}$$

since all the lines must have become extinct independently. Some numerical results are shown in Table 12.2.

Two points stand out. The first is the great variability in the time to extinction. There is a 50% chance that a line started by one individual will be extinct within two generations, but to have a 95% chance of extinction we must wait 40 generations; the distribution has the peculiar property that, although ultimate extinction is certain, the expected time to extinction is infinite! The second point is that, to a good approximation,

$$q(at,a) \cong q(t,1) \tag{22}$$

Bell (1988) suggests that the approximate time for one turn of the ratchet,

Table 12.2. The probability $q(t, a)$ of extinction of the optimal class by t generations starting with a individuals initially.

t	$q(t, 1)$	$q(5t, 5)$	$q(10t, 10)$	$q(20t, 20)$
1	0.368	0.210	0.179	0.160
2	0.531	0.422	0.400	0.386
5	0.732	0.689	0.681	0.676
10	0.842	0.825	0.822	0.821
20	0.912	0.907	0.906	0.905
40	0.954	0.952	0.952	0.951
100	0.981	0.980	0.980	0.980

starting with a individuals, can be defined as

$$T_a \cong 10a \tag{23}$$

which is the time by which the probability of extinction is about 80%. The approximate time to complete k turns of the ratchet, starting each time with about a individuals in the optimal class and ignoring the time spent in the establishment phase, can be taken as

$$T_{a,k} \cong 10ak \tag{24}$$

(A more detailed investigation confirms that this is a good approximation.)

To assess the importance of the ratchet, assume that at the beginning of the extinction phase the number in the optimal class is given by the deterministic equilibrium, so that the time for one turn of the ratchet is approximately

$$10Ne^{-U/s} \tag{25}$$

and the time for k turns is k times this quantity. The ratchet will be most effective when N is small, U large, and s small. Weak selection (s small) means that only weakly deleterious mutations accumulate, since strongly deleterious ones are eliminated quickly by selection. Large total mutation rate (U) means that organisms or organelles (such as mitochondria) with very small genomes may be immune from the ratchet. Small population size (N) means that unicellular organisms with very large population sizes may be immune.

We have already seen that for *Drosophila* $\exp(-U/s) \cong 3 \times 10^{-7}$ so that if the population size were 10^{10} the number of individuals in the least loaded class would be about 3000 and the ratchet would (very approximately) turn once every 30,000 generations. One hundred turns of the ratchet would reduce viability to $(1-s)^{100} = 0.975^{100} = 0.08$ and would happen in 3 million generations, which is a short span in geological time. Thus the ratchet would seem to be effective unless the population size is substantially larger than this, or the genome size much smaller.

Muller's ratchet shows that it would be very difficult for an asexual population to survive indefinitely unless it has a very large population size or a very small genome, because of the accumulation of deleterious mutations. The ratchet should also lead to genetic deterioration in any nonrecombining part of a sexual genome and is thought to be responsible for the genetic inertness of the male-determining Y chromosome, which does not recombine because it does not have a partner with which to pair (Charlesworth, 1978; Rice, 1994). There is however a problem in using the ratchet as an explanation for the maintenance of sex. Suppose that an asexual mutation occurs in a sexual population of *Drosophila* of size 10^{10}. At first it will be at a twofold advantage, and it will take about 800,000 generations, corresponding to 27 turns of the ratchet, before the average fitness of the asexual form is reduced by half, so canceling out this twofold advantage. During this time the asexual form will maintain an advantage, varying from twofold at the beginning to a negligible advantage toward the end, so that it seems virtually certain that the sexual form will have gone extinct before Muller's ratchet has completely overcome the twofold advantage of the asexual form. It is for this reason that the ratchet is classified as a long-term-advantage model for the maintenance of sex.

Kondrashov's synergistic-fitness theory

> "By my count, 20 hypotheses trying to explain the evolution of sex have been proposed to date. ... It is time to realize that the controversy about the factors involved in the evolution of sex and several other phenomena will not be resolved by theoretical arguments. It is clear what data are necessary, and experimentalists should not pass a unique opportrunity to make 19 hypotheses obsolete at a single stroke."
>
> —Alexey Kondrashov (1994)

Muller's ratchet assumes that the deleterious genes act independently on fitness, so that the overall fitness is the product of the fitnesses at the individual loci. It depends for its effect on random genetic drift in populations of finite size. Kondrashov (1982) suggested a mechanism that would give an advantage to sex in an infinite population if the deleterious genes act synergistically, so that a mutant would be more harmful if there are many other harmful mutants in the genome than if there are only a few. Figure 12.3 shows three possible relationships between the number of mutants and fitness. The thin line is the fitness curve

$$w(i) = 0.9^{i} \tag{26}$$

representing multiplicative gene action, where $w(i)$ is the fitness of an individual with i mutant genes; the thick curve is

$$w(i) = 0.995^{i^2} \tag{27}$$

representing moderate synergism; and the thick dashed line shows truncation selection, individuals with fewer than 20 mutant loci having maximum fitness and those with more being lethal, which is an extreme form of synergism.

Suppose as before that there are l loci with mutation rate u per locus

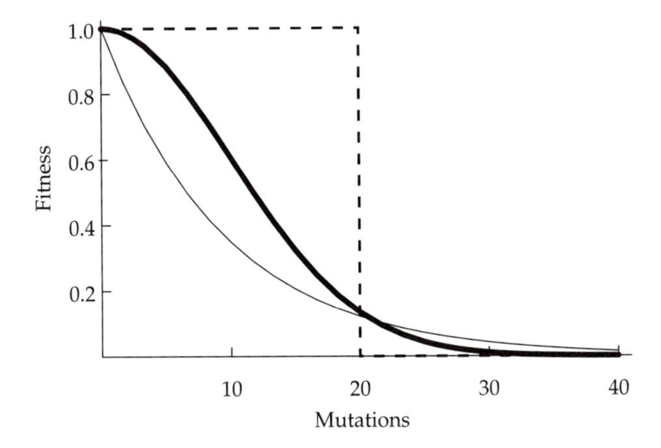

Figure 12.3. Multiplicative and synergistic fitness functions.

and write $U = lu$ for the mutation rate per genome. Consider truncation selection with $w = 1$ for $i < k$ and zero otherwise. In the asexual population all individuals will have $k - 1$ mutations after selection, provided the process has been running for some time. If the number of new mutations follows a Poisson distribution with mean U, the probability that no mutation occurs is e^{-U}, and this is the average viability (fitness) of their offspring. Therefore, in the asexual population

$$\overline{w}_{\text{asexual}} = e^{-U} \tag{28}$$

This is the same as the average fitness under multiplicative selection (Equation 17) and is in fact true for an infinite asexual population under any form of gene action (Kimura and Maruyama, 1966; Crow, 1970). It is also true for a sexual population under multiplicative selection, but under synergistic gene action a sexual population will have a higher mean fitness.

Consider a sexual population under truncation selection and assume that alleles at all loci are independently distributed after recombination (though this will not be exactly true). If the average number of mutations after recombination but before selection is λ, then the frequency of individuals with i mutations follows a Poisson distribution with this mean. The average number of mutations will now be reduced by selection but will then be restored in the next generation by the occurrence of fresh mutations. The average number of mutations after selection (which eliminates all individuals with k or more mutations) is the mean, μ, of the truncated Poisson distribution

$$\mu = \frac{\sum_{i=0}^{k-1} i\lambda^i / i!}{\sum_{i=0}^{k-1} \lambda^i / i!} \tag{29}$$

The average number of mutations in the next generation after fresh mutation is $\mu + U$, which at equilibrium must equal λ:

$$\lambda = \mu + U \tag{30}$$

This can be solved numerically as an equation in λ. The probability of survival of the offspring is the fraction with fewer than k mutations, so that the fitness in the sexual population is

$$\overline{w}_{\text{sexual}} = e^{-\lambda} \sum_{i=0}^{k-1} \lambda^i / i! \tag{31}$$

The ratio $\overline{w}_{\text{sexual}} / \overline{w}_{\text{asexual}}$ gives the relative advantage of sex, which is shown in Table 12.3 for different values of U and k.

The mechanism of the effect is that selection with synergistic action decreases the variability of the number of mutants. Recombination restores this variability, so that selection acts more efficiently in the sexual population in eliminating mutants. Truncation selection is extreme, but the mechanism works with other forms of synergistic action. Charlesworth (1990b) has shown under a more realistic model of synergism that a twofold relative advantage of sex can be generated if $U > 1.4$. Thus an asexual mutant introduced into a sexual population will eventually equilibrate at a mean fitness less than half that of its sexual parent, thus counterbalancing its twofold advantage, provided that U exceeds about 1.4 and that there is a moderate degree of synergism in fitness. The asexual

Table 12.3. Relative advantage of sex under truncation selection.

	k		
U	5	20	80
0.5	1.40	1.55	1.61
2.0	3.24	5.30	6.47
8.0	29.82	366.80	1320.00

mutant will initially increase in frequency because of its twofold advantage, but Charlesworth (1990b) estimates from numerical studies that the average fitness of the asexual form will decline to its equilibrium value over a period of time of the order of a hundred generations. It seems unlikely that the sexual form will have gone extinct over such a short time, so that this provides a potential short-term explanation for the maintenance of sex. However, it remains to be shown empirically that the total mutation rate per generation is substantially larger than the estimate of 0.39 given above for *Drosophila* and that there is adequate synergistic gene action on fitness for this model to work.

Another factor that should be taken into account is that the germ-line mutation rate is higher in males than in females because of the larger number of cell divisions in spermatogenesis than in oogenesis. In consequence, male gametes carry on average more deleterious mutations than female gametes, so that the cost of male mutations can exceed the benefit of sexual recombination, causing females to produce fitter progeny by parthenogenesis than by mating (Redfield, 1994). It is still possible to obtain a twofold advantage of sex despite the additional cost to a female of contaminating her genome with an inferior male genome, but the conditions required are more severe (Kondrashov, 1994).

The maintenance of sex by group selection

Muller's ratchet and Kondrashov's model both favor sex by lessening the impact of deleterious mutations on the population, but they differ in their mechanism. Muller's ratchet depends on random genetic drift and is only effective in finite populations (though this may include very large populations), while Kondrashov's model is effective in an infinite population if deleterious genes act synergistically. Another important difference is that Kondrashov's model provides a short-term advantage to the sexual form which, if it is strong enough, might counterbalance the twofold cost of sex each generation, whereas Muller's ratchet only offers a long-term advantage to sex. Under the ratchet an asexual mutant would initially have a twofold advantage but would ultimately, in the long term, be doomed to extinction through the accumulation of deleterious mutations, whereas the sexual form can purge itself of deleterious mutations and could survive indefinitely. But once an asexual mutant arises it will for many generations have a twofold advantage over its sexual ancestor and will almost certainly drive it to extinction before itself going extinct through accumulating deleterious mutations. Sex can only be maintained under these circumstances by a process of group selection that has been modeled by Nunney (1989) as follows.

Suppose that there are S sexual and A asexual species in some taxon, with $S + A = N$. The model has three components: (i) An asexual form arises at rate p in a sexual species (this is the mutation rate multiplied by the population size); when an asexual form arises it replaces the sexual form almost instantaneously because of its twofold advantage. (ii) Asexual species go extinct faster than sexual species, either because of Muller's ratchet or because sexual species can evolve more rapidly as discussed in the next subsection; write e_s and e_a for the extinction rate of sexual and asexual species with $e_a > e_s$. (iii) To balance extinction and ensure that the total number of species N remains constant, assume that when an extinction occurs one of the remaining lineages is chosen at random as the source of a new species. These three assumptions lead to the differential equation

$$\frac{dS}{dt} = \frac{S}{N}[(e_a - e_s)(N - S) - pN] \tag{32}$$

which has a unique stable equilibrium defined by

$$\hat{S} = N\left(1 - \frac{p}{e_a - e_s}\right) \tag{33}$$

provided that

$$p < e_a - e_s \tag{34}$$

For Muller's ratchet, in which it is assumed that sexual species are immortal, this gives the equilibrium

$$\hat{S} = N(1 - p/e_a) \tag{35}$$

provided that $p < e_a$.

Thus it is possible for sex to be maintained by a mechanism, such as Muller's ratchet, involving group selection. The taxonomic distribution of parthenogenesis supports this view, since it suggests that asexual forms become extinct quite rapidly; they usually occur as isolated species within an otherwise sexual genus, or sometimes as an asexual genus within a sexual family, but very rarely throughout a larger phylogenetic group. (The best known exception to this generalization are the Bdelloid rotifers all of which are believed to be asexual.) But this interpretation requires that the likelihood of viable asexual mutants arising is extraordinarily low. In many taxa the frequency of parthenogenetic forms is about 1%. If we ignore the extinction rate of sexual species, Equation 35 requires that $p \cong 0.01e_a$; that is to say, only one such mutation occurs in a sexual species in the time it takes 100 asexual species to go extinct. This seems rather implausible unless there is some mechanism making the origin of asexual mutants extremely rare.

Nunney (1989) suggests the following mechanism that might bring this about. If the value of p varies between lineages, then one of the effects of group selection is to favor groups (i.e., species) with the lowest values of p, because lines that convert to asexuality (because of a high p) go extinct quickly, and in the long term only those that do not convert to asexuality (because of a low p) survive. To formalize this argument let the number of sexual lines with the ith value of p be S_i and call that value p_i. The analogue of Equation 33 incorporating this extension to the model is

$$\frac{\mathrm{d}S_i}{\mathrm{d}t} = \frac{S_i}{N}[(e_{\mathrm{a}} - e_{\mathrm{s}})(N - S) - p_i N] \tag{36}$$

where $S = \Sigma S_i$. If p_1 is the smallest of the p_i's, the unique stable equilibrium is

$$\begin{aligned} \hat{S}_1 &= N\left(1 - \frac{p_1}{e_{\mathrm{a}} - e_{\mathrm{s}}}\right) \\ \hat{S}_i &= 0 \qquad\qquad i > 1 \end{aligned} \tag{37}$$

with all of the sexual lines except those with the smallest mutation rate to asexuality going extinct.

Ecological models

Muller's ratchet and Kondrashov's model both favor sex by lessening the impact of deleterious mutations on the population. A number of models have been suggested that might give sex an advantage in a varying environment. I shall discuss these models very briefly; the brevity of this review does not imply that I consider these models unimportant, but that they are less amenable to a simple theoretical treatment.

Fisher (1930) and Muller (1932) suggested that sex and recombination speed up the progress of evolution because they allow two or more beneficial mutations that arise in different individuals to be united in the same genotype. Suppose that the population is fixed at two loci for alleles A_1 and B_1, and that a change in the physical or biotic environment makes the alternative alleles A_2 and B_2 beneficial. An asexual population will, on average, take longer to become fixed for both of these new alleles than a sexual population, since both alleles must arise in the same lineage, whereas in a sexual population the B_2 allele may arise in an A_1 individual and subsequently be united with the A_2 allele in another individual by recombination. Theoretical investigations show that the Fisher-Muller process will work, but the advantage for sex depends on the second beneficial mutation arising while the first is on its way to fixation (Felsenstein, 1974; Pamilo et al., 1987). It seems unlikely that evolution is fast enough to make this plausible.

A variant on the Fisher-Muller hypothesis suggests that advantageous mutations are selected more quickly in a sexual than an asexual population in the presence of a background of deleterious mutations (Fisher, 1930; Manning and Thompson, 1984; Peck, 1994). Suppose that a locus is fixed for the allele A_1 and that A_2 becomes advantageous. In an asexual population a new mutant A_2 must arise in an individual with a very small number of deleterious mutations to have an appreciable chance of becoming fixed, while this is not the case for sexual populations, since the new mutant is shuffled around between different genetic backgrounds through recombination. Suppose for example that A_2 has a selective advantage 0.01 over A_1, that deleterious mutations occur at rate $U = 0.25$ per genome per generation, and that each deleterious mutation has selective disadvantage $s = 0.02$. Peck (1994) shows that a new mutant A_2 has a probability of only 7×10^{-8} of becoming established in an asexual population, whereas it has a probability of 0.02 of establishment in a sexual population of the same size. Thus an asexual population must wait, on average, about 3×10^6 times as long as a sexual population before A_2 becomes established. This variant seems more plausible than the original Fisher-Muller hypothesis because deleterious mutations are always present, and it does not require that more than one beneficial mutation is on its way to establishment at any one time. Peck (1994) also shows that a little bit of sex (say, reproducing sexually 10% of the time) is almost as good as obligate sexuality in favoring the fixation of favorable mutations.

The Fisher-Muller model and its variants rely on group selection to maintain sex since they depend on a long-term advantage of sexual over asexual forms. Ecological models that attempt to find a short-term advantage of sex to balance its twofold cost are the host–parasite coevolution model (often called the Red Queen model) and various forms of the sib-competition model.

The sib-competition model supposes that there is less competition between sexual sibs than between asexual sibs, because the former are genetically different while the latter are genetically identical. Thus a mother who produces five daughters and five sons sexually may have more surviving daughters than one who produces 10 genetically identical daughters asexually. The model has two variants, the lottery model (Williams, 1975) and the elbow-room model (Young, 1981), which is similar to the tangled-bank model of Bell (1982).

In the lottery model, several mothers lay eggs in a patch, and the offspring compete with each other for survival. There is only one survivor, the individual with the best genotype, but because of environmental variability in space and time the ranking of genotypes varies from patch to patch and from year to year in an unpredictable way. In these circumstances it is better to produce genetically variable offspring sexually (since each of them has an independent chance of winning) than to produce genetically identical offspring. As Williams (1975) remarks, an asexual parent is like a man who buys 100 tickets in a lottery and finds that they all have the same number, whereas a sexual parent may be able to buy fewer tickets, but at least they all have different numbers. (Bulmer [1980] and Barton and Post [1986] develop the theory underlying this model.)

The elbow-room model also depends on competition being less severe between genetically variable individuals than between genetically identical individuals, but the relaxation of competition arises less directly either through differential resource utilization or through negative frequency-dependent selection leading to a minority advantage of rare genotypes. (This situation has been modeled by Price and Waser [1982]). Either of these types of sib competition is capable in theory of generating an advantage capable of offsetting the twofold cost of sex. Kelley et al. (1988) have shown experimentally that there is less competition between sexually generated sibs than between asexually generated sibs in sweet vernal grass, leading to a substantial selective advantage for sexual reproduction, but the mechanism underlying this effect is unknown.

The host–parasite coevolution model views sexual reproduction as an adaptation on the part of the host to resist parasites (Hamilton et al., 1990). Consider a parasite with a gene for virulence with two alleles, v_1 and v_2, and a host with a corresponding gene for resistance to the parasite with two alleles, r_1 and r_2, such that r_1 hosts are resistant to v_1 but not to v_2 parasites, and likewise r_2 hosts are resistant to v_2 but not to v_1 parasites; thus a host will want to match the parasite, but the parasite will want to avoid matching the host. These competing selection pressures will set up cyclical oscillations in the two gene frequencies, rather like a prey–predator oscillation (see Chapter 7). Hamilton et al. (1990) simulated this process with a host subject to infection by several parasite species, each with its own set of virulence genes to which the host had corresponding resistance genes. The host also had a gene for sexuality versus parthenogenesis, and in many simulations it was found that the sexual form increased from a low initial frequency to fixation despite the twofold cost of sex. The explanation offered for its counterbalancing advantage is that sex stores genes that are currently bad but that have promise for reuse when the cycle swings in their favor, whereas asexual reproduction eliminates them so that the population must wait for a mutation to occur if they are needed again. There are several interacting factors involved in the maintenance of sex by this mechanism (Seger and Hamilton, 1988; Nee, 1989; Hamilton, 1993). Simulation studies show that it can work, but it is not easy to articulate exactly how it works. Howard and Lively (1994) have proposed a model in which sex is maintained through the interacting effects of parasites and Muller's ratchet.

Empirical support for the maintenance of sex by parasitism has been found by Lively (1987) in a New Zealand freshwater snail that has both sexual and parthenogenetic populations. The frequency of sexual reproduction, measured by the frequency of males, was positively correlated with the degree of parasitism by trematodes in the population, suggesting that sexuality is a response to parasite loads, though skeptics argue that sex might cause parasitism through transfer of parasites between sexual partners or that sex and parasitism might both be consequences of an unidentified third factor. Lively et al. (1990) have found increased parasitism by trematode larvae in clonal strains of the topminnow *Poeciliopsis* than in their sexual counterparts, suggesting that genetic similarity in a clone may expose it to higher disease levels. Parasitism may be an important factor in the maintenance of sex, acting either through the host–parasite coevolution model or as the agent of frequency-dependent selection in the elbow-room variant of the sib-competition model.

THE EVOLUTION OF GENDER

> "There is much difficulty in understanding why hermaphrodite plants should ever have been rendered dioecious."
>
> —Charles Darwin, *The Different Forms of Flowers on Plants of the Same Species* (1877)

At the beginning of this chapter we discussed the evolution of anisogamy, leading to the development of two specialist types of gamete that unite to form the zygote—small, motile microgametes and large, sessile macrogametes; they are the sperm and ova of animals, the pollen and ovules of plants. Three types of individual are possible: males who produce only microgametes, females who produce only macrogametes, and hermaphrodites who produce both types of gamete. Most higher animals have separate sexes, individuals being either male or female; this is called *dioecy* (from the Greek for "two houses," because the two types of gamete live in separate houses). Most flowering plants are *hermaphrodite*, the pollen and ovules occurring either in the same flower or on different flowers of the same plant. What evolutionary pressures account for this difference?

A resource-allocation model

We consider first a model based on the efficiency of resource allocation to male and female function (Charnov et al., 1976). Suppose that a female produces F macrogametes and a male M microgametes, while a self-incompatible hermaphrodite that allocates a proportion x of its resources to female and $1 - x$ to male function can produce $\alpha(x)F$ macrogametes and $\beta(x)M$ microgametes; as x goes from 0 to 1, $\alpha(x)$ will increase from 0 to 1, and $\beta(x)$ will decrease from 1 to 0.

Consider a rare hermaphrodite in a dioecious population with an unbiased sex ratio. Define fitness as number of daughters plus number of females inseminated by sons. Assuming the same viability of all offspring, the fitness of a female is F, as is that of a male (since the sex ratio is unbiased). The fitness of the rare hermaphrodite is

$$\alpha(x)F + \beta(x)M\frac{F}{M} = [\alpha(x) + \beta(x)]F \tag{38}$$

The hermaphrodite will be unable to invade, so that dioecy will persist, if

$$\alpha(x) + \beta(x) \leq 1 \qquad \text{for all } x \tag{39}$$

Verbally, this means that dioecy will be stable if having separate sexes is a more efficient way of producing micro- and macrogametes than hermaphroditism, a reassuringly obvious conclusion.

Consider next a hermaphrodite population. We must first determine the value of x that will evolve. Suppose that most of the population has a value x while a rare mutant has value y. The fitnesses of the wild-type and mutant strategies are

$$\begin{aligned} w_{\text{wild type}} &= \alpha(x)F + \beta(x)M\frac{\alpha(x)F}{\beta(x)M} \\ &= 2\alpha(x)F \\ w_{\text{mutant}} &= \alpha(y)F + \beta(y)M\frac{\alpha(x)F}{\beta(x)M} \\ &= \left[\alpha(y) + \beta(y)\frac{\alpha(x)}{\beta(x)}\right]F \end{aligned} \tag{40}$$

Apart from a factor of 2, the relative fitness of the mutant is

$$w(y,x) = \frac{\alpha(y)}{\alpha(x)} + \frac{\beta(y)}{\beta(x)} \tag{41}$$

so that, from Equation 10.12, the optimal value x^* satisfies

$$\frac{\alpha'(x^*)}{\alpha(x^*)} + \frac{\beta'(x^*)}{\beta(x^*)} = 0 \tag{42}$$

This is the criterion for finding the value of x that maximizes the product $\alpha(x)\beta(x)$. This is the product theorem of MacArthur (1965), which holds for many sex-ratio problems.

Suppose now that the hermaphrodite population has evolved to this optimal value, and consider whether a pure female or male phenotype can invade. The fitness of a mutant female phenotype is F, and the fitness of the hermaphrodite wild type is $2\alpha(x^*)F$, so that the condition for hermaphroditism to persist is that

$$\alpha(x^*) \geq \tfrac{1}{2} \tag{43}$$

The female phenotype can only invade if it produces more than twice as many macrogametes as the hermaphrodite. The fitness of a mutant male phenotype is

$$M\frac{\alpha(x^*)F}{\beta(x^*)M} = \frac{\alpha(x^*)}{\beta(x^*)}F \tag{44}$$

so that the condition for hermaphroditism to persist is

$$\beta(x^*) \geq \tfrac{1}{2} \tag{45}$$

The male phenotype can only invade if it produces more than twice as many microgametes as the hermaphrodite. Thus hermaphroditism is stable if the hermaphrodite is more efficient than having separate sexes in the sense that it does each job at least half as well as the corresponding specialist.

The general conclusion is that dioecy is stable if it is more efficient than hermaphroditism in the sense of Equation 39, while hermaphroditism is stable if it is more efficient than dioecy in the sense of Equations 43 and 45. In general one might expect a specialist to be more efficient than a generalist, which may explain the prevalence of dioecy in animals. There are two special features of many plants that might favor hermaphroditism (Maynard Smith, 1978b). First, the flowers that attract insects to insect-pollinated plants serve both male and female function, since the insects bring pollen to the plant to fertilize its ovules and carry pollen from it to fertilize other plants. There is therefore an economy in having one structure to serve both functions. Secondly, since plants are sessile, there may be a diminishing return on investment in pollen production. If a plant can only pollinate its near neighbors, doubling pollen production will not double the number of successful pollinations, since some of the pollen will be competing with itself. In other words, $\beta(1 - x)$, which measures the return on investment in male function, will not increase linearly with $1 - x$ but will be concave. (This is the same argument of diminishing returns on investment in males that leads to a female-biased sex ratio under local mate competition.)

The evolution of gynodioecy

In a few plant species some individuals are female and the rest hermaphrodite, a condition called gynodioecy. It seems likely that hermaphroditism is the primitive condition in plants, from which both dioecy and gynodioecy evolved; a plausible view is that dioecy evolved via gynodioecy, some hermaphrodite plants first becoming male-sterile (female), the rest of the population later becoming male.

I begin by discussing a simple scenario based on the model of the preceding subsection. Consider a hermaphrodite plant population and suppose that $\alpha(x^*) < 0.5$, so that females will increase in frequency. In consequence the allocation to female function by hermaphrodites, x, will decrease. It may be that the frequency of females will increase and that the allocation to female function by hermaphrodites will decrease until $x = 0$, so that the hermaphrodites have become males and the population has become dioecious, or it may happen that the coevolution of these two quantities will stop short at a position where $x = x^{**}$ $(0 < x^{**} < x^*)$. In the latter case the population will be gynodioecious, having a proportion p of females and a proportion $q = 1 - p$ of hermaphrodites. At this equilibrium the fitness of females and hermaphrodites must be the same. Write $\alpha(x^{**}) = \alpha$ and $\beta(x^{**}) = \beta$ for brevity. The fitness of females is F. The fitness of hermaphrodites is

$$\alpha F + \beta M \frac{pF + q\alpha F}{q\beta M} = \left(2\alpha + \frac{p}{q}\right)F \tag{46}$$

Equating to F yields

$$p = \frac{1 - 2\alpha}{2(1 - \alpha)} \tag{47}$$

In this equation α is the seed production of hermaphrodites relative to that of females. The result is often expressed in terms of $f = 1/\alpha$, the seed production of females relative to hermaphrodites, giving

$$p = \frac{f-2}{2(f-1)} \tag{48}$$

This result was first obtained by a simple population genetic argument by Lewis (1941). Consider a dominant mutation that renders hermaphrodites male-sterile but at the same time increases their seed production by a factor f. Thus *aa* individuals are hermaphrodite, producing say M pollen grains and F ovules, while *Aa* individuals produce fF ovules but no pollen. Write p for the relative frequency of females in generation t. All the pollen is of type a, while there are $\frac{1}{2}fFp$ A ovules out of a total number of $F(fp + 1 - p)$ ovules. Hence the recurrence relation for p is

$$p' = \frac{\frac{1}{2}fp}{1+(f-1)p} \tag{49}$$

When females are rare, the condition for their initial increase is that $f > 2$. If this is true, their frequency will increase until an equilibrium is reached with $p' = p$, at which point p satisfies Equation 48.

There are however three problems with this scenario: (i) it is difficult to find a pair of functions, $\alpha(x)$ and $\beta(x)$, which give gynodioecy rather than dioecy or hermaphroditism as an ESS; (ii) in some gynodioecious species the frequency of females is much higher than that predicted in Equation 48 from their relative advantage in female function (Gouyon and Couvet, 1987); and (iii) there is no explanation for the fact that androdioecy (with a population consisting of males and hermaphrodites) is much rarer in plants than gynodioecy. These problems may arise from ignoring two important biological facts: (i) hermaphrodites may be self-compatible, and selfed seed may suffer from inbreeding depression; and (ii) male sterility may be caused by cytoplasmic rather than nuclear genes. I shall now consider in turn the consequences of these facts.

Gynodioecy and dioecy in plants have often been seen as mechanisms to promote outcrossing in partially self-compatible hermaphrodites and so to avoid the cost of inbreeding depression. The theory has been developed by Lloyd (1974, 1975, 1976) and by Charlesworth and Charlesworth (1978). Consider the population genetic model of Lewis (1941), but suppose that a fraction ϕ of the hermaphrodite ovules are selfed and that they have viability $1 - \delta$ compared with outcrossed ovules, where δ expresses the degree of inbreeding depression. If p is the frequency of females in generation t, there are as before $\frac{1}{2}fFp$ A ovules (giving *Aa* plants in the next generation), but the a ovules that are selfed must be devalued by $1 - \delta$, so that the total number of ovules is

$$F\{fp + (1-p)[1-\phi+\phi(1-\delta)]\} \tag{50}$$

Hence,

$$p' = \frac{\frac{1}{2}fp}{fp + (1-p)(1-\phi\delta)} \tag{51}$$

The condition for initial spread of the female genotype is

$$f > 2(1-\phi\delta) \tag{52}$$

This is less stringent than the condition $f > 2$ in the absence of selfing, particularly when the selfing rate and inbreeding depression are substantial. At equilibrium

$$\begin{aligned} p &= \frac{f-2(1-\phi\delta)}{2[f-(1-\phi\delta)]} \\ &> \frac{f-2}{2(f-1)} \end{aligned} \tag{53}$$

The evolution of androdioecy (males and hermaphrodites) can be investigated in a similar way by considering a dominant gene for female sterility, having male function m times that of hermaphrodites. The condition for the initial spread of this gene is

$$m > 2(1-\phi\delta)/(1-\phi) \tag{54}$$

which is more stringent than in the absence of selfing. This provides an explanation for the rarity of androdioecy in plants. However, once gynodioecy due to a nuclear male sterility gene has begun to spread, it becomes easier for a gene causing female sterility in the hermaphrodite to invade. Imagine that a gene for male sterility begins to spread because Equation 52 is satisfied, and that it goes to an equilibrium frequency. It may be that at this point a gene for female sterility in the hermaphrodite still cannot invade, so that gynodioecy is stable; or it may be that it can invade, leading to dioecy if it goes to fixation. This model for the evolution of dioecy from gynodioecy is examined further by Charlesworth and Charlesworth (1978). Kjellberg et al. (1987) discuss its application to the evolution of dioecy in figs.

The other complication is that male sterility can be caused by a cytoplasmic (usually mitochondrial) gene that suppresses pollen formation. Since cytoplasmic genes are maternally inherited, all the offspring of female plants will be female since they carry the male-sterility gene in their cytoplasm, and all the offspring of hermaphrodite plants will be hermaphrodite. Thus the frequency of females will increase if $f > 1 - \phi\delta$. However, a conflict of interest will arise between nuclear and cytoplasmic genes, so that the presence of a cytoplasmic male-sterility gene in the population will create a selection pressure in favor of a nuclear restorer gene to suppress it. The outcome of this interaction will depend in a complex way on its detailed mechanism and on the population structure (Gouyon and Couvet, 1987; Frank, 1989; Couvet et al., 1990). Dioecy could also evolve from a system with a cytoplasmic male sterility gene and a nuclear restorer gene if a female sterility gene closely linked to the restorer arose, converting hermaphrodites into males (Gouyon and Couvet, 1987).

In conclusion, two factors favor the evolution of gynodioecy rather than androdioecy. First, inbreeding depression in a partially self-fertile hermaphrodite favors the spread of a nuclear gene causing male-sterility but disfavors the spread of a nuclear gene causing female-sterility. Second, female inheritance of cytoplasmic genes greatly facilitates the spread of a cytoplasmic gene causing male-sterility, though this may establish counterselection in favor of a nuclear restorer gene. The establishment of gynodioecy paves the way for the evolution of dioecy. However, the large majority of flowering plants remain hermaphrodite for reasons discussed earlier in this section.

FURTHER READING

Maynard Smith (1978b) and Bell (1982) are single author books on the evolution of sex; Stearns (1987) and Michod and Levin (1988) are multi-author reviews; the September/October 1993 issue of the *Journal of Heredity* (84:321–424) is a symposium issue on this topic.

EXERCISES

1. (a) Consider the survival function in Figure 12.1b. Find the optimal value of m_1 for fixed m_2 as a function of m_2, with the constraint that $m_1, m_2 \geq 0.01$. Hence show that the Nash equilibria are $\{0.01, 1.10\}$ and $\{1.10, 0.01\}$. (You may assume without loss of generality that $M = 100$.) (b) Repeat this exercise for the survival function in Figure 12.1a and show that there is a unique equilibrium at $\{0.01, 0.01\}$.

2. Verify Equation 5. Find the equilibrium values and determine their

local stability. Investigate the behavior of the system by simulation.

3. Find the mating table for the 16 types of gamete union for the model of cytoplasmic inheritance leading to Equation 8. Hence verify Equation 8 and show that the system behaves in the way described in the text.

4. The model for the cost of sex discussed in the text supposes that *aa* females are sexual whereas *Aa* females are parthenogenetic. This model can be generalized by supposing that *Aa* females reproduce asexually with probability p (called the *penetrance* of the gene) and sexually with probability $1 - p$. Write x for the relative frequency of *Aa* among females and y for the corresponding frequency among males, both supposed small. Find the mating table and hence find the transition matrix of the linearized recurrence relationship for x and y (recall the methodology developed in Chapters 9 and 10). Obtain an algebraic expression for the eigenvalues of this matrix. Show that the dominant eigenvalue with complete penetrance ($p = 1$) is $1/(1 - r)$, in agreement with the result found in the text. Linearize the eigenvalues about $p = 0$ and show that the dominant eigenvalue for low penetrance is approximately $1 + p/2(1 - r)$. (The factor of 2 in the divisor is probably due to the fact that *Aa* is heterozygous.)

5. Verify the recurrence relationship in Equation 15 and find the eigenvalues of the matrix.

6. Write computer programs to verify the results in (a) Table 12.1 and (b) Table 12.2.

7. Write a program to calculate the advantage of sex under Kondrashov's model.

8. (a) Verify that Equation 32 follows from the assumptions in the text and has an equilibrium value given by Equation 33.
(b) Verify that Equation 37 gives the unique stable equilibrium to Equation 36 when there are two different mutation rates to asexuality, with $e_a = 10^{-4}$, $e_s = 0$, $p_1 = 10^{-6}$, $p_2 = 2 \times 10^{-6}$, and $N = 100$.

9. Consider the functions $\alpha(x) = x$, $\beta(x) = 1.582[1 - \exp(x - 1)]$ as a model of resource allocation to female and male function showing diminishing returns on resource allocation to male function. Plot $\beta(1 - x)$ against x to verify this statement. Plot $\alpha(x) + \beta(x)$ against x and hence

show from Equation 39 that dioecy is unstable. Find the ESS value of x^*, evaluate α and β at this point, and hence show from Equations 43 and 45 that hermaphroditism is stable.

10. (a) Verify the recurrence relation in Equation 51 and hence deduce Equations 52 and 53.
(b) Investigate the evolution of androdioecy in the same way and hence verify Equation 54.

APPENDIXES

Some Mathematical Tools for Population Biologists

"During the three years which I spent at Cambridge my time was wasted, as far as the academical studies were concerned, as completely as at Edinburgh and at school. I attempted mathematics, and even went during the summer of 1828 with a private tutor (a very dull man) to Barmouth, but I got on very slowly. The work was repugnant to me, chiefly from my not being able to see any meaning in the early steps in algebra. This impatience was very foolish, and in after years I have deeply regretted that I did not proceed far enough at least to understand something of the great leading principles of mathematics; for men thus endowed seem to have an extra sense. But I do not believe that I should ever have succeeded beyond a very low grade."

—Charles Darwin, *Autobiography* (Barlow, 1958)

I assume a basic knowledge of elementary algebra and calculus. This appendix describes, in a nonrigorous way, some other mathematical tools that are frequently used in the development of ecological and evolutionary models. Some simple worked examples are shown. My expectation is that a computer will often be used to do more complex calculations, but to understand what the computer has done it is helpful to have done a similar (though simpler) calculation yourself with pencil and paper. The sections of this appendix are referenced in the text as Appendix A, Appendix B, and so on.

A. TAYLOR SERIES EXPANSIONS

If $f(x)$ is a function of x, it is often useful to linearize it about a value c that is of particular interest. We can write

$$f(x) \cong f(c) + (x - c)f'(c) \quad \text{(A1)}$$

where $f'(c)$ is the derivative of $f(x)$ evaluated at $x = c$. This approximation is valid for values of x sufficiently close to c. It relies on the function being locally linear; the derivative is the slope of the curve at c, so that the right-hand side of Equation A1 is the best linear approximation to the curve. Consider for example the logarithmic function, $f(x) = \log(x)$, with derivative $f'(x) = 1/x$. Taking $c = 1$, we find the approximation

$$\log(x) \cong \log(1) + \frac{(x-1)}{1} = x - 1 \quad \text{(A2)}$$

valid for values of x near 1; for $x = 1.07$ this gives the approximate value 0.07 compared with the exact value $\log(1.07) = 0.06766$.

The linear approximation in Equation A1 depends on the linearity of the function near c, and it will break down as soon as x moves far enough from c for any curvature to become appreciable. One might hope that a better approximation could be obtained by taking into account higher powers of $(x - c)$, leading to a power-series expansion of the form

$$f(x) = a_0 + a_1(x-c) + a_2(x-c)^2 + a_3(x-c)^3 + \cdots \quad \text{(A3)}$$

For the power series to have the same derivatives as $f(x)$ at $x = c$ requires that

$$a_r = \frac{f^{(r)}(c)}{r!} \quad \text{(A4)}$$

where $f^{(r)}(c)$ denotes the rth derivative of $f(x)$ evaluated at $x = c$. Thus the power series is

$$f(x) = f(c) + (x-c)f'(c) + \frac{(x-c)^2}{2!}f''(c) + \frac{(x-c)^3}{3!}f'''(c) + \cdots \quad \text{(A5)}$$

This is known as a Taylor series expansion. By taking more terms into account, one expects to get a better and better approximation. It is left as an exercise to show that the Taylor series expansion for the logarithmic function about $x = 1$ is

$$\log(x) = (x-1) - \frac{(x-1)^2}{2} + \frac{(x-1)^3}{3} - \cdots \quad \text{(A6)}$$

Table A1. Accuracy of the Taylor series expansion for log(1.07).

Terms used	Approximation
Linear	0.07
Quadratic	0.06755
Cubic	0.0676643
Exact	0.0676586

The accuracy of the approximation for log(1.07) is shown in Table A1.

Four important Taylor series expansions about $c = 0$ are those for the exponential, sine and cosine functions, and the geometric series expansion:

$$\begin{aligned} \exp(x) &= 1 + x + \frac{x^2}{2!} + \frac{x^3}{3!} + \cdots \\ \sin(x) &= x - \frac{x^3}{3!} + \frac{x^5}{5!} - \cdots \\ \cos(x) &= 1 - \frac{x^2}{2!} + \frac{x^4}{4!} - \cdots \\ \frac{1}{1-x} &= 1 + x + x^2 + x^3 + \cdots \end{aligned} \tag{A7}$$

These results can be verified by calculating the appropriate derivatives. The geometric series can be obtained in an easier way as follows. Write

$$S_n = 1 + x + x^2 + \cdots + x^{n-1} \tag{A8}$$

Then

$$S_n x = x + x^2 + \cdots + x^{n-1} + x^n \tag{A9}$$

so that

$$\begin{aligned} S_n - S_n x &= S_n(1-x) = 1 - x^n \\ S_n &= \frac{1-x^n}{1-x} \end{aligned} \tag{A10}$$

If $|x| < 1$, then $x^n \to 0$ as $n \to \infty$, so that

$$1 + x + x^2 + x^3 + \cdots \equiv \sum_{n=0}^{\infty} x^n = \frac{1}{1-x} \tag{A11}$$

The technique of linear approximation in Equation A1 can be extended to a function of several variables, $f(x_1, x_2, \ldots, x_n)$, near the point $\mathbf{c} = \{c_1, c_2, \ldots, c_n\}$. The best linear approximation is

$$f(x_1, x_2, \cdots, x_n) \cong f(c_1, c_2, \cdots, c_n) + \sum_{i=1}^{n} (x_i - c_i) \frac{\partial f}{\partial x_i} \qquad \text{(A12)}$$

where $\partial f / \partial x_i$ is the partial derivative of f with respect to x_i evaluated at $\mathbf{c}$. Consider for example the function $f(x_1, x_2) = \log(x_1/x_2)$ near $\{x_1, x_2\} = \{1, 1\}$. We find

$$\begin{aligned} f(1,1) &= \log(1) = 0 \\ \partial f/\partial x_1 &= x_2/x_1 = 1 \\ \partial f/\partial x_2 &= -1/x_2 = -1 \end{aligned} \qquad \text{(A13)}$$

so that

$$\log(x_1/x_2) \cong (x_1 - 1) - (x_2 - 1) = x_1 - x_2 \qquad \text{(A14)}$$

B. COMPLEX NUMBERS

The quadratic equation

$$ax^2 + bx + c = 0 \qquad \text{(B1)}$$

has two solutions

$$x = \frac{-b \pm \sqrt{b^2 - 4ac}}{2a} \qquad \text{(B2)}$$

When $b^2 < 4ac$, this presents the problem of finding the square root of a negative number. A similar problem can occur in solving higher-order polynomial equations and in other contexts. It is convenient to resolve this problem by inventing an "imaginary" number i whose square is –1, in the same vein as earlier mathematicians invented negative numbers. Thus the solutions of the equation

$$x^2 - 2x + 5 = 0 \qquad \text{(B3)}$$

are

$$x = 1 \pm 2i \qquad \text{(B4)}$$

A number like $1 + 2i$, with a real part (1) and an imaginary part (2) is called a *complex number*.

Complex numbers provide a consistent system when manipulated by the ordinary laws of algebra, provided that i^2 is replaced by –1 whenever it occurs. Define

$$\begin{aligned} z_1 &= 1 + 2i \\ z_2 &= 2 - i \end{aligned} \tag{B5}$$

Then,

$$\begin{aligned} z_1 + z_2 &= 3 + i \\ z_1 z_2 &\equiv (1 + 2i)(2 - i) = 2 + 3i - 2i^2 = 4 + 3i \end{aligned} \tag{B6}$$

The ratio $z_3 = z_1/z_2$ is the number such that $z_2 z_3 = z_1$. In the above example, if $z_3 = x + iy$, then

$$(2 - i)(x + iy) = (2x + y) + (-x + 2y)i = 1 + 2i \tag{B7a}$$

so that

$$\begin{aligned} 2x + y &= 1 \\ -x + 2y &= 2 \end{aligned} \tag{B7b}$$

Hence,

$$x = 0 \qquad y = 1 \tag{B7c}$$

that is to say,

$$z_3 = i \tag{B7d}$$

A complex number has two parts, real and imaginary, which can be represented in a two-dimensional diagram (the Argand diagram) with the real part along the horizontal axis and the imaginary part along the vertical axis. The number $z_1 = 1 + 2i$ is shown in the Argand diagram in Figure B1.

It is sometimes convenient to represent a complex number z by its polar coordinates in the Argand diagram, that is to say its *modulus* or *absolute value*, $|z| = r$ (its distance from the origin), and its *argument*, $\arg(z) = \theta$ (its direction from the origin). These quantities are defined for any complex number $z = x + iy$ as

$$\begin{aligned} r &= \sqrt{x^2 + y^2} \\ \theta &= \arctan(y/x) \end{aligned} \tag{B8}$$

For $z_1 = 1 + 2i$,

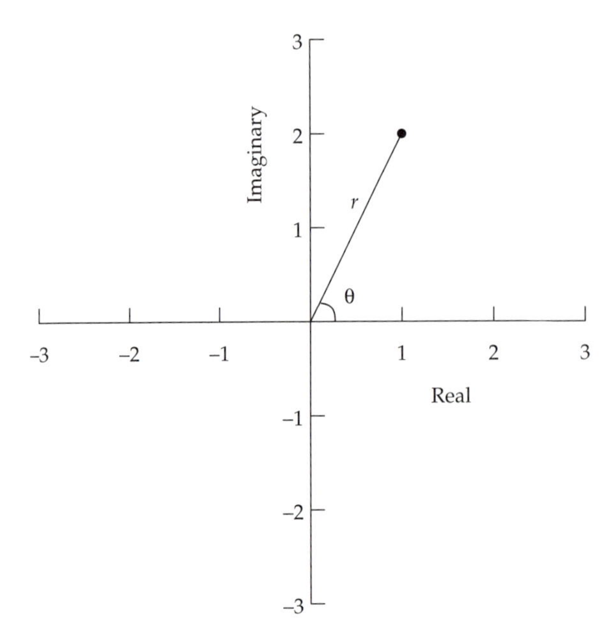

Figure B1. Representation of the number $z = 1 + 2i$ in an Argand diagram. The number may be represented in polar coordinates by its *absolute value*, $r = 2.236$ and its *argument*, $\theta = 63.4°$ or 1.107 radians.

$$
\begin{aligned}
r &= \sqrt{5} = 2.236 \\
\theta &= \arctan 2 = 63.4° \text{ or } 1.107 \text{ radians}
\end{aligned}
\tag{B9}
$$

Contrariwise, we can go from polar to Cartesian coordinates. It is clear from the Argand diagram that

$$
\begin{aligned}
x &= \text{real part} = r\cos\theta \\
y &= \text{imaginary part} = r\sin\theta
\end{aligned}
\tag{B10}
$$

Thus

$$z = x + iy = r(\cos\theta + i\sin\theta) \tag{B11}$$

It is possible to define functions of a complex variable by analogy with the corresponding definition for a real variable. Of particular importance is the exponential function, which can be defined by the series expansion

$$\exp(z) = 1 + z + \frac{z^2}{2!} + \frac{z^3}{3!} + \cdots \tag{B12}$$

by analogy with Equation A7. If z_1 and z_2 are complex numbers, it can be verified from Equation B12 that

$$\exp(z_1 + z_2) = \exp(z_1)\exp(z_2) \tag{B13}$$

a well-known property for real variables. In particular, if $z = x + iy$

$$\exp(z) = \exp(x + iy) = \exp(x)\exp(iy) \tag{B14}$$

The exponential of an imaginary number is

$$\begin{aligned}\exp(iy) &= 1 + iy + \frac{i^2y^2}{2!} + \frac{i^3y^3}{3!} + \cdots \\ &= (1 - \frac{y^2}{2!} + \frac{y^4}{4!} - \cdots) + i(y - \frac{y^3}{3!} + \frac{y^5}{5!} - \cdots) \\ &= \cos y + i \sin y\end{aligned} \tag{B15}$$

Hence

$$z = r(\cos\theta + i\sin\theta) = r\exp(i\theta) \tag{B16}$$

It follows that

$$z_1z_2 = r_1\exp(i\theta_1)r_2\exp(i\theta_2) = r_1r_2\exp[i(\theta_1 + \theta_2)] \tag{B17}$$

Thus the rule for multiplying complex numbers is simple in polar coordinates: multiply the absolute values and add the arguments.

C. ELEMENTARY VECTOR OPERATIONS

A vector of length l is an ordered list of l numbers (or symbols that behave like numbers). Thus

$$\mathbf{x}_1 = \{1, 2, -1\} \tag{C1}$$

and

$$\mathbf{x}_2 = \{4, 6, 3\} \tag{C2}$$

are vectors of length 3. (It is conventional to denote vectors by bold lowercase letters. Note that I will use braces for defining a vector.)[1]

Addition of two vectors of the same length is defined by adding their corresponding elements:

---

[1] Some readers may be used to distinguishing between row vectors and column vectors. I do not use these concepts. In my terminology a vector is simply an ordered list of numbers. A row vector is not a vector in this sense but a matrix with a single row; likewise a column vector is a matrix with a single column.

$$\mathbf{x}_1 + \mathbf{x}_2 = \{1+4, 2+6, -1+3\} = \{5, 8, 2\} \tag{C3}$$

Multiplication of a vector by a scalar (an ordinary number) is defined by multiplying each element by that number:

$$2\mathbf{x}_1 \equiv 2\{1, 2, -1\} = \{2, 4, -2\} \tag{C4}$$

There are two ways of multipling a pair of vectors, the inner product and the outer product. The inner product of two vectors of the same length is a scalar obtained by multiplying corresponding elements and then summing these products:

$$\mathbf{x}_1\mathbf{x}_2 = 1\times 4 + 2\times 6 - 1\times 3 = 13 \tag{C5}$$

The outer product of two vectors **u** and **v**, of lengths m and n , written $\mathbf{u} \times \mathbf{v}$ is a matrix of order $m \times n$ whose ijth element is the product of the ith element of **u** and the jth element of **v**:

$$\mathbf{x}_1 \times \mathbf{x}_2 \equiv \{1, 2, -1\} \times \{4, 6, 3\} = \begin{bmatrix} 4 & 6 & 3 \\ 8 & 12 & 6 \\ -4 & -6 & -3 \end{bmatrix} \tag{C6}$$

A vector can be represented geometrically either as a point in l-dimensional space or as a directed line. Suppose that there are 20 prey and 10 predators per unit area; this defines the density vector {20, 10} that can be represented as a point in the phase plane (Figure C1a). Suppose now that the prey density is increasing at a rate of 3 per year while the predator density is decreasing at a rate of 1 per year. This defines the rate vector {3, –1}, that can be represented by a directed line with components +3 in the horizontal (prey) direction and –1 in the vertical (predator) direction (Figure C1b, arrow at lower left); this line has magnitude (distance from the origin)

$$\sqrt{3^2 + (-1)^2} = 3.16 \tag{C7}$$

and is in a direction making an angle with the horizontal axis of

$$\theta = \arctan\left(-\tfrac{1}{3}\right) = -18.4^\circ \tag{C8}$$

This directed line can be moved (Figure C1b, arrow at upper right), with unchanged magnitude and direction, to have its origin at {20, 10} to represent the rate of increase in the phase plane at this population density. This is the meaning of the velocity vectors shown in the vector field diagrams in Chapters 2 and 3.

The magnitude of a vector **v** is denoted |**v**| and can be calculated as the square root of the inner product of the vector with itself:

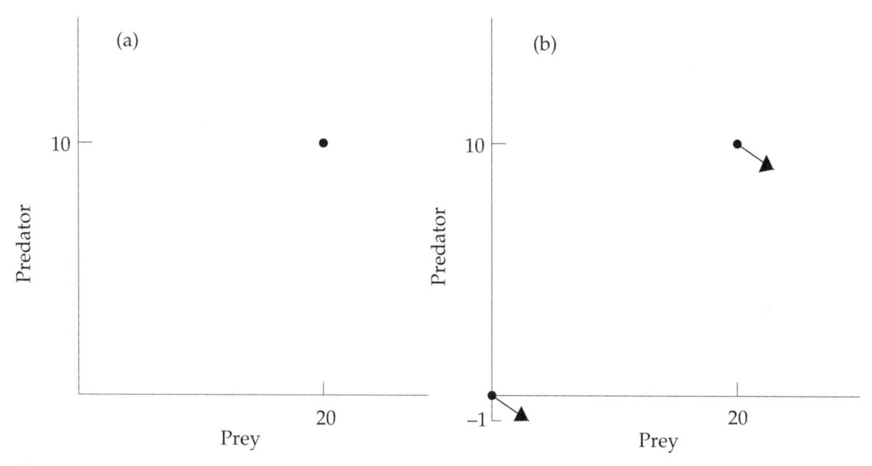

Figure C1. Geometric representation of vectors.

$$|\mathbf{v}| = \sqrt{\mathbf{vv}} = \sqrt{\mathbf{v}^2} \tag{C9}$$

The angle θ between two vectors **u** and **v** can be calculated from the formula

$$\cos\theta = \mathbf{uv} / |\mathbf{u}|\,|\mathbf{v}| \tag{C10}$$

For the vectors $\mathbf{x}_1$ and $\mathbf{x}_2$ defined in Equations 1 and 2,

$$\begin{aligned} \cos\theta &= 13/\sqrt{6 \cdot 61} = 0.68 \\ \theta &= 47° \end{aligned} \tag{C11}$$

D. ELEMENTARY MATRIX OPERATIONS

Matrix algebra was invented by William Cayley in the last century to provide a compact way of writing simultaneous linear equations. A matrix of order $r \times c$ is a rectangular array of numbers (or number-like symbols) with r rows and c columns. Thus

$$\mathbf{A}_1 = \begin{bmatrix} 1 & 3 & 4 \\ 9 & -1 & 6 \end{bmatrix} \tag{D1}$$

and

$$\mathbf{A}_2 = \begin{bmatrix} 3 & -1 & 2 \\ 10 & 3 & 4 \end{bmatrix} \tag{D2}$$

are matrices of order 2×3. (It is conventional to denote matrices by bold uppercase letters.)[2]

Addition of two matrices of the same order is defined by adding their corresponding elements:

$$\mathbf{A}_1 + \mathbf{A}_2 = \begin{bmatrix} 4 & 2 & 6 \\ 19 & 2 & 10 \end{bmatrix} \tag{D3}$$

The transpose of a matrix $\mathbf{A}$, denoted $\mathbf{A}^{\mathrm{T}}$, has the rows and columns transposed:

$$\mathbf{A}_1^{\mathrm{T}} = \begin{bmatrix} 1 & 9 \\ 3 & -1 \\ 4 & 6 \end{bmatrix} \tag{D4}$$

Multiplication of a matrix by a scalar is defined by multiplying each element by the scalar:

$$2\mathbf{A}_1 = \begin{bmatrix} 2 & 6 & 8 \\ 18 & -2 & 12 \end{bmatrix} \tag{D5}$$

Matrix multiplication is an extension of the inner product of two vectors. Suppose that $\mathbf{A}$ and $\mathbf{B}$ are matrices of orders $m \times n$ and $n \times p$, respectively. Their product $\mathbf{AB}$ is defined as the matrix with dimensions $m \times p$, whose ijth element is the inner product of the ith row of $\mathbf{A}$ and the jth column of $\mathbf{B}$:

$$\mathbf{A}_1^{\mathrm{T}}\mathbf{A}_2 \equiv \begin{bmatrix} 1 & 9 \\ 3 & -1 \\ 4 & 6 \end{bmatrix} \begin{bmatrix} 3 & -1 & 2 \\ 10 & 3 & 4 \end{bmatrix} = \begin{bmatrix} 93 & 26 & 38 \\ -1 & -6 & 2 \\ 72 & 14 & 32 \end{bmatrix} \tag{D6}$$

Notice that the matrix product $\mathbf{AB}$ is only defined when the number of columns of $\mathbf{A}$ is equal to the number of rows of $\mathbf{B}$. If $\mathbf{A}$ and $\mathbf{B}$ are square matrices of the same order, both $\mathbf{AB}$ and $\mathbf{BA}$ are defined, but they are not usually the same; that is to say, matrix multiplication is not commutative:

$$\begin{aligned} \begin{bmatrix} 1 & 3 \\ 2 & -1 \end{bmatrix} \begin{bmatrix} 4 & 6 \\ -2 & 8 \end{bmatrix} &= \begin{bmatrix} -2 & 30 \\ 10 & 4 \end{bmatrix} \\ \begin{bmatrix} 4 & 6 \\ -2 & 8 \end{bmatrix} \begin{bmatrix} 1 & 3 \\ 2 & -1 \end{bmatrix} &= \begin{bmatrix} 16 & 6 \\ 14 & -14 \end{bmatrix} \end{aligned} \tag{D7}$$

[2] The rows of a matrix are lists of numbers (vectors), so that we may think of a matrix as a list of lists: $\mathbf{A}_1 = \{\{1, 3, 4\}, \{9, -1, 6\}\}$. This is how matrices are represented in *Mathematica*.

The product of a vector **v** of length l with a matrix **A** of order $l \times m$ is a vector of length m whose ith element is the inner product of **v** and the ith column of **A**:

$$\{1,3\}\begin{bmatrix} 1 & 3 & 4 \\ 9 & -1 & 6 \end{bmatrix} = \{28,0,22\} \tag{D8}$$

The product of a matrix **A** of order $m \times n$ with a vector **v** of length n is a vector of length m whose ith element is the inner product of the ith row of **A** and **v**:[3]

$$\begin{bmatrix} 1 & 3 & 4 \\ 9 & -1 & 6 \end{bmatrix}\{1,2,-1\} = \{3,1\} \tag{D9}$$

The motivation for defining matrix multiplication like this is that it leads to useful applications. Consider the problem of solving three linear equations in three unknowns:

$$\begin{aligned} &3x_1 + 4x_2 - x_3 = 4 \\ &4x_1 + x_2 + 12x_3 = 12 \\ &9x_1 - 3x_2 + x_3 = 6 \end{aligned} \tag{D10}$$

This can be written more compactly in matrix terminology as

$$\mathbf{Ax} = \mathbf{c} \tag{D11}$$

where

$$\begin{aligned} &\mathbf{A} = \begin{bmatrix} 3 & 4 & -1 \\ 4 & 1 & 12 \\ 9 & -3 & 1 \end{bmatrix} \\ &\mathbf{x} = \{x_1, x_2, x_3\} \\ &\mathbf{c} = \{4,12,6\} \end{aligned} \tag{D12}$$

It is not only more convenient to use this compact terminology, but it also leads to new methodology, such as solution of the system of equations by finding the inverse of the matrix **A**, as will be explained in the next section.

[3] Readers familiar with row and column vectors will recognize these rules as meaning that a vector behaves like a row vector when it premultiples a matrix and like a column vector when it postmultiplies a matrix.

E. DETERMINANTS, LINEAR INDEPENDENCE, AND MATRIX INVERSION

Suppose that **A** is a square matrix of order $n \times n$ with elements a_{ij}. Its determinant is a scalar defined as

$$\det \mathbf{A} = \sum \pm a_{1j_1} a_{2j_2} \dots a_{nj_n} \tag{E1}$$

Each term contains one element from each row and from each column. The column subscripts in each term are a permutation of the integers {1, 2, … , n}, and the sum (with a "±" sign) is taken over all possible permutations, of which there are n! A permutation is classified as odd or even depending on whether it can be obtained from the natural ordering by an odd or an even number of reciprocal interchanges, a "+" sign being given to even permutations and a "−" sign to odd ones. The determinant of a 2 × 2 matrix is

$$\det \mathbf{A} = a_{11}a_{22} - a_{12}a_{21} \tag{E2}$$

The determinant of the 3 × 3 matrix defined in Equation D12 is

$$\begin{aligned}\det \mathbf{A} &= (3\times1\times1)-(3\times12\times-3)-(4\times4\times1)\\ &\quad+(4\times12\times9)+(-1\times4\times-3)-(-1\times1\times9)\\ &=548\end{aligned} \tag{E3}$$

The following properties follow almost immediately from the definition of the determinant. If all the elements of a row (or column) are multiplied by a constant, the determinant is multiplied by the same constant. If two rows (or columns) are interchanged, the sign of the determinant changes (because even permutations become odd and vice versa), but its absolute value is unchanged. If two rows (or columns) are identical, the determinant is zero (because interchanging them leaves the matrix the same). Hence for any square matrix, adding one row (or column) to another leaves the determinant unchanged. It follows that if the rows are linearly dependent the determinant is zero. (Suppose that there is a linear dependence involving the first row, for example. Then by adding multiples of the other rows to it, it is possible, without changing the determinant, to make this row zero; a matrix with a zero row has zero determinant.) The converse is also true: if the rows are independent then the determinant is nonzero. Thus there is a one-to-one correspondence between the dependence or independence of the rows (and columns) and whether the determinant is zero or nonzero. A square matrix that has linear relationships among its rows or columns (and hence has zero determinant) is called *singular*. The matrix

$$\begin{bmatrix} 1 & 2 & 4 \\ 3 & 1 & 2 \\ 7 & 4 & 8 \end{bmatrix} \tag{E4}$$

is singular because the third row is equal to twice the second row plus the first row; it is left as an exercise to show that its determinant is zero. A square matrix without linear relationships among its rows or columns (and hence with nonzero determinant) is called *nonsingular*. Testing for singularity is one of the most important uses of the determinant.

The identity matrix of order n, denoted $\mathbf{I}_n$, is a square matrix of order $n \times n$ with 1 on the leading diagonal and 0 everywhere else:

$$\mathbf{I}_2 = \begin{bmatrix} 1 & 0 \\ 0 & 1 \end{bmatrix}$$

$$\mathbf{I}_3 = \begin{bmatrix} 1 & 0 & 0 \\ 0 & 1 & 0 \\ 0 & 0 & 1 \end{bmatrix} \tag{E5}$$

The identity matrix plays the same role in matrix algebra as unity in scalar arithmetic. In particular, if $\mathbf{A}$ is any matrix, square or rectangular, and $\mathbf{I}$ is the appropriate identity matrix, then

$$\begin{aligned} \mathbf{AI} &= \mathbf{A} \\ \mathbf{IA} &= \mathbf{A} \end{aligned} \tag{E6}$$

(Verify this for $\mathbf{A}_1$ defined in Equation D1; the appropriate identity matrix is $\mathbf{I}_3$ for $\mathbf{A}_1\mathbf{I}$ and $\mathbf{I}_2$ for $\mathbf{IA}_1$.)

If $\mathbf{A}$ is a nonsingular square matrix it has a unique inverse, denoted $\mathbf{A}^{-1}$, such that

$$\mathbf{AA}^{-1} = \mathbf{A}^{-1}\mathbf{A} = \mathbf{I} \tag{E7}$$

For example, the inverse of the matrix defined in Equation D12 is

$$\mathbf{A}^{-1} = \frac{1}{548}\begin{bmatrix} 37 & -1 & 49 \\ 104 & 12 & -40 \\ -21 & 45 & -13 \end{bmatrix} \tag{E8}$$

(Verify that this satisfies Equation E7 above.) The calculation of the inverse is in general a complicated procedure that is best left to a computer program, but there is a simple formula for the inverse of a 2×2 matrix. If

$$\mathbf{A} = \begin{bmatrix} a & b \\ c & d \end{bmatrix} \tag{E9}$$

then

$$\mathbf{A}^{-1} = \frac{1}{ad - bc}\begin{bmatrix} d & -b \\ -c & a \end{bmatrix} \tag{E10}$$

(Verify this assertion by multiplying together the right-hand sides of Equations E9 and E10.)

Note that the divisor in Equation E10 is the determinant of **A**, so that the result is only meaningful for a nonsingular matrix. In fact a singular matrix cannot have an inverse. Suppose that $\mathbf{AB} = \mathbf{I}$, write $\mathbf{r}_i$ for the *i*th row of **A** and $\mathbf{c}_j$ for the *j*th column of **B**, and suppose also that $\mathbf{r}_3 = \mathbf{r}_4 + 2\mathbf{r}_5$. Then

$$\begin{aligned} &\mathbf{r}_3\mathbf{c}_3 = 1 \\ &\mathbf{r}_4\mathbf{c}_3 + 2\mathbf{r}_5\mathbf{c}_3 = (\mathbf{r}_4 + 2\mathbf{r}_5)\mathbf{c}_3 = 0 \end{aligned} \tag{E11}$$

which contradicts the assumption that $\mathbf{r}_3 = \mathbf{r}_4 + 2\mathbf{r}_5$. Thus a matrix with any linear dependencies among its rows cannot have an inverse.

We return to the problem of solving *n* linear equations in *n* unknowns, considered at the end of the last section, which can be expressed in matrix notation as

$$\mathbf{Ax} = \mathbf{c} \tag{E12}$$

where **A** is a matrix of known coefficients, **x** is the vector of unknowns, and **c** is a vector of known constants.

Consider first the nonhomogeneous equation with $\mathbf{c} \neq \mathbf{0}$. If **A** is nonsingular it has a unique inverse; premultiplying both sides by this inverse we find the unique solution

$$\mathbf{x} = \mathbf{A}^{-1}\mathbf{c} \tag{E13}$$

For the system defined in Equation D12

$$\mathbf{x} = \frac{1}{548}\begin{bmatrix} 37 & -1 & 49 \\ 104 & 12 & -40 \\ -21 & 45 & -13 \end{bmatrix}\{4,12,6\} = \frac{1}{274}\{215,160,189\} \tag{E14}$$

If **A** is singular, there are linear relationships among its rows. If these relationships do not hold among the c's, the equations are inconsistent and have no solution; if the relationships among the rows of **A** also happen to hold among the c's, a range of solutions exists. For example, the equations

$$\begin{aligned} x_1 + x_2 &= 1 \\ 2x_1 + 2x_2 &= 3 \end{aligned} \tag{E15}$$

have no solution whereas the equations

$$\begin{aligned} x_1 + x_2 &= 1 \\ 2x_1 + 2x_2 &= 2 \end{aligned} \tag{E16}$$

are compatible with any values of x_1 and x_2 on the line $x_2 = 1 - x_1$.

Consider finally the homogeneous equation

$$\mathbf{Ax} = \mathbf{0} \tag{E17}$$

If **A** is nonsingular, premultiplication by its inverse shows that the only solution is the trivial solution $\mathbf{x} = \mathbf{0}$. If **A** is singular, then any linear relationship among the rows of **A** automatically holds on the right-hand side so that a range of nontrivial solutions exists. For example, the equations

$$\begin{aligned} x_1 + x_2 &= 0 \\ 2x_1 + 2x_2 &= 0 \end{aligned} \tag{E18}$$

are satisfied by any values on the line $x_2 = -x_1$.

F. EIGENVALUES AND EIGENVECTORS

If **A** is a square matrix of order $n \times n$ and **x** is a vector of length n, then **Ax** is also a vector of length n. Thus **A** can be regarded as defining a linear transformation on points in n-dimensional space. Consider a simple example in two-dimensional space:

$$\mathbf{A}_3 = \begin{bmatrix} 1 & 2 \\ 3 & 2 \end{bmatrix} \tag{F1}$$

Then $\mathbf{A}_3\{0, 1\} = \{2, 2\}$ while $\mathbf{A}_3\{1, 1\}=\{3, 5\}$. Thus $\mathbf{A}_3$ has transformed the point {0, 1} into the point {2, 2} and the point {1, 1} into {3, 5}.

We shall now consider an important tool in studying linear transformations. If λ is a scalar and **e** a nonzero vector such that

$$\mathbf{Ae} = \lambda\mathbf{e} \tag{F2}$$

then λ is called an eigenvalue of **A** and **e** is the corresponding eigenvector. This equation can be written

$$(\mathbf{A} - \lambda\mathbf{I})\mathbf{e} = \mathbf{0} \tag{F3}$$

where **0** is the null vector, all of whose elements are zero. This homogeneous equation can only have a nonzero solution if

$$\det(\mathbf{A} - \lambda\mathbf{I}) = 0 \tag{F4}$$

The determinant defines a polynomial of degree n in λ, and the equation will in general have n solutions, though there may be fewer if there are repeated roots. In practice there are usually n distinct eigenvalues, and for simplicity we shall from now on assume that this is the case. For the 2 × 2 matrix defined in Equation F1,

$$\begin{aligned} \mathbf{A}_3 - \lambda\mathbf{I} &= \begin{bmatrix} 1-\lambda & 2 \\ 3 & 2-\lambda \end{bmatrix} \\ \det(\mathbf{A}_3 - \lambda\mathbf{I}) &= (1-\lambda)(2-\lambda) - 6 \end{aligned} \tag{F5}$$

Equating the determinant to zero gives a quadratic equation in λ whose roots are $\lambda = -1$ and $\lambda = 4$. These are the eigenvalues of $\mathbf{A}_3$.

To each eigenvalue there corresponds an eigenvector $\mathbf{e} = \{e_1, e_2\}$. Substituting $\lambda = -1$ in Equation F3 gives

$$\begin{aligned} 2e_1 + 2e_2 &= 0 \\ 3e_1 + 3e_2 &= 0 \end{aligned} \tag{F6}$$

One of these equations is redundant (this is ensured by the fact that λ is an eigenvalue), so that we may choose e_2 arbitrarily. If $e_2 = 1$, then $e_1 = -1$; if $e_2 = 2$, then $e_1 = -2$, and so on. Thus $\mathbf{e} = \{-1, 1\}$ or any multiple thereof is an eigenvector corresponding to $\lambda = -1$. It is a general rule that eigenvectors are only determined up to a multiplicative constant. It can be shown in the same way that the eigenvector corresponding to $\lambda = 4$ is $\{0.667, 1\}$ or any multiple thereof.

Eigenvalues may be real or complex. Complex eigenvalues occur as conjugate pairs, and the corresponding eigenvectors are also complex conjugates. Consider for example the matrix

$$\begin{bmatrix} 1 & 2 \\ -2 & 1 \end{bmatrix} \tag{F7}$$

It is left as an exercise to show that the eigenvalues are $1 - 2i$ and $1 + 2i$

with corresponding eigenvectors $\{i, 1\}$ and $\{-i, 1\}$, respectively.

A real eigenvector can be thought of as a direction in n-dimensional space that is conserved under the linear transformation **A** while the magnitude of the vector is multiplied by λ. Suppose that $\mathbf{n}_t$ is a vector of population sizes (or gene frequencies or similar variables) at time t satisfying the linear difference equation

$$\mathbf{n}_{t+1} = \mathbf{A}\mathbf{n}_t \tag{F8}$$

If the population vector **n** lies in the direction of a particular eigenvector this year, it will lie in the same direction next year but will be multiplied in magnitude by λ; that is to say, all the elements of **n** will be multiplied by the same constant λ. Linear difference equations will be considered in more detail in Appendix J.

So far we have been considering *right* eigenvalues and eigenvectors. One can define in the same way *left* eigenvalues and eigenvectors satisfying

$$\mathbf{eA} = \lambda\mathbf{e} \tag{F9}$$

The left eigenvalues satisfy the same determinantal equation as the right eigenvalues and are therefore the same, but the corresponding eigenvectors are different. The left eigenvector for $\mathbf{A}_3$ defined in Equation F1 corresponding to $\lambda = -1$ satisfies

$$2e_1 + 3e_2 = 0 \tag{F10}$$

Thus the left eigenvector is $\{-1.5, 1\}$ or any multiple thereof. It is convenient to standardize a left eigenvector so that its inner product with the corresponding right eigenvector is unity. If we take the right eigenvector as $\{-1, 1\}$, the standardized left eigenvector is $\{-0.6, 0.4\}$. It is left as an exercise to show that a left eigenvector corresponding to the eigenvalue $\lambda = 4$ is $\{1, 1\}$; if the right eigenvector is taken as $\{0.667, 1\}$, the standardized left eigenvector is $\{0.6, 0.6\}$. I shall denote a standardized left eigenvector by $\boldsymbol{\varepsilon}$.

Eigenvalues and eigenvectors will usually be calculated by using a computer package. Many packages only find right eigenvectors. It is useful to note that the left eigenvectors of a matrix **A** are the right eigenvectors of its transpose, $\mathbf{A}^{\mathrm{T}}$ since $\mathbf{eA} = \mathbf{A}^{\mathrm{T}}\mathbf{e}$.

An important fact is that a left and a right eigenvector corresponding to different eigenvalues are orthogonal; geometrically this means that they are at right angles and algebraically that their inner product is zero. Suppose that $\mathbf{e}_1$ is a right eigenvector corresponding to λ_1 and that $\boldsymbol{\varepsilon}_2$ is a left eigenvector corresponding to λ_2. Then,

$$\begin{aligned} \boldsymbol{\varepsilon}_2\mathbf{A}\mathbf{e}_1 &= \boldsymbol{\varepsilon}_2(\mathbf{A}\mathbf{e}_1) = \boldsymbol{\varepsilon}_2(\lambda_1\mathbf{e}_1) = \lambda_1\boldsymbol{\varepsilon}_2\mathbf{e}_1 \\ \boldsymbol{\varepsilon}_2\mathbf{A}\mathbf{e}_1 &= (\boldsymbol{\varepsilon}_2\mathbf{A})\mathbf{e}_1 = (\lambda_2\boldsymbol{\varepsilon}_2)\mathbf{e}_1 = \lambda_2\boldsymbol{\varepsilon}_2\mathbf{e}_1 \end{aligned} \tag{F11}$$

It follows that $\boldsymbol{\varepsilon}_2\mathbf{e}_1 = 0$ if λ_1 differs from λ_2. Define **R** as the matrix of right eigenvectors stacked on top of each other (so that the *i*th row of **R** is the *i*th right eigenvector) and **L** as the corresponding matrix of standardized left eigenvectors. For example, for the matrix $\mathbf{A}_3$

$$\mathbf{R} = \begin{bmatrix} -1 & 1 \\ 0.667 & 1 \end{bmatrix}$$

$$\mathbf{L} = \begin{bmatrix} -0.6 & 0.4 \\ 0.6 & 0.6 \end{bmatrix} \tag{F12}$$

It follows from the orthogonality of left and right eigenvectors corresponding to different eigenvalues and from the standardization of left eigenvectors that

$$\mathbf{L}\mathbf{R}^{\mathrm{T}} = \mathbf{I} \tag{F13}$$

In other words, **L** is the inverse of the transpose of **R**. Since a left inverse is also a right inverse

$$\mathbf{R}^{\mathrm{T}}\mathbf{L} = \mathbf{I} \tag{F14}$$

It is left as an exercise to verify these results for the matrices in Equation F12.

Finally, we consider the effect of a small perturbation of the elements of a matrix **A** on an eigenvalue λ with right and left eigenvectors **e** and $\boldsymbol{\varepsilon}$. A perturbation from **A** to **A** + d**A** will change λ to λ + dλ and **e** to **e** + d**e** such that

$$\begin{aligned} \mathbf{A}\mathbf{e} &= \lambda\mathbf{e} \\ (\mathbf{A} + \mathrm{d}\mathbf{A})(\mathbf{e} + \mathrm{d}\mathbf{e}) &= (\lambda + \mathrm{d}\lambda)(\mathbf{e} + \mathrm{d}\mathbf{e}) \end{aligned} \tag{F15}$$

Ignoring second-order terms

$$\mathrm{d}\mathbf{A}\mathbf{e} + \mathbf{A}\mathrm{d}\mathbf{e} = \lambda\mathrm{d}\mathbf{e} + \mathrm{d}\lambda\mathbf{e} \tag{F16}$$

Premultiplying both sides by $\boldsymbol{\varepsilon}$ and remembering that $\boldsymbol{\varepsilon}\mathbf{A} = \lambda\boldsymbol{\varepsilon}$ and that $\boldsymbol{\varepsilon}\mathbf{e} = 1$,

$$\boldsymbol{\varepsilon}\mathrm{d}\mathbf{A}\mathbf{e} = \mathrm{d}\lambda \tag{F17}$$

If only one element in **A** is perturbed, say a_{ij} to a_{ij} + da_{ij}, then

$$\mathrm{d}\lambda = \varepsilon_i e_j \mathrm{d}a_{ij} \tag{F18}$$

where ε_i and e_j are the *i*th and *j*th elements of $\boldsymbol{\varepsilon}$ and **e**, respectively. This result is useful in life-history theory.

G. SPECTRAL DECOMPOSITION OF A MATRIX

It is often useful to represent a matrix in a particular way known as its *spectral decomposition*. We first observe that

$$\mathbf{A}\mathbf{R}^{\mathrm{T}} = \mathbf{R}^{\mathrm{T}}\boldsymbol{\Lambda} \tag{G1}$$

where $\mathbf{R}$ is the matrix of right eigenvectors defined in Appendix F and $\boldsymbol{\Lambda}$ is a diagonal matrix with the eigenvalues on the leading diagonal and zero everywhere else. To prove this, note that the ith column of $\mathbf{A}\mathbf{R}^{\mathrm{T}}$ is $\mathbf{A}\mathbf{e}_i = \lambda_i\mathbf{e}_i$; this is also the ith column of $\mathbf{R}^{\mathrm{T}}\boldsymbol{\Lambda}$. Postmultiplying Equation G1 by $\mathbf{L}$ and using Equation F14, we find that

$$\mathbf{A} = \mathbf{R}^{\mathrm{T}}\boldsymbol{\Lambda}\mathbf{L} \tag{G2}$$

Write e_{ij} for the ijth element of $\mathbf{R}$ (the jth element of the ith right eigenvector) and ε_{ij} for the corresponding element of $\mathbf{L}$ (the jth element of the ith standardized left eigenvector). The ith row of $\mathbf{R}^{\mathrm{T}}$ is

$$\{e_{1i}, e_{2i}, \cdots, e_{ni}\} \tag{G3}$$

while the jth column of $\boldsymbol{\Lambda}\mathbf{L}$ is

$$\{\lambda_1\varepsilon_{1j}, \lambda_2\varepsilon_{2j}, \cdots, \lambda_n\varepsilon_{nj}\} \tag{G4}$$

Equation G3 says that the inner product of these two vectors is the ijth element of $\mathbf{A}$:

$$a_{ij} = \lambda_1 e_{1i}\varepsilon_{1j} + \lambda_2 e_{2i}\varepsilon_{2j} + \cdots + \lambda_n e_{ni}\varepsilon_{nj} \tag{G5}$$

We now define

$$\mathbf{T}_k = \mathbf{e}_k \times \boldsymbol{\varepsilon}_k \tag{G6}$$

the outer product of the kth right and left eigenvectors. The ijth element of $\mathbf{T}_k$ is $e_{ki}\varepsilon_{kj}$ so that we finally obtain

$$\mathbf{A} = \lambda_1\mathbf{T}_1 + \lambda_2\mathbf{T}_2 + \cdots + \lambda_n\mathbf{T}_n = \sum_{k=1}^{n}\lambda_k\mathbf{T}_k \tag{G7}$$

This is called the *spectral decomposition* of $\mathbf{A}$.

For the matrix $\mathbf{A}_3$ defined in Equation F1,

$$\begin{aligned} \mathbf{T}_1 &= \mathbf{e}_1 \times \boldsymbol{\varepsilon}_1 = \{-1, 1\} \times \{-0.6, 0.4\} = \begin{bmatrix} 0.6 & -0.4 \\ -0.6 & 0.4 \end{bmatrix} \\ \mathbf{T}_2 &= \mathbf{e}_2 \times \boldsymbol{\varepsilon}_2 = \{0.667, 1\} \times \{0.6, 0.6\} = \begin{bmatrix} 0.4 & 0.4 \\ 0.6 & 0.6 \end{bmatrix} \end{aligned} \tag{G8}$$

so that the spectral decomposition is

$$\mathbf{A}_3 = -\begin{bmatrix} 0.6 & -0.4 \\ -0.6 & 0.4 \end{bmatrix} + 4\begin{bmatrix} 0.4 & 0.4 \\ 0.6 & 0.6 \end{bmatrix} \tag{G9}$$

The reader may verify that the right-hand side of this equation reproduces the original matrix.

The significance of the spectral decomposition is that the matrices $\mathbf{T}_k$ have the special properties

$$\begin{aligned} \mathbf{T}_k\mathbf{T}_k &= \mathbf{T}_k \\ \mathbf{T}_j\mathbf{T}_k &= \mathbf{0} \qquad j \neq k \end{aligned} \tag{G10}$$

(To prove this, note that $\mathbf{T}_j\mathbf{T}_k = (\mathbf{e}_j \times \boldsymbol{\varepsilon}_j)(\mathbf{e}_k \times \boldsymbol{\varepsilon}_k)$ and that $\boldsymbol{\varepsilon}_j\mathbf{e}_k = 1$ if $j = k$ and 0 otherwise. Verify these properties for the matrices in Equation G8.)

Thus the pth power of $\mathbf{A}$ is

$$\mathbf{A}^p = \left(\sum_k \lambda_k \mathbf{T}_k\right)^p = \sum_k \lambda_k^p \mathbf{T}_k \tag{G11}$$

This is useful for determining the behavior of large matrix powers. If λ_1 dominates the other eigenvalues in the sense that it is larger than any of them in absolute value, then Equation G11 will be dominated by this eigenvalue when p is large so that

$$\mathbf{A}^p \cong \lambda_1^p \mathbf{T}_1 \tag{G12}$$

For example, the 50th power of $\mathbf{A}_3$ can be approximated as

$$\mathbf{A}_3^{50} \cong 4^{50}\begin{bmatrix} 0.4 & 0.4 \\ 0.6 & 0.6 \end{bmatrix} \tag{G13}$$

This idea can be extended to other functions of matrices. The exponential of a matrix, exp($\mathbf{A}$), can be defined by the power series appropriate for the exponential of a real variable (see Appendix A):

$$\exp(\mathbf{A}) \equiv \mathbf{I} + \mathbf{A} + \frac{\mathbf{A}^2}{2!} + \frac{\mathbf{A}^3}{3!} + \cdots \tag{G14}$$

It can be shown from this definition that the matrix exponential shares the properties of the ordinary exponential function. In particular

$$\exp(\mathbf{A} + \mathbf{B}) = \exp(\mathbf{A})\exp(\mathbf{B}) \tag{G15}$$

We shall often be interested in exp($\mathbf{A}t$) where t is time. The derivative of this matrix is

$$\frac{d}{dt}\exp(\mathbf{A}t) = \frac{d}{dt}\left(\mathbf{I} + \mathbf{A}t + \frac{\mathbf{A}^2t^2}{2!} + \frac{\mathbf{A}^3t^3}{3!} + \cdots\right)$$
$$= \mathbf{A} + \mathbf{A}^2t + \frac{\mathbf{A}^3t^2}{2!} + \cdots \qquad \text{(G16)}$$
$$= \mathbf{A}\left(\mathbf{I} + \mathbf{A}t + \frac{\mathbf{A}^2t^2}{2!} + \cdots\right)$$
$$= \mathbf{A}\exp(\mathbf{A}t)$$

which is the analogue of the scalar derivative

$$\frac{d}{dt}\exp(at) = a\exp(at) \qquad \text{(G17)}$$

(Differentiating a matrix with respect to a scalar variable means differentiating each of the terms of the matrix.)

Using Equation G11 we can find a closed expression for the matrix exponential:

$$\exp(\mathbf{A}) = \sum_k (1 + \lambda_k + \frac{\lambda_k^2}{2!} + \frac{\lambda_k^3}{3!} + \cdots)\mathbf{T}_k = \sum_k \exp(\lambda_k)\mathbf{T}_k \qquad \text{(G18)}$$

For example

$$\exp(\mathbf{A}_3) = \exp(-1)\mathbf{T}_1 + \exp(4)\mathbf{T}_2 = \begin{bmatrix} 22.06 & 21.69 \\ 32.54 & 32.91 \end{bmatrix} \qquad \text{(G19)}$$

We shall often be interested in $\exp(\mathbf{A}t)$:

$$\exp(\mathbf{A}t) = \sum_k \exp(\lambda_k t)\mathbf{T}_k \qquad \text{(G20)}$$

H. LINEAR DIFFERENTIAL EQUATIONS AND LOCAL STABILITY ANALYSIS IN CONTINUOUS TIME

Linear differential equations

The linear differential equation

$$dx/dt = ax \qquad \text{(H1)}$$

expresses the model that the rate of change of x is a constant multiple a of its current value. Starting with the value $x(0)$ at time $t = 0$, the solution is

$$x(t) = x(0)\exp(at) \tag{H2}$$

It is straightforward to verify that this satisfies both the differential equation and the initial condition.

Consider now the set of n linear differential equations in n variables:

$$\begin{aligned} dx_1/dt &= a_{11}x_1 + a_{12}x_2 + \cdots + a_{1n}x_n \\ dx_2/dt &= a_{21}x_1 + a_{22}x_2 + \cdots + a_{2n}x_n \\ &\vdots \\ dx_n/dt &= a_{n1}x_1 + a_{n2}x_2 + \cdots + a_{nn}x_n \end{aligned} \tag{H3}$$

which can be expressed in matrix notation as

$$d\mathbf{x}/dt = \mathbf{Ax} \tag{H4}$$

Starting with the value $\mathbf{x}(0)$ at time $t = 0$, the solution is

$$\mathbf{x}(t) = \exp(\mathbf{A}t)\mathbf{x}(0) \tag{H5}$$

which satisfies the differential equations (compare Equation G16) and the initial conditions. From Equation G20 we can write

$$\mathbf{x}(t) = \sum_k \exp(\lambda_k t)\mathbf{T}_k\mathbf{x}(0) \tag{H6}$$

We now define

$$c_k = \boldsymbol{\varepsilon}_k\mathbf{x}(0) \tag{H7}$$

the inner product of the kth standardized left eigenvector and the vector of initial values, and we note that

$$\mathbf{T}_k\mathbf{x}(0) = \mathbf{e}_k \times \boldsymbol{\varepsilon}_k\mathbf{x}(0) = c_k\mathbf{e}_k \tag{H8}$$

Hence,

$$\mathbf{x}(t) = \sum_k c_k \exp(\lambda_k t)\mathbf{e}_k \tag{H9}$$

The vector of constants $\mathbf{c} = \{c_1, c_2, \ldots, c_n\}$ can also be expressed in terms of the right eigenvectors since

$$\mathbf{c} = \mathbf{Lx}(0) = \mathbf{x}(0)\mathbf{L}^{\mathrm{T}} = \mathbf{x}(0)\mathbf{R}^{-1}$$

from Equation F14.

I shall now consider two examples. First consider the equations

$$\begin{aligned} dx_1/dt &= x_1 + 2x_2 \\ dx_2/dt &= 3x_1 + 2x_2 \end{aligned} \tag{H10}$$

with initial values $x_1(0) = x_2(0) = 1$. The matrix **A** is $\mathbf{A}_3$, defined in Equation F1, with eigenvalues –1 and 4, right eigenvectors {–1, 1} and {0.67, 1}, and with standardized left eigenvectors {–0.6, 0.4} and {0.6, 0.6}. Hence,

$$\begin{aligned} c_1 &= \{-0.6, 0.4\}\{1, 1\} = -0.2 \\ c_2 &= \{0.6, 0.6\}\{1, 1\} = 1.2 \\ \mathbf{x}(t) &= -0.2\exp(-t)\{-1, 1\} + 1.2\exp(4t)\{0.67, 1\} \\ &= \{0.2\exp(-t) + 0.8\exp 4t, -0.2\exp(-t) + 1.2\exp 4t\} \end{aligned} \tag{H11}$$

In other words,

$$\begin{aligned} x_1(t) &= 0.2\exp(-t) + 0.8\exp(4t) \\ x_2(t) &= -0.2\exp(-t) + 1.2\exp(4t) \end{aligned} \tag{H12}$$

The behavior of the system after a long time is determined by the term in $\exp(4t)$ and may be represented as

$$\begin{aligned} x_1(t) &\cong 0.8\exp(4t) \\ x_2(t) &\cong 1.2\exp(4t) \end{aligned} \tag{H13}$$

Both x_1 and x_2 increase exponentially at a rate determined by the largest eigenvalue, $\lambda = 4$, in a direction determined by the corresponding right eigenvector, {0.667, 1}. The constant c multiplying both terms (in this case $c_2 = 1.2$) is a weighted sum of the vector of initial values, the weights being the terms in the corresponding standardized left eigenvector. Thus the elements of this eigenvector reflect the relative importance of the different initial values in determining the ultimate values. In the present case this eigenvector is {0.6, 0.6), which means that the two initial values are equally important in determining the ultimate values.

The second example will illustrate what happens with complex eigenvalues. Consider the equations

$$\begin{aligned} dx_1/dt &= x_1 + 2x_2 \\ dx_2/dt &= -2x_1 + x_2 \end{aligned} \tag{H14}$$

with initial values $x_1(0) = x_2(0) = 1$. The matrix **A** is defined in Equation F7, with eigenvalues $1 - 2i$ and $1 + 2i$, right eigenvectors $\{i, 1\}$, and $\{-i, 1\}$, and standardized left eigenvectors $0.5\{-i, 1\}$ and $0.5\{i, 1\}$. (The calculation of the left eigenvectors is left as an exercise.) Hence

$$\begin{aligned} c_1 &= 0.5\{-i,1\}\{1,1\} = 0.5(1-i) \\ c_2 &= 0.5\{i,1\}\{1,1\} = 0.5(1+i) \\ \mathbf{x}(t) &= 0.5(1-i)\exp(t-2ti)\{i,1\} + 0.5(1+i)\exp(t+2ti)\{-i,1\} \end{aligned} \tag{H15a}$$

so that

$$\begin{aligned} x_1(t) &= 0.5(1+i)\exp(t)\exp(-2ti) + 0.5(1-i)\exp(t)\exp(2ti) \\ x_2(t) &= 0.5(1-i)\exp(t)\exp(-2ti) + 0.5(1+i)\exp(t)\exp(2ti) \end{aligned} \tag{H15b}$$

We now use the relationships

$$\begin{aligned} \exp(-2ti) &= \cos(-2t) + i\sin(-2t) = \cos(2t) - i\sin(2t) \\ \exp(2ti) &= \cos(2t) + i\sin(2t) \end{aligned} \tag{H16}$$

to obtain the final result

$$\begin{aligned} x_1(t) &= \exp(t)(\cos 2t + \sin 2t) \\ x_2(t) &= \exp(t)(\cos 2t - \sin 2t) \end{aligned} \tag{H17}$$

Note that imaginary numbers appear in the intermediate calculations but not in the final answer.

The following general conclusions can be drawn from these results. A real eigenvalue, say $\lambda = a$, gives rise to an exponential term $\exp(at)$ in the solution, which will increase without limit or converge to zero depending on whether a is positive or negative. A pair of complex conjugate eigenvalues, say $\lambda = a \pm ib$, gives rise to an oscillatory term $\exp(at)$ $[k_1\cos(bt) + k_2\sin(bt)]$, where k_1 and k_2 depend on the initial conditions. This term represents a sinusoidal oscillation with period $2\pi/b$ and with amplitude proportional to $\exp(at)$ which will increase without limit or converge to zero depending on whether a is positive or negative. After a sufficiently long time, as $t\to\infty$, the solution will be dominated by the eigenvalue with the largest real part. If this eigenvalue is real, there will be exponential increase when it is positive and exponential convergence to zero when it is negative. If the dominant eigenvalue is a pair of complex conjugates, there will eventually be oscillations whose amplitude will increase exponentially or converge to zero depending on whether its real part is positive or negative. When the dominant eigenvalue is real, the system will eventually diverge or converge exponentially in a direction determined by the corresponding right eigenvector, and the terms in the corresponding left eigenvector represent the weights to be attached to the different initial values in determining the ultimate values. This interpretation of the dominant eigenvectors only holds when the dominant eigenvalue is real.

Local stability analysis of nonlinear differential equations

An important application of these results is to the local stability analysis of a system of differential equations. Consider the system of k equations in k variables

$$\mathrm{d}n_i/\mathrm{d}t = f_i(n_1, n_2, \dots, n_k) \tag{H18}$$

for $i = 1, \dots, k$. These variables may be the population sizes of different species, or of age classes within the same species, or they may be the frequencies of different alleles at a genetic locus. Consider an equilibrium value at which all the variables are stationary at $\hat{\mathbf{n}} = \{\hat{n}_1, \hat{n}_2, \dots, \hat{n}_k\}$, so that $f_i\{\hat{n}_1, \hat{n}_2, \dots, \hat{n}_k\} = 0$ for all i. Expanding the function f_i in a Taylor series about the equilibrium value we find

$$f_i(\mathbf{n}) = f_i(\hat{\mathbf{n}}) + \sum_j (n_j - \hat{n}_j)\partial f_i/\partial n_j = \sum_j (n_j - \hat{n}_j)\partial f_i/\partial n_j \tag{H19}$$

and similarly for the other functions, where the derivatives are to be evaluated at $\hat{\mathrm{n}}$. Define a vector of small perturbations from the equilibrium

$$\mathbf{p} = \{p_1, p_2, \dots, p_k\} = \mathbf{n} - \hat{\mathbf{n}} \tag{H20}$$

Since

$$\mathrm{d}n_i/\mathrm{d}t = \mathrm{d}(\hat{n}_i + p_i)/\mathrm{d}t = \mathrm{d}p_i/\mathrm{d}t \tag{H21}$$

the equations for the perturbations, as long as they remain small enough that nonlinear terms can be ignored, are

$$\mathrm{d}p_i/\mathrm{d}t = \sum_j p_j \partial f_i/\partial n_j \tag{H22}$$

for $i = 1, \dots, k$. These equations can be written in matrix terminology as

$$\mathrm{d}\mathbf{p}/\mathrm{d}t = \mathbf{Jp} \tag{H23}$$

where the ijth element of the Jacobian matrix $\mathbf{J}$ is $\partial f_i/\partial n_j$ evaluated at $\mathbf{n} = \hat{\mathbf{n}}$.

The long-term behavior of the system depends on the real parts of the eigenvalues of the matrix **J**. The dominant eigenvalue, which dominates the ultimate behavior of the system, is the one with the largest real part. If the dominant eigenvalue is real, the system will ultimately converge to or diverge from zero exponentially depending on whether this eigenvalue is negative or positive. If the dominant eigenvalue is complex (in which case it will be one of a pair of complex conjugates), the system will ultimately oscillate with an amplitude that either decreases or increases exponentially depending on whether the real part of this eigenvalue is negative or positive. In general the system will ultimately tend toward zero if all the eigenvalues have negative real parts; otherwise it will increase without limit. These conclusions are of course only valid as long as the perturbations remain small. If the system is locally unstable, further investigation is required to determine whether the perturbations will in fact increase without limit or whether there will be a stable limit cycle or, possibly, chaotic behavior.

With two variables there is a simple criterion for the local stability of the system. Write the Jacobian matrix as

$$\mathbf{J} = \begin{bmatrix} a & b \\ c & d \end{bmatrix} \tag{H24}$$

and define

$$\begin{aligned} T &= \text{trace}(\mathbf{J}) = a + d \\ D &= \det(\mathbf{J}) = ad - bc \end{aligned} \tag{H25}$$

The necessary and sufficient condition that both eigenvalues are negative (if real) or have negative real parts (if complex) is that

$$T < 0 \text{ and } D > 0 \tag{H26}$$

Proof: The eigenvalues can be expressed as

$$\frac{1}{2}\left(T \pm \sqrt{T^2 - 4D}\right) \tag{H27}$$

If $4D > T^2$, they are complex conjugates with real part T. If $4D < T^2$, they are real, and the larger of them is obtained by taking the "+" sign in Equation H27; for this quantity to be negative requires that Equation H26 is true.

J. LINEAR DIFFERENCE EQUATIONS AND LOCAL STABILITY ANALYSIS IN DISCRETE TIME

Linear difference equations

The linear difference equation

$$x(t+1) = ax(t) \tag{J1}$$

expresses the model that the value of x next time is a constant multiple a of its value this time. Starting with the value $x(0)$ at time zero, we find

$$\begin{aligned} x(1) &= ax(0) \\ x(2) &= ax(1) = a^2x(0) \end{aligned} \tag{J2}$$

and so on. The solution for any time t is clearly

$$x(t) = a^t x(0) \tag{J3}$$

Consider now the set of n linear difference equations in n variables:

$$\begin{aligned} x_1(t+1) &= a_{11}x_1(t) + a_{12}x_2(t) + \cdots + a_{1n}x_n(t) \\ x_2(t+1) &= a_{21}x_1(t) + a_{22}x_2(t) + \cdots + a_{2n}x_n(t) \\ &\vdots \\ x_n(t+1) &= a_{n1}x_1(t) + a_{n2}x_2(t) + \cdots + a_{nn}x_n(t) \end{aligned} \tag{J4}$$

which can be expressed in matrix notation as

$$\mathbf{x}(t+1) = \mathbf{A}\mathbf{x}(t) \tag{J5}$$

where $\mathbf{x}(t)$ is the vector $\{x_1(t), x_2(t), \ldots, x_n(t)\}$ and $\mathbf{A}$ is the matrix of coefficients a_{ij}. We want to solve this set of equations starting from the initial value $\mathbf{x}(0)$ at time zero. By the argument used in the univariate case, the solution is found to be

$$\mathbf{x}(t) = \mathbf{A}^t\mathbf{x}(0) \tag{J6}$$

From Equations G11 and H8 we can write

$$\mathbf{x}(t) = \sum_k \lambda_k^t \mathbf{T}_k \mathbf{x}(0) = \sum_k c_k \lambda_k^t \mathbf{e}_k \tag{J7}$$

where c_k is defined in Equation H7. (Compare Equation J7 with the result for a set of linear differential equations in Equation H9.)

I shall now consider two examples analogous to the differential-equation examples in the previous section. First consider the equations

$$\begin{aligned} x_1(t+1) &= x_1(t) + 2x_2(t) \\ x_2(t+1) &= 3x_1(t) + 2x_2(t) \end{aligned} \tag{J8}$$

with initial values $x_1(0) = x_2(0) = 1$. (Compare Equation H10.) The matrix **A** is the same as in that example, so that

$$\begin{aligned} c_1 &= -0.2 & c_2 &= 1.2 \\ \lambda_1 &= -1 & \lambda_2 &= 4 \\ \mathbf{e}_1 &= \{-1, 1\} & \mathbf{e}_2 &= \{0.667, 1\} \end{aligned} \tag{J9}$$

Hence,

$$\begin{aligned} x_1(t) &= 0.2 \times (-1)^t + 0.8 \times 4^t \\ x_2(t) &= -0.2 \times (-1)^t + 1.2 \times 4^t \end{aligned} \tag{J10}$$

(Compare Equation H12.) After a long time, the term in 4^t will ultimately dominate the term in $(-1)^t$, and the system will lie in the direction of the dominant right eigenvector, increasing fourfold each year.

Consider next the equations

$$\begin{aligned} x_1(t+1) &= x_1(t) + 2x_2(t) \\ x_2(t+1) &= -2x_1(t) + x_2(t) \end{aligned} \tag{J11}$$

with initial values $x_1(0) = x_2(0) = 1$. (Compare Equation H14.) As in that example

$$\begin{aligned} c_1 &= 0.5(1-i) & c_2 &= 0.5(1+i) \\ \lambda_1 &= 1-2i & \lambda_2 &= 1+2i \\ \mathbf{e}_1 &= \{i, 1\} & \mathbf{e}_2 &= \{-i, 1\} \end{aligned} \tag{J12}$$

The solution in Equation J7 involves the tth power of the eigenvalues, which is easier to deal calculate if they are expressed in polar coordinates, r and θ. In the present case $r = \sqrt{5}$, $\theta = \arctan(\pm 2) = \pm 1.107$ radians, so that

$$\begin{aligned} \lambda_1^t &= \left(\sqrt{5}\right)^t [\cos(1.107t) - i\sin(1.107t)] \\ \lambda_2^t &= \left(\sqrt{5}\right)^t [\cos(1.107t) + i\sin(1.107t)] \end{aligned} \tag{J13}$$

Hence,

$$\begin{aligned} x_1(t) &= \left(\sqrt{5}\right)^t [\cos(1.107t) + \sin(1.107t)] \\ x_2(t) &= \left(\sqrt{5}\right)^t [\cos(1.107t) - \sin(1.107t)] \end{aligned} \tag{J14}$$

The following general conclusions can be drawn from these results. A real eigenvalue, say $\lambda = a$, gives rise to a geometric term a^t in the solution, which will increase without limit or converge to zero depending on whether $|a|$ is positive or negative; there will be a change of sign in alternate years if $a < 0$. A pair of complex conjugate eigenvalues, say $\lambda = a \pm ib$, is best expressed in polar coordinates

$$\begin{aligned} r &= |\lambda| = \sqrt{a^2 + b^2} \\ \theta &= \arctan b/a \end{aligned} \tag{J15}$$

and gives rise to an oscillatory term $r^t[k_1 \cos(\theta t) + k_2 \sin(\theta t)]$, where k_1 and k_2 depend on the initial conditions. This term represents a sinusoidal oscillation with period $2\pi/\theta$ and with amplitude proportional to r^t which will increase without limit or converge to zero depending on whether r is positive or negative. After a sufficiently long time, as $t \to \infty$, the solution will be dominated by the eigenvalue with the largest absolute value. If this eigenvalue is real, the behavior of the system is as follows:

$\lambda > 1$	geometric increase
$0 < \lambda < 1$	geometric decrease
$-1 < \lambda < 0$	geometric decrease, with alternation of sign
$\lambda < -1$	geometric increase, with alternation of sign

If the dominant eigenvalue is a pair of complex conjugates, there will be oscillations whose amplitude will increase geometrically or converge to zero depending on whether its absolute value is greater or less than 1.

Local stability analysis of nonlinear difference equations

An important application of these results is to the local stability analysis of a system of difference equations. Consider the system of k equations in k variables

$$n_i(t+1) = F_i[n_1(t), n_2(t), \ldots, n_k(t)] \tag{J16}$$

for $i = 1, \ldots, k$. As in the continuous-time case considered in the previous section, these variables may be the population sizes of different species, or of age classes within the same species, and so on. Consider an equilibrium value at which all the variables are stationary at $\hat{\mathbf{n}} = \{\hat{n}_1, \hat{n}_2, \ldots, \hat{n}_k\}$; this implies that $F_i(\hat{\mathbf{n}}) = \hat{\mathbf{n}}$ for all i. Expanding the function F_i in a Taylor series about the equilibrium value we find

$$F_i(\mathbf{n}) = F_i(\hat{\mathbf{n}}) + \sum_j (n_j - \hat{n}_j)\partial F_i/\partial n_j = \hat{\mathbf{n}} + \sum_j p_j \partial F_i/\partial n_j \tag{J17}$$

where the derivatives are to be evaluated at $\hat{\mathbf{n}}$ and $p_j = (n_j - \hat{n}_j)$ is a small perturbation in the jth variable. Hence,

$$n_i(t+1) = \hat{n}_i + p_i(t+1) \cong \hat{n}_i + \sum_j p_j(t)\partial F_i/\partial n_j \tag{J18a}$$

so that

$$p_i(t+1) \cong \sum_j p_j(t)\partial F_i/\partial n_j \tag{J18b}$$

with similar equations for the other perturbations. The set of approximate difference equations for the perturbations can be written in matrix terminology as

$$\mathbf{p}(t+1) = \mathbf{J}\mathbf{p}(t) \tag{J19}$$

where the ijth element of the Jacobian matrix $\mathbf{J}$ is $\partial f_i/\partial n_j$ evaluated at $\mathbf{n} = \hat{\mathbf{n}}$.

The long-term behavior of the system depends on the eigenvalues of the matrix $\mathbf{J}$ and can be determined as follows. (We suppose that these eigenvalues are distinct.) Rank the eigenvalues by their *absolute value*. There are two common situations. (i) There may be a real eigenvalue λ greater in absolute value than any of the others. The system will ultimately increase or decrease exponentially depending on whether $|\lambda| > 1$ or $|\lambda| < 1$; it will alternate in sign or not depending on whether λ is negative or positive. (ii) There may be a pair of complex conjugate eigenvalues greater in absolute value than any of the others. The system will ultimately oscillate with increasing or decreasing amplitude depending on whether $|\lambda| > 1$ or $|\lambda| < 1$, with period $2\pi/\theta$ where θ is the argument of λ. This analysis determines the local stability of the system, as long as the perturbations remain small. The equilibrium is locally stable if all the eigenvalues are less than 1 in absolute value, and it is unstable otherwise.

With two variables there is a simple criterion for the local stability of the system. The necessary and sufficient condition that both eigenvalues are less than 1 in absolute value is that

$$2 > 1 + D > |T| \tag{J20}$$

where D is the determinant and T is the trace of the matrix $\mathbf{J}$ defined in Equation H25.

Proof: The eigenvalues are

$$\frac{1}{2}\left(T \pm \sqrt{T^2 - 4D}\right) \tag{J21}$$

Suppose first that

$$4D > T^2 \tag{J22}$$

so that the roots are complex. Then

$$|\lambda|^2 = \tfrac{1}{4}(T^2 + 4D - T^2) = D \tag{J23}$$

so that the criterion for stability is that $D < 1$. Thus Equation J20 implies stability. Also Equation J22 implies that

$$D + 1 > \tfrac{1}{4}T^2 + 1 = [(\tfrac{1}{2}|T| - 1)^2 + |T|] > |T| \tag{J24}$$

so that stability implies Equation J20.

Suppose now that

$$4D < T^2 \tag{J25}$$

so that the roots are real. The dominant eigenvalue has absolute value

$$|\lambda| = \frac{1}{2}\left(|T| + \sqrt{T^2 - 4D}\right) \tag{J26}$$

The fact that $|\lambda| < 1$ implies that

$$|T| < 2 \text{ and } |T| < D + 1 \tag{J27}$$

which together with Equation J25 imply Equation J20. Furthermore, Equation J20 obviously implies Equation J27, which in turn implies that $|\lambda| < 1$.

This completes the proof that Equation J20 is a necessary and sufficient condition for stability whether the eigenvalues are real or complex.

K. THE POISSON DISTRIBUTION

The Poisson distribution was first derived by S.D. Poisson in 1837 and has many applications in biology. It is a discrete probability distribution with the formula

$$P(x) = e^{-\mu}\mu^x/x! \qquad x = 0, 1, 2, \ldots \tag{K1}$$

$P(x)$ represents the probability that some random event occurs exactly x times. It depends on the parameter μ that is the mean of the distribution, representing the average number of occurences of the event; a characteristic of the Poisson distribution is that its mean is also its variance.

The Poisson distribution is a limiting case of the binomial distribution,. which gives the probability of getting x "successes" and $n - x$ "failures" in n trials if each trial results in one of two outcomes with a constant probability p of success at each trial, which is independent of the outcomes of previous trials. For example, the number of boys in families of fixed size is to a good approximation binomial, because the probability of a male birth is very nearly the same for all families and independent of birth order. The formula for this distribution is

$$P(x) = \frac{n!}{x!(n-x)!} p^x (1-p)^{n-x} \qquad x = 0,1,2,\ldots,n \tag{K2}$$

The reason is that the probability of getting x successes and $n - x$ failures in a particular order is $p^x(1 - p)^{n-x}$ and there are $n!/x!(n - x)!$ ways in which this can happen.

The Poisson distribution is the limiting case of this distribution when the probability p becomes very small but the number of trials n becomes very large. Write $\mu = np$ for the mean of the binomial distribution (this is clearly its mean, since p is the probability of success in each trial, whence np is the expected number of successes in n trials) and let n increase and p decrease with μ held constant. Then with μ constant and $n \to \infty$,

$$\begin{aligned} &\frac{n!}{x!(n-x)!} \to \frac{n^x}{x!} \\ &p^x = \frac{\mu^x}{x!} \\ &(1-p)^{n-x} = (1-\mu/n)^{n-x} \cong (1-\mu/n)^n \to e^{-\mu} \end{aligned} \tag{K3}$$

Hence,

$$P(x) \to e^{-\mu} \mu^x / x! \tag{K4}$$

which is the Poisson distribution with parameter μ.

A classic example of the Poisson distribution is the numbers of deaths from horse kicks in the Prussian army between 1875 and 1894. Data on 14 army corps were collected by von Bortkiewicz (1898) for each of these years, giving 280 observations in all on the number of deaths from this cause in a corps in a year. The data are shown in Table K1, together with the Poisson distribution with the same mean, $\bar{x} = 0.7$. There is good agreement between the observed and predicted frequencies. One would expect these data to follow a Poisson distribution since there is a very large number of men at risk, each of whom has a very small chance of being killed by a horse in a year; otherwise the conditions for the binomial distribution hold.

The Poisson distribution can be derived by an independent argument

Table K1. The numbers of deaths from horse kicks in the Prussian army between 1875 and 1894.

	Number of deaths (x)						
	0	1	2	3	4	5+	Total
Observed frequency:	144	91	32	11	2	0	280
Poisson probability, $P(x)$:	0.497	0.348	0.122	0.029	0.005	0.000	1
Expected frequency, $280P(x)$:	139	97	34	8	1	0	280

as the distribution of the number of random events occurring in a fixed time. Suppose that in a short time interval dt there is a chance λdt that an event occurs, which does not depend on the number of events that have already occurred. (This is what we mean by saying that events occur at random.) Write $P(n, t)$ for the probability that exactly n events have occurred by time t, starting with zero events at time zero. Now n events can occur by time t + dt in two ways: (1) n events occur by time t and none between t and t + dt, (2) $n - 1$ events occur by time t and another event occurs between t and t + dt. (The second possibility only applies when $n \geq 1$.) Thus

$$P(n,t+\mathrm{d}t) = P(n,t)(1-\lambda\mathrm{d}t) + P(n-1,t)\lambda\mathrm{d}t \tag{K5}$$

whence

$$\begin{aligned} \mathrm{d}P(n,t)/\mathrm{d}t &= \lambda[P(n-1,t) - P(n,t)] \qquad n \geq 1 \\ \mathrm{d}P(0,t)/\mathrm{d}t &= -\lambda P(0,t) \end{aligned} \tag{K6}$$

with initial conditions

$$\begin{aligned} P(n,0) &= 0 \qquad n \geq 1 \\ P(0,0) &= 1 \end{aligned} \tag{K7}$$

It is straightforward to verify that these differential equations are satisfied by the Poisson distribution with parameter $\mu = \lambda t$.

The properties of the Poisson distribution are most easily investigated by calculating its probability generating function (pgf), $G(s)$. The pgf of a discrete random variable X with probability function

$$\Pr(X = x) = P(x) \tag{K8}$$

is defined as

$$G(s) = E(s^X) = \sum_x s^x P(x) \tag{K9}$$

For the Poisson distribution,

$$G(s) = \sum_{x=0}^{\infty} e^{-\mu} (s\mu)^x / x! = e^{-\mu} e^{s\mu} = e^{(s-1)\mu} \tag{K10}$$

If we differentiate the pgf r times with respect to s and then set $s = 1$, we obtain

$$G^{(r)}(s)\Big|_{s=1} = E[X(X-1) \cdots (X-r+1)] \tag{K11}$$

These are called the factorial moments of the distribution, and they can be used to calculate the ordinary moments. For the Poisson distribution,

$$G^{(r)}(s)\Big|_{s=1} = \mu^r \tag{K12}$$

The first factorial moment is the mean of the distribution, which is in this case μ. The second factorial moment can be used to find the variance

$$\begin{aligned} \mathrm{Var}(X) &= E(X^2) - E^2(X) = E[X(X-1)] + E(X) - E^2(X) \\ &= \mu^2 + \mu - \mu^2 = \mu \end{aligned} \tag{K13}$$

Thus for the Poisson distribution, the variance is equal to the mean.

If we differentiate the pgf r times with respect to s and then set $s = 0$, we obtain

$$G^{(r)}(s)\Big|_{s=0} = P(r)/r! \tag{K14}$$

This enables the probability distribution to be calculated from the pgf, which is useful if the pgf has been obtained indirectly without direct knowledge of the probability distribution.

Another useful property is that the pgf of the sum of two independent random variables is equal to the product of their pgf's. Suppose X and Y are independent random variables with pgf's $G_X(s)$ and $G_Y(s)$; the pgf of their sum, $X + Y$, is

$$G_{X+Y}(s) = E(s^{X+Y}) = E(s^X s^Y)$$
$$= E(s^X)E(s^Y) \text{ (under independence)} \tag{K15}$$
$$= G_X(s)G_Y(s)$$

If X and Y are independent Poisson variates with means μ_1 and μ_2, respectively, the pgf of their sum is

$$e^{(s-1)\mu_1} e^{(s-1)\mu_2} = e^{(s-1)(\mu_1+\mu_2)} \tag{K16}$$

which is the pgf of a Poisson variate with mean $\mu_1 + \mu_2$. Hence the sum of two independent Poisson variates is itself a Poisson variate, with mean equal to the sum of the means.

The sum of n identically and independently distributed random variables X_i with pgf $G_X(s)$, say $Y = X_1 + X_2 + \cdots + X_n$, has pgf $G_Y(s) = G_X^n(s)$. Suppose however that that the number of X_i's is not fixed but is itself a random variable N with probability distribution $P_N(n)$ and pgf $G_N(s)$. The pgf of $Y = X_1 + X_2 + \cdots + X_N$ is

$$\begin{aligned} G_Y(s) &= P_N(0) + P_N(1)G_X(s) + P_N(2)G_X^2(s) + \cdots \\ &= G_N(G_X(s)) \end{aligned} \tag{K17}$$

For example, suppose an individual has a Poisson number of offspring with mean l and then dies, and that each of these offspring has a Poisson number of offspring with the same mean. Then the number of grandoffspring of the original individual has pgf $G(G(s))$, where $G(s)$ is the pgf of the Poisson distribution

$$\begin{aligned} G(s) &= \exp(s-1) \\ G[G(s)] &= \exp[\exp(s-1)-1] \end{aligned} \tag{K18}$$

This can be used to find the probability distribution of the number of grandoffspring from Equation K14, shown in Table K2.

This process is called a Poisson branching process. If $G(s)$ is the Poisson pgf, the pgf of the descendants of a single individual after t generations is

$$G_t(s) = G(G(\cdots G(s) \cdots)) \tag{K19}$$

where there are t G's on the right-hand side; this can be calculated from the recurrence relationship

$$G_t(s) = G[G_{t-1}(s)] \tag{K20}$$

Table K2. Probability distribution of the number of grandoffspring of a single individual if the offspring distribution in each generation is a Poisson with mean 1.

x	$P(x)$, offspring	$P(x)$, grandoffspring
0	0.368	0.531
1	0.368	0.196
2	0.184	0.134
3	0.061	0.073
4	0.015	0.036
5	0.003	0.017
6	0.001	0.008
7		0.003
8		0.001
9		0.001

In particular, if q_t is the probability that the line has gone extinct by t generations, which is the probability that there are no descendants, then $q_t = G_t(0)$, which satisfies the recurrence

$$q_t = G(q_{t-1}) \tag{K21}$$

If $G(s)$ is the pgf of a Poisson distribution with mean 1, this becomes

$$q_t = \exp(q_{t-1} - 1) \tag{K22}$$

with the initial condition that $q_0 = 0$. This equation can be solved recursively with the results shown in Table K3.

Table K3. Probability of extinction of a line by t generations, starting with a single individual in generation 0.

t:	1	2	5	10	50	100	250
q_t:	0.368	0.531	0.732	0.842	0.962	0.981	0.992

Answers to Exercises

CHAPTER 1

1. Young fledged are {3.00, 3.50, 3.375, 4.25, 4.125, 5.00, 4.50} for clutch sizes {4, 5, ... , 10}. Young fledged are {3.75, 4.25, 3.75} for genotypes {*aa*, *Aa*, *AA*}.

CHAPTER 2

2. The best-fitting line is $\log n_t = 1.25 + 1.04t$.
5. (b) $K = 200$, $r = 0.68$.
6. $K = 540$, $r = 0.73$ (taking the same cutoff point).
7. $f'(n) = r(1 - 2n/K)$. Hence $f'(0) = r, f'(n) = -r$.
8. $f'(n) = b/(1 + n)^2 - d$. Hence $f'(0) = b - d, f'(\text{k}) = -d(b - d)/b$ at $n = (b - d)/d$.
9. $F(0) = 0, F(K) = K; F'(0) = e^r, F'(K) = 1 - r$.
11. For Great Tits, $x_{t+1} = 2.42 + 0.31x_t$, corrected slope is 0.35 ± 0.14 (see text); for Blue Tits, $x_{t+1} = 1.17 + 0.68x_t$, corrected slope is 0.75 ± 0.11.

CHAPTER 3

2. The Jacobian matrices at the four equilibria are:

(a) $$\begin{bmatrix} r_1 & 0 \\ 0 & r_2 \end{bmatrix}$$

(b) $$\begin{bmatrix} -r_1 & -r_1\gamma_{12} \\ 0 & \frac{r_2}{K_2}(K_2 - \gamma_{21}K_1) \end{bmatrix}$$

(c) $$\begin{bmatrix} \frac{r_1}{K_1}(K_1 - \gamma_{12}K_2) & 0 \\ -r_2\gamma_{21} & -r_2 \end{bmatrix}$$

(d) $$\begin{bmatrix} \frac{-r_1(K_1 - \gamma_{12}K_2)}{K_1(1-\gamma_{12}\gamma_{21})} & \frac{-r_1\gamma_{12}(K_1 - \gamma_{12}K_2)}{K_1(1-\gamma_{12}\gamma_{21})} \\ \frac{-r_2\gamma_{21}(K_2 - \gamma_{21}K_1)}{K_2(1-\gamma_{12}\gamma_{21})} & \frac{-r_2(K_2 - \gamma_{21}K_1)}{K_2(1-\gamma_{12}\gamma_{21})} \end{bmatrix}$$

3. (a) $T = r_1 + r_2$; never stable.

   (b) $D = -r_1r_2(1 - \gamma_{21}K_1/K_2)$ is positive if $K_2 < \gamma_{21}K_1$; in this case $T = -r_1 + r_2(1 - \gamma_{21}K_1/K_2) < 0$ so that the equilibrium is stable.

   (c) Reversing the subscripts, this is stable if $K_1 < \gamma_{12}K_2$.

   (d) $D = \dfrac{r_1r_2(K_1 - \gamma_{12}K_2)(K_2 - \gamma_{21}K_1)}{K_1K_2(1-\gamma_{12}\gamma_{21})}$

   For a feasible equilibrium, with $(K_1 - \gamma_{12}K_2)$ and $(K_2 - \gamma_{21}K_1)$ having the same sign, D is positive when both these terms are positive, which implies that the denominator is also positive. In this case

   $$T = \frac{-r_1K_2(K_1 - \gamma_{12}K_2) - r_2K_1(K_2 - \gamma_{21}K_1)}{K_1K_2(1-\gamma_{12}\gamma_{21})} < 0$$

   so that the equilibrium is stable.

4. The eigenvalues are $\pm i\sqrt{r_1r_2}$. There is neutral stability. We expect oscillations with period $2\pi\sqrt{r_1r_2}$.

5. (a) $\hat{n}_1 = 200, \hat{n}_2 = 200(1-200/K_1)$. The eigenvalues are real and negative when $200 < K_1 < 257$, complex with negative real part when $257 < K_1 < 600$, and complex with positive real part when $K_1 > 600$.

6. For the first equilibrium with $\hat{x} = N$, $\hat{y} = 0$, the Jacobian matrix is

   $$\begin{bmatrix} -d & (D - d - \beta N) \\ 0 & k \end{bmatrix}$$

   where $k = \beta N - (D+\gamma)$. The trace is $-d + k$ and the determinant is $-dk$. If $k < 0$, the trace is negative and the determinant positive, proving sta-

bility; otherwise the determinant is negative, proving instability.
For the second equilibrium with $\hat{x} = (D + \gamma)/\beta$, $\hat{y} = d(N - \hat{x})/(d + \gamma)$, the Jacobian matrix is

$$\begin{bmatrix} -(d+K) & -(d+\gamma) \\ K & 0 \end{bmatrix}$$

where $K = dk/(d + \gamma)$ which has the same sign as k. The trace is $-(d + K)$, and the determinant is $(d + \gamma)K$. If $K > 0$, the trace is negative and the determinant positive, proving stability; otherwise, the determinant is negative, proving instability.

7. $\hat{p} = 0.100$, $\hat{q} = 0.00026$. The eigenvalues are $-0.07 \pm 2.53i$, giving weakly damped oscillations with a period of 2.48 years.

8. (a) $\hat{n}_1 = R \log R/a(R-1)$, $\hat{n}_2 = \log R/a$.

9. (a) $\hat{n}_1 = \dfrac{R}{R-1}\left(\dfrac{\log R}{Q}\right)^2$ $\hat{n}_2 = \left(\dfrac{\log R}{Q}\right)^2$

CHAPTER 4

3. $\bar{x} = 1.907$, $q = 0.524$, $p = 0.476$. Expected numbers are {90.2, 42.9, 20.4, 9, 7, 4.6, 4.2}; chi-square = 2.27 with four degrees of freedom (having fitted one parameter from the data); not significant.

4. (b) {1, –0.5 + 0.866i, –0.5 – 0.866i}; all have absolute value 1.

5. 2.7 years.

CHAPTER 5

2. $P_1 = 0.25$ in a population of annuals for all values of b^*, whereas P_1 varies with b^* in a population of perennials, and is 0.22, 0.21, and 0.20 for $b^* = 70$, 77, and 85. Hence perennials can invade a population of annuals when $b^* = 77$ or 85, and annuals can invade a population of perennials when $b^* = 70$ or 77. Simulations will show an equilibrium with annuals only when $b^* = 70$, with perennials only when $b^* = 85$, and with both forms present when $b^* = 77$.

3. The geometric growth rates of the two types are $P_0 m_1$ and $\sqrt{P_1 P_2 m_2}$, while the lifetime reproductive successes are $P_0 m_1$ and $P_0 P_1 m_2$. The

condition for the second to outgrow the first is that $P_1 m_2 > P_0 m_1^2$; with the numerical values given, this gives $P_1 m_2 > 1.6$, 2.0, and 2.4, respectively, whereas the condition for the lifetime reproductive success of the second type to exceed that of the first is $P_1 m_2 > 2.0$ in all three cases. Lifetime reproductive success gives the same criterion as geometric growth rate when the first population is stationary ($P_0 = 0.5$), but it is too severe in a declining population ($P_0 = 0.4$) and not severe enough in an increasing population ($P_0 = 0.6$).

4. (a) Make queens during the last four weeks of the season.
   (b) Make queens during the last 27 days of the season.

5. Make only workers up to week 16, make workers and queens in week 17 (with 41.4% of resources being allocated to making workers), and make only queens after then.

6. The sensitivities to changes in $\log P_a$ for $a = \{0, \ldots, 6\}$ are {1.11, 0.79, 0.53, 0.32, 0.19, 0.11, 0.05}.

   The sensitivities to changes in m_a for $a = \{1, \ldots, 7\}$ are {0.25, 0.12, 0.09, 0.06, 0.04, 0.02, 0.02}.

8. (b) The optimal germination rates (g) are:

$$\begin{array}{cc} & \overbrace{\begin{array}{ccc} 0.10 & 0.25 & 0.75 \end{array}}^{p} \\ d\left\{\begin{array}{c} 0.01 \\ 0.10 \\ 0.50 \end{array}\right. & \begin{bmatrix} 0.04 & 0.07 & 0.20 \\ 0.09 & 0.17 & 0.47 \\ 0.10 & 0.24 & 0.69 \end{bmatrix} \end{array}$$

The numbers of surviving seeds (s) are:

$$\begin{array}{cc} & \overbrace{\begin{array}{ccc} 0.10 & 0.25 & 0.75 \end{array}}^{p} \\ d\left\{\begin{array}{c} 0.01 \\ 0.10 \\ 0.50 \end{array}\right. & \begin{bmatrix} 15.8 & 5.2 & 1.4 \\ 57.9 & 9.7 & 1.7 \\ 13{,}211 & 74 & 2.5 \end{bmatrix} \end{array}$$

CHAPTER 6

1. The rewards for different values of n are {0.44, 0.59, 0.65, 0.66, 0.66, 0.65, 0.62, 0.59, 0.56, 0.53}, with a maximum at position 4 or 5. The marginal values for different values of n are {1.02, 0.93, 0.83, 0.74, 0.65, 0.55, 0.46,

0.36, 0.27, 0.17}; there is approximate equality of the reward with the marginal value at position 5. The bees are doing the right thing.

3. The expected utilities under the two setups are 0.57 and 0.6875, respectively; the second is preferable.

4. The optimal habitats for $x = \{1, 2, 3, 4, 5\}$ are {3, 1, 1, 1, 1} on day 99, {3, 3, 1, 1, 1} on day 98, {3, 3, 2, 1, 1} on days 95 through 97 and {3, 3, 2, 2, 1} on days 94 and before.

6. 220,000. The yields under open access and sole ownership are 2880 and 7920 whales per year, compared with a maximum sustainable yield of 8000 per year.

CHAPTER 7

4. (a) The dominant eigenvalues for the three equilibria (with absolute values in parentheses when complex) are (i) $1 \pm 0.096i$ (1.005), 1.11, 1.43; (ii) $0.997 \pm 0.096i$ (1.002), 1.11, 1.43; (iii) $0.990 \pm 0.096i$ (0.995), 1.11, 1.41. The internal equilibrium is a stable spiral for the third mutation rate and an unstable spiral for the first two; the boundary equilibria are unstable nodes.

5. For the reward function of Equation 21, $n_{11} = 2n_{22} - 20$, $n_{12} = 30 - n_{22}$, and $n_{21} = 80 - 2n_{22}$, with $10 < n_{22} < 30$. For the reward function of Equation 22, $n_{11} = 40$, $n_{12} = 0$, $n_{21} = 20$, and $n_{22} = 30$.

8. If

$$G(t) = \int_t^b g(t,u)\,du$$

then

$$\frac{dG}{dt} = -g(t,t) + \int_t^b \frac{dg(t,u)}{dt}\,du$$

The first term reflects the effect of a small change in the lower limit of integration, the second that of a small change in the integrand. Hence

$$\begin{aligned}\frac{dw}{dt} &= -\frac{f(t)}{m^*(t)} + \lambda\int_t^b \exp[-\lambda(u-t)]\frac{f(u)}{m^*(u)}\,du \\ &= -\frac{f(t)}{m^*(t)} + \lambda w(t) \\ &= -\frac{f(t)}{m^*(t)} + \lambda \qquad \text{if } m(t) > 0 \quad \text{(from Equation 32)}\end{aligned}$$

Equating dw/dt to zero leads to Equation 33, whence

$$f'(t) = \lambda \frac{dm^*(t)}{dt} \qquad \text{if } m(t) > 0$$

By a similar argument (though with t as the upper rather than the lower limit of integration),

$$\begin{aligned} \frac{dm^*(t)}{dt} &= m(t) - \lambda m^*(t) \\ &= m(t) - f(t) \qquad \text{if } m(t) > 0 \end{aligned}$$

from which Equation 34 follows.

9. The mean and variance of the female emergence date are 18.82 and 19.49, the corresponding figures for the male emergence date being 12.42 and 18.22. The predicted truncation point for male emergence is $t^* = 17.4$, and the predicted mean and variance are 13.5 and 7.3, respectively.

CHAPTER 8

1. To show that Equation 6 is a Nash equilibrium, first calculate $E_A(1, \mathbf{p}_B) = E_A(2, \mathbf{p}_B) = 0.67$, $E_A(3, \mathbf{p}_B) = E_A(4, \mathbf{p}_B) = -2$; thus any reply by A involving only his first two strategies has a payoff of 0.67 and is a best reply to $\mathbf{p}_B$. Similarly, $E_B(\mathbf{p}_A, 1) = E_B(\mathbf{p}_A, 2) = -0.67$, so that any reply by B has the same payoff of –0.67 and is a best reply to $\mathbf{p}_A$.
2. (i) It is a Nash equilibrium because the fitnesses of hawk and dove against this strategy are the same.
   (ii) For the dynamic stability evaluate $dp/dt = f(p)$ from Equations 7.5 and 8.11, check that there is an equilibrium at $p = 0.5$, and find that

   $$f'(p)\big|_{p=0.5} = -0.5$$

   (iii) Verify the ESS calculations in Equations 16 and 17.
3. Repeat the above calculations for this situation.
4. (i) For the dynamic stability analysis the eigenvalues of the Jacobian matrix are complex with real part –0.167.

   (ii) $E(\mathbf{p}, \mathbf{q}) = 1.833$, $E(\mathbf{q}, \mathbf{q}) = 2.5$.
5. It is a Nash equilibrium since $E(1, \mathbf{p}) = E(2, \mathbf{p}) = E(3, \mathbf{p}) = 0.33$. It is not

an evolutionarily stable state since the eigenvalues of the Jacobian matrix are complex with real part 0.33. It is not an ESS since it can be invaded by any pure strategy; for example, $E(\mathbf{p}, 1) = 0.33$, $E(1, 1) = 1$. Each of the frequencies oscillates with increasing period and amplitude, getting closer and closer to the boundaries.

7. The payoff matrix is

$$\begin{bmatrix} 0 & 4 \\ -2 & 8 \end{bmatrix}$$

The Nash equilibria are {1, 0}, {0.67, 0.33}, and {0, 1}, of which the first and third are evolutionarily stable states and strategies. TFT will go to fixation if its initial frequency exceeds one-third; otherwise it will go extinct.

8. Write $\mathbf{p} = \{0.5, 0, 0, 0.5\}$, $\mathbf{q} = \{0, 1, 0, 0\}$. Then $\mathbf{p}$ is a Nash equilibrium because $E(i, \mathbf{p}) = 0.5$ where i is any one of the four pure strategies; it is not an ESS because $E(\mathbf{p}, \mathbf{q}) = 0.5$ but $E(\mathbf{q}, \mathbf{q}) = 1$.

11. $v^* = 0.25$.

CHAPTER 9

1. $aa \times aa$ matings [probability $(1-x)^2 \cong 1$] produce $(S + s + b - c)$ offspring with genotype aa; $Aa \times aa$ or $aa \times Aa$ matings (probability $\cong 2x$) produce the offspring types shown below.

	Offspring types from Aa × aa matings (elder sib first)			
	aa, aa	*aa, Aa*	*Aa, aa*	*Aa, Aa*
Frequency	$\frac{1}{4}$	$\frac{1}{4}$	$\frac{1}{4}$	$\frac{1}{4}$
Aa survivors	0	$s + b$	S	$S + s$

Hence,

$$x' = \frac{1}{4}\left[\frac{(s+b)+S+(S+s)}{S+s+b-c}\right]2x$$

$$= \left[1 - \frac{(\frac{1}{2}b-c)}{S+s+b-c}\right]x$$

5. (a) $\frac{3}{4}$; (b) $\frac{3}{8}$; (c) $\frac{3}{8}$; (d) $\frac{3}{16}$.
6. The dominant left eigenvectors (corresponding to the eigenvalue 1) are (a) {2, 1} and (b) $\{1 + p, 1\}$.

CHAPTER 10

5. Equilibrium genotype frequencies in females (*aa*, *Aa*, *AA*) and males (*a*, *A*) are: (a) {0, 0, 1} and {0, 1}; (b) {0.557, 0.386, 0.057} and {0.677, 0.323}; (c) {0.25, 0.5, 0.25} and {0.5, 0.5}. Equilibrium sex ratios are: (a) 0.4; (b) 0.5; (c) 0.425.
6. (a) The average coefficients of relatedness of a foundress to male and female larvae are $\frac{1}{2}$ and $\frac{1}{4}$, respectively, so that the life-for-life coefficients are both $\frac{1}{4}$; an unbiased sex ratio is predicted under queen control. The average coefficients of relatedness of a worker to male and female larvae are $\frac{1}{4}$ and $\frac{3}{8}$, so that the life-for-life coefficients are $\frac{1}{8}$ and $\frac{3}{8}$; the predicted sex ratio is $\frac{1}{4}$ under worker control.

   (b) The average coefficients of relatedness of a foundress to male and female larvae are $\frac{7}{8}$ and $\frac{7}{16}$, respectively, so that the life-for-life coefficients are both $\frac{7}{16}$; an unbiased sex ratio is predicted under queen control. The average coefficients of relatedness of a worker to male and female larvae are $\frac{7}{16}$ and $\frac{15}{32}$, so that the life-for-life coefficients are $\frac{7}{32}$ and $\frac{15}{32}$; the predicted sex ratio is $\frac{7}{22} = 0.32$ under worker control.
11. The optimal sex ratios for the three-patch model are all males in patch 1, 75% males in patch 2, and all females in patch 3.

CHAPTER 11

1. (a) {0, 0, 1}
   (b) {0, 0, 1} and {0, 1, 0}
   (c) {0, 0, 1} and {1, 0, 0}
   (d) {0, 0, 1}, {0, 1, 0}, and {1, 0, 0}.

CHAPTER 12

1. (a) The optimal response is about 1 for $m_2 < 0.0728$ and switches to 0.01 above this critical value. The optimal response to 0.01 is 1.1042, and the optimal response to 1.1042 is 0.01, which together form an

(asymmetric) Nash equilibrium.

(b) The optimal response to any value of m_2 is 0.01.

2. The equilibrium values are 0, $2 - 1/s$ (relevant only when $0.5 \le s \le 1$), and 1. When $0 < s < 0.5$, 0 is unstable and 1 is stable; the system goes to fixation at $p = 1$. When $0.5 < s < 1$, 0 and 1 are both stable and $2 - 1/s$ is unstable; the system goes to fixation at 0 or 1 depending on whether it starts below or above the internal equilibrium.

3. The mating table is as follows:

Gamete unions					**Gametic output**			
Female		Male	**Frequency**	**Zygote**	*IU*	*IB*	*NU*	***NB***
IU	×	*IU*	x_1^2	*I*/*UU*	1	0	0	0
IU	×	*IB*	x_1x_2	*I*/*UB*	$\frac{1}{2}$	$\frac{1}{2}$	0	0
IU	×	*NU*	x_1x_3	*I*/*UU*	1	0	0	0
IU	×	*NB*	x_1x_4	*I*/*UB*	$\frac{1}{2}$	$\frac{1}{2}$	0	0
IB	×	*IU*	x_1x_2	*I*/*UB*	$\frac{1}{2}$	$\frac{1}{2}$	0	0
IB	×	*IB*	x_2^2	*I*/*BB*	0	1	0	0
IB	×	*NU*	x_2x_3	*I*/*UB*	$\frac{1}{2}$	$\frac{1}{2}$	0	0
IB	×	*NB*	x_2x_4	*I*/*BB*	0	1	0	0
NU	×	*IU*	x_1x_3	*N*/*UU*	0	0	1	0
NU	×	*IB*	x_2x_3	*I*/*UB*	$\frac{1}{2}$	$\frac{1}{2}$	0	0
NU	×	*NU*	x_3^2	*N*/*UU*	0	0	1	0
NU	×	*NB*	x_3x_4	*N*/*UB*	0	0	$\frac{1}{2}$	$\frac{1}{2}$
NB	×	*IU*	x_1x_4	*N*/*UB*	0	0	$\frac{1}{2}$	$\frac{1}{2}$
NB	×	*IB*	x_2x_4	*I*/*BB*	0	1	0	0
NB	×	*NU*	x_3x_4	*N*/*UB*	0	0	$\frac{1}{2}$	$\frac{1}{2}$
NB	×	*NB*	x_4^2	*N*/*BB*	0	0	0	1

4. The mating table is as follows:

		Female offspring		**Male offspring**	
Mating type	**Frequency**	*Aa*	*aa*	*Aa*	*aa*
$Aa \times aa$	x	$p+\frac{1}{2}(1-p)(1-r)$	$\frac{1}{2}(1-p)(1-r)$	$\frac{1}{2}(1-p)r$	$\frac{1}{2}(1-p)r$
$aa \times Aa$	y	$\frac{1}{2}(1-r)$	$\frac{1}{2}(1-r)$	$\frac{1}{2}r$	$\frac{1}{2}r$
$aa \times aa$	$1-x-y$	0	$1-r$	0	r

The transition matrix is

$$\begin{bmatrix} p/(1-r)+\frac{1}{2}(1-p) & \frac{1}{2} \\ \frac{1}{2}(1-p) & \frac{1}{2} \end{bmatrix}$$

8. (b) The equilibria are {0, 0}, {99, 0}, and {0, 98}. The eigenvalues, multiplied by 10^6, are {98, 99}, {–99, –1}, and {–98, 1}.

9. $x^* = 0.557$.

References

Abraham, R.H. and Shaw, C.D. 1992. *Dynamics: The Geometry of Behavior*, 2nd Ed. Addison-Wesley, Redwood City, CA.

Aitchison, J. and Brown, J.A.C. 1957. *The Lognormal Distribution, with Special Reference to Its Uses in Economics*. Cambridge University Press, Cambridge.

Akçakaya, H.R. 1992. Population cycles of mammals: evidence for a ratio-dependent predation hypothesis. Ecol. Monogr. 62:119–142.

Alexander, R.M. 1982. *Optima for Animals*. Arnold, London.

Allee, W.C., Emerson, A.E., Park, O., Park, T. and Schmidt, K. 1949. *Principles of Animal Ecology*. W.B. Saunders, Philadelphia.

Allen, J.A. 1988. Frequency-dependent selection by predators. Phil. Trans. Roy. Soc. Lond. B 319:485–503.

Anderson, D.J. 1990. Evolution of obligate siblicide in boobies. 1. A test of the insurance egg hypothesis. Amer. Natur. 135:334–350.

Anderson, R.M. 1981. Population ecology of infectious disease agents. In R.M. May, ed., *Theoretical Ecology*, Blackwell, Oxford, pp. 318–355.

Anderson, R.M. and May, R.M. 1991. *Infectious Diseases of Humans*. Oxford University Press, Oxford.

Andersson, M. 1982. Female choice selects for extreme tail length in a widowbird. Nature 299:818–820.

Andersson, M. 1984. The evolution of eusociality. Ann. Rev. Ecol. Syst. 15: 165–189.

Andersson, M. 1994. Sexual Selection. Princeton University Press, Princeton.

Arditi, R. and Ginzburg, L.R. 1989. Coupling in predator–prey dynamics: ratio-dependence. J. Theor. Biol. 139:311–326.

Axelrod, R. 1984. *The Evolution of Cooperation*. Basic Books, New York.

Axelrod, R. and Dion, D. 1988. The further evolution of cooperation. Science 242:1385–1390.

Axelrod, R. and Hamilton, W.D. 1981. The evolution of cooperation. Science 211:1390–1396.

Ayala, F.J. 1972. Competition between species. Amer. Scientist 60:348–357.

Bailey, N.T.J. 1975. *The Mathematical Theory of Infectious Diseases*. Griffin, London.

Barlow, N., ed. 1958. *The Autobiography of Charles Darwin, 1809–1882: With Original Omissions Restored. Edited with Appendix and Notes by his Grand-daughter*. Collins, London.

Barton, N.H. and Post, R.J. 1986. Sibling competition and the advantage of mixed families. J. Theor. Biol. 120:381–387.

Bell, G. 1982. *The Masterpiece of Nature: The Evolution and Genetics of Sexuality*. Croom Helm, New York.

Bell, G. 1988. *Sex and Death in Protozoa*. Cambridge University Press, Cambridge.
Benford, F.A. 1978. Fisher's theory of the sex ratio applied to the social hymenoptera. J. Theor. Biol. 72:701–727.
Bernardelli, H. 1941. Population waves. J. Burma Res. Soc. 31:1–18.
Best, L.S. and Bierzychudek, P. 1982. Pollinator foraging on foxglove (*Digitalis purpurea*): a test of a new model. Evolution 36:70–79.
Bishop, D.T. and Cannings, C. 1978. A generalised war of attrition. J. Theor. Biol. 70:85–124.
Boomsma, J.J. 1989. Sex-investment ratios in ants: Has female bias been systematically overestimated? Amer. Natur. 133:517–532.
Boomsma, J.J. and Grafen, A. 1991. Colony-level sex ratio selection in the eusocial Hymenoptera. J. Evol. Biol. 4:383–407.
Bourke, A.F.G. 1988. Worker reproduction in the higher eusocial Hymenoptera. Quart. Rev. Biol. 63:291–311.
Boyce, M. and Perrins, C.M. 1987. Optimizing Great Tit clutch size in a fluctuating environment. Ecology 68:142–153.
Bradbury, J.W. and Andersson, M.B., eds. 1987. *Sexual Selection: Testing the Alternatives*. Wiley, Chichester.
Breden, F. 1990. Partitioning of covariance as a method of studying kin selection. Trends Ecol. Evol. 5:224–228.
Brown, J.L. 1987. *Helping and Communal Breeding in Birds*. Princeton University Press, Princeton.
Bull, J.J. 1983. *The Evolution of Sex Determining Mechanisms*. Benjamin/Cummings, Menlo Park, CA.
Bull, J.J. and Charnov, E.L. 1985. Irreversible evolution. Evolution 39:1149–1155.
Bulmer, M.G. 1974. A statistical analysis of the 10-year cycle in Canada. J. Anim. Ecol. 43:701–718.
Bulmer, M.G. 1975. Phase relations in the ten-year cycle. J. Anim. Ecol. 44: 609–621.
Bulmer, M.G. 1976. The theory of prey–predator oscillations. Theor. Pop. Biol. 9:137–150.
Bulmer, M.G. 1977. Periodical insects. Amer. Natur. 111:1099–1117.
Bulmer, M.G. 1980. The sib competition model for the maintenance of sex and recombination. J. Theor. Biol. 82:335–345.
Bulmer, M.G. 1983a. Sex ratio evolution in social hymenoptera under worker control with behavioral dominance. Amer. Natur. 121:899–902.
Bulmer, M.G. 1983b. Sex ratio theory in social insects with swarming. J. Theor. Biol. 100:329–339.
Bulmer, M.G. 1983c. Models for the evolution of protandry in insects. Theor. Pop. Biol. 23:314–322.
Bulmer, M.G. 1984. Delayed germination of seeds: Cohen's model revisited. Theor. Pop. Biol. 26:367–377.
Bulmer, M. G. 1989. Structural stability of models of sexual selection. Theor. Pop. Biol. 35:195–206.
Bulmer, M.G. and Perrins, C.M. 1973. Mortality in the Great Tit *Parus major*. Ibis 115:277–281.
Burgess, J.W. 1976. Social spiders. Sci. Amer. 234(3):100–106.
Burley, N. 1986. Sexual selection for aesthetic traits in species with biparental care. Amer. Natur. 127:415–445.
Burley, N., Krantzberg, G. and Radman, P. 1982. Influence of colour-banding on the conspecific preferences of Zebra Finches. Anim. Behav. 30:444–455.

Caraco, T., Blanckenhorn, W.U., Gregory, G.M., Newman, J.A., Recer, G.M. and Zwicker, S.M. 1990. Risk-sensitivity: Ambient temperature affects foraging choice. Anim. Behav. 39:338–345.

Caraco, T., Martindale, S. and Whittam, T.S. 1980. An empirical demonstration of risk-sensitive foraging preferences. Anim. Behav. 28:820–830.

Caswell, H. 1989. *Matrix Population Models*. Sinauer, Sunderland, MA.

Charlesworth, B. 1978. Model for evolution of Y chromosomes and dosage compensation. Proc. Nat. Acad. Sci. USA 75:5618–5622.

Charlesworth, B. 1980. The cost of sex in relation to mating system. J. Theor. Biol. 84:655–671.

Charlesworth, B. 1990a. Optimization models, quantitative genetics, and mutation. Evolution 44:520–538.

Charlesworth, B. 1990b. Mutation–selection balance and the evolutionary advantage of sex and recombination. Genet. Res. 55:199–221.

Charlesworth, B. 1994. *Evolution in Age-Structured Populations*, 2nd Ed. Cambridge University Press, Cambridge.

Charlesworth, D. and Charlesworth, B. 1978. A model for the evolution of dioecy and gynodioecy. Amer. Natur. 112:975–997.

Charnov, E.L. 1976. Optimal foraging, the marginal value theorem. Theor. Pop. Biol. 9:129–136.

Charnov, E.L. 1978. Sex-ratio selection in eusocial Hymenoptera. Amer. Natur. 112:715–734.

Charnov, E.L. 1979. Natural selection and sex change in Pandalid shrimp: test of a life history theory. Amer. Natur. 113:715–734.

Charnov, E.L. 1982. *The Theory of Sex Allocation*. Princeton University Press, Princeton.

Charnov, E.L., Maynard Smith, J. and Bull, J.J. 1976. Why be an hermaphrodite? Nature 263:125–126.

Charnov, E.L. and Schaffer, W.M. 1973. Life-history consequences of natural selection: Cole's result revisited. Amer. Natur. 107:791–792.

Clark, A.B. 1978. Sex ratio and local resource competition in a prosimian primate. Science 201:163–165.

Clark, C. 1990. *Mathematical Bioeconomics: The Optimal Management of Renewable Resources*, 2nd Ed. Wiley, New York.

Clarke, B.C. and Partridge, L., eds. Frequency-dependent selection. Phil. Trans. Roy. Soc. Lond. B 319:457–640.

Clayton, D.H. 1991. The influence of parasites on host sexual selection. Parasitol. Today 7:329–334.

Clutton-Brock, T.H., Albon, S.D. and Guinness, F.E. 1984. Maternal dominance, breeding success and birth sex ratios in red deer. Nature 308:358–360.

Cockburn, A. 1991. *An Introduction to Evolutionary Ecology*. Blackwell, Oxford.

Cohen, D. 1966. Optimizing reproduction in a randomly varying environment. J. Theor. Biol. 12:110–129.

Cohen, D. 1971. Maximizing final yield when growth is limited by time or limiting resources. J. Theor. Biol. 33:299–307.

Cole, L.C. 1954. The population consequences of life history phenomena. Quart. Rev. Biol. 29:103–137.

Colwell, R.K. 1981. Group selection is implicated in the evolution of female-biased sex ratios. Nature 290:401–404.

Conover, D.O. 1984. Adaptive significance of temperature-dependent sex determination in a fish. Amer. Natur. 123:297–313.

Conover, D.O. and Heins, S.W. 1987. Adaptive variation in environmental and genetic sex determination in a fish. Nature 326:496–498.

Cooper, W.S. and Kaplan, R.H. 1982. Adaptive "coin-flipping": a decision-theoretic examination of natural selection for random individual variation. J. Theor. Biol. 94:135–151.

Cosmides, L.M. and Tooby, J. 1981. Cytoplasmic inheritance and intragenomic conflict. J. Theor. Biol. 89:83–129.

Couvet, D., Atlan, A., Belhassen, E., Gliddon, C., Gouyon, P.-H. and Kjellberg, F. 1990. Co-evolution between two symbionts: the case of cytoplasmic male-sterility in higher plants. Oxford Surv. Evol. Biol. 7:225–247.

Craig, R. 1980. Sex investment ratios in social hymenoptera. Amer. Natur. 116:311–323.

Crow, J.F. 1970. Genetic loads and the cost of natural selection. In K. Kojima, ed., *Mathematical Topics in Population Genetics*, Springer-Verlag, Berlin, pp. 128–177.

Curtsinger, J.W. 1986. Stay times in *Scatophaga* and the theory of evolutionarily stable strategies. Amer. Natur. 128:130–136.

Darwin, C. 1859. *On the Origin of Species by Means of Natural Selection.* John Murray, London.

Darwin, C. 1871. *The Descent of Man and Selection in Relation to Sex.* John Murray, London; 2nd Ed., 1874.

Darwin, C. 1877. *The Different Forms of Flowers on Plants of the Same Species.* John Murray, London.

Davies, N.B. 1978. Territorial defence in the speckled wood butterfly (*Pararge aegeria*): The resident always wins. Anim. Behav. 26:138–147.

Edelstein-Keshet, L. 1988. *Mathematical Models in Biology.* Random House, New York.

Elton, C. and Nicholson, M. 1942. The ten-year cycle in numbers of the lynx in Canada. J. Anim. Ecol. 11:215–244.

Emlen, J.M. 1973. *Ecology: An Evolutionary Approach.* Addison-Wesley, Reading, MA.

Emlen, J.M. 1984. *Population Biology. The Coevolution of Population Dynamics and Behavior.* Macmillan, New York.

Emlen, S.T. 1990. White-fronted Bee-eaters: helping in a colonially nesting species. In P.B. Stacey and W.D. Koenig, eds., *Cooperative Breeding in Birds,* Cambridge University Press, Cambridge, pp. 489–526.

Emlen, S.T. 1991. Evolution of cooperative breeding in birds and mammals. In J.R. Krebs and N.B. Davies, eds., *Behavioural Ecology: An Evolutionary Approach,* 3rd Ed., Blackwell, Oxford, pp. 301–337.

Eshel, I. 1983. Evolutionary and continuous stability. J. Theor. Biol. 103:99–111.

Eshel, I. and Feldman, M.W. 1982. On evolutionary genetic stability of the sex ratio. Theor. Pop. Biol. 21:430–439.

Evans, H.E. 1977. Extrinsic versus intrinsic factors in the evolution of insect sociality. BioScience 27:613–617.

Felsenstein, J. 1974. The evolutionary advantage of recombination. Genetics 78:737–756.

Fisher, R.A. 1930. *The Genetical Theory of Natural Selection.* Oxford University Press, Oxford.

Fitzgibbon, C.D. and Fanshaw, J.H. 1988. Stotting in Thompson's gazelles: an honest signal of condition. Behav. Ecol. Sociobiol. 23:69–74.

Forsyth, A. 1981. Sex ratio and parental investment in an ant population. Evolution 35:1252–1253.
Frank, S.A. 1989. The evolutionary dynamics of cytoplasmic male sterility. Amer. Natur. 133:345–376.
Fretwell, S.D. and Lucas, H.J. 1970. On territorial behavior and other factors influencing habitat distribution in birds. Acta Biotheor. 19:16–36.

Gadgil, M. and Bossert, W.H. 1970. Life historical consequences of natural selection. Amer. Natur. 104:1–24.
Gause, G.F. 1934. *The Struggle for Existence*. Williams and Wilkins, Baltimore.
Gibson, R.M. and Bachman, G.C. 1992. The costs of female choice in a lekking bird. Behav. Ecol. 3:300–309.
Gilpin, M.E. and Ayala, F.J. 1973. Global models of growth and competition. Proc. Nat. Acad. Sci. USA 70:3590–3593.
Gilpin, M.E. and Justice, K.E. 1972. Reinterpretation of the invalidation of the principle of competitive exclusion. Nature 236:273–274, 299–301.
Godfray, H.C.J. 1987. The evolution of clutch size in parasitic wasps. Amer. Natur. 129:221–233.
Godfray, H.C.J. 1991. Signalling of need by offspring to their parents. Nature 352:328–330.
Godfray, H.C.J. 1994. *Parasitoids: Behavioral and Evolutionary Ecology*. Princeton University Press, Princeton.
Godfray, H.C.J. and Grafen, A. 1988. Unmatedness and the evolution of eusociality. Amer. Natur. 131:303–305.
Godfray, H.C.J. and Harper, A.B. 1990. The evolution of brood reduction by siblicide in birds. J. Theor. Biol. 145:163–175.
Godfray, H.C.J., Partridge, L. and Harvey, P.H. 1991. Clutch size. Ann. Rev. Ecol. Syst. 22:409–429.
Gould, S.J. and Lewontin, R.C. 1979. The spandrels of San Marco and the Panglossian paradigm: a critique of the adaptationist programme. Proc. Roy. Soc. Lond. B 205:581–598.
Gouyon, P.-H. and Couvet, D. 1987. A conflict between two sexes, females and hermaprodites. In S.C. Stearns, ed., *The Evolution of Sex and Its Consequences*, Birkhäuser, Basel, pp. 245–261.
Grafen, A. 1987. The logic of divisely asymmetric contests: respect for ownership and the desperado effect. Anim. Behav. 35:462–467.
Grafen, A. 1990a. Sexual selection unhandicapped by the Fisher process. J. Theor. Biol. 144:473–516.
Grafen, A. 1990b. Biological signals as handicaps. J. Theor. Biol. 144:517–546.

Haigh, J. 1975. Game theory and evolution. Adv. Appl. Prob. 7:8–11.
Haigh, J. 1978. The accumulation of deleterious genes in a population—Muller's ratchet. Theor. Pop. Biol. 14:251–267.
Hamilton, W.D. 1964. The genetical evolution of social behaviour. J. Theor. Biol. 7:1–52.
Hamilton, W.D. 1967. Extraordinary sex ratios. Science 156:477–488.
Hamilton, W.D. 1970. Selfish and spiteful behaviour in an evolutionary model. Nature 228:1218–1220.
Hamilton, W.D. 1971. Geometry for the selfish herd. J. Theor. Biol. 31:295–311.
Hamilton, W.D. 1972. Altruism and related phenomena, mainly in social insects. Ann. Rev. Ecol. Syst. 3:193–232.

Hamilton, W.D. 1993. Haploid dynamic polymorphism in a host with matching parasites: effects of mutation/subdivision, linkage, and patterns of selection. J. Hered. 84:328–338.

Hamilton, W.D., Axelrod, R. and Tanese, R. 1990. Sexual reproduction as an adaptation to resist parasites. Proc. Nat. Acad. Sci. USA 87:3566–3573.

Hamilton, W.D. and May, R.M. 1977. Dispersal in stable habitats. Nature 269: 578–581.

Hamilton, W.D. and Zuk, M. 1982. Heritable true fitness and bright birds: a role for parasites? Science 218:384–387.

Hanson, J. and Green, L. 1989. Foraging decisions: prey choice by pigeons. Anim. Behav. 37:429–443.

Harvey, P.H., Partridge, L. and Southwood, T.R.E., eds. 1991. *The Evolution of Reproductive Strategies*. Phil. Trans. Roy. Soc. Lond. B 332:1–104.

Hassell, M.P. 1978. *The Dynamics of Arthropod Predator–Prey Systems*. Princeton University Press, Princeton.

Hassell, M.P. 1981. Arthropod predator–prey systems. In R.M. May, ed., *Theoretical Ecology*, Blackwell, Oxford, pp. 105–131.

Hassell, M.P., Lawton, J.H. and May, R.M. 1976. Patterns of dynamical behaviour in single species populations. J. Anim. Ecol. 45:471–486.

Hassell, M.P. and Varley, G.C. 1969. New inductive population model for insect parasites and its bearing on population control. Nature 223:1133–1136.

Hastings, A., Hom, C.H., Ellner, S., Turchin, P. and Godfray, H.C.J. 1993. Chaos in ecology: Is mother nature a strange attractor? Ann. Rev. Ecol. Syst. 24:1–33.

Hastings, A. and Powell, T. 1991. Chaos in a three-species food chain. Ecology 72:896–903.

Hastings, I.M. 1992. Population genetic aspects of deleterious cytoplasmic genomes and their effect on the evolution of sexual reproduction. Genet. Res. 59:215–225.

Herre, E.A. 1985. Sex ratio adjustment in fig wasps. Science 228:896–900.

Hill, G.E. 1990. Female House Finches prefer colourful males: sexual selection for a condition-dependent trait. Anim. Behav. 40:563–572.

Hill, G.E. 1991. Plumage coloration is a sexually selected indicator of male quality. Nature 350:337–339.

Hoekstra, R.F. 1987. The evolution of sexes. In S.C. Stearns, ed., *The Evolution of Sex and Its Consequences*, Birkhäuser, Basel, pp. 59–91.

Hoelzer, G.A. 1989. The good parent process of sexual selection. Anim. Behav. 38:1067–1078.

Holling, C.S. 1965. The functional response of predators to prey density and its role in mimicry and population regulation. Mem. Entomol. Soc. Can. 45:1–60.

Hori, M. 1993. Frequency-dependent natural selection in the handedness of scale-eating cichlid fish. Science 260:216–219.

Houston, A., Clark, C., McNamara, J.M. and Mangel, M. 1988. Dynamic models in behavioural and evolutionary ecology. Nature 332:29–34.

Houston, A.I. and McNamara, J.M. 1988. The ideal free distribution when competitive abilities differ: an approach based on statistical mechanics. Anim. Behav. 36:166–174.

Howard, R.D. 1978a. Factors influencing early embryo mortality in bullfrogs. Ecology 59:789–798.

Howard, R.D. 1978b. The evolution of mating strategies in bullfrogs *Rana catesbeiana*. Evolution 32:850–871.

Howard, R.S. and Lively, C.M. 1994. Parasitism, mutation accumulation and the maintenance of sex. Nature 367:554–557.

Hughes, K.A. and Charlesworth, B. 1994. A genetic analysis of senescence in *Drosophila*. Nature 367:64–66.
Hurst, L.D. 1992. Intragenomic conflict as an evolutionary force. Proc. Roy. Soc. Lond. B 248:135–140.
Hurst, L.D. and Hamilton, W.D. 1992. Cytoplasmic fusion and the nature of sexes. Proc. Roy. Soc. Lond. B 247:189–194.

Intriligator, M.D. 1971. *Mathematical Optimization and Economic Theory*. Prentice-Hall, Englewood Cliffs, NJ.
Iwasa, Y. and Haccou, P. 1994. ESS emergence pattern of male butterflies in stochastic environments. Evol. Ecol. (*in press*).
Iwasa, Y., Odendaal, F.J., Murphy, D.D., Ehrlich, P.R. and Launer, A. 1983. Emergence patterns in male butterflies: a hypothesis and a test. Theor. Pop. Biol. 23:363–379. (*in press*).
Iwasa, Y., Pomiankowski, A. and Nee, S. 1991. The evolution of costly mate preferences. II. The "handicap" principle. Evolution 45:1431–1442.

Johnson, C.N. 1988. Dispersal and the sex ratio at birth in primates. Nature 332:726–728.

Karlin, S. and Lessard, S. 1986. *Theoretical Studies on Sex Ratio Evolution*. Princeton University Press, Princeton.
Kelley, S.E., Antonovics, J. and Schmitt, J. 1988. A test of the short-term advantage of sexual reproduction. Nature 331:714–716.
Kimura, M. and Maruyama, T. 1966. The mutational load with epistatic gene interactions in fitness. Genetics 54:1337–1351.
King, D. and Roughgarden, J. 1982a. Multiple switches between vegetative and reproductive growth in annual plants. Theor. Pop. Biol. 21:194–204.
King, D. and Roughgarden, J. 1982b. Graded allocation between vegetative and reproductive growth for annual plants in growing seasons of random length. Theor. Pop. Biol. 22:1–16.
Kirkpatrick, M. 1982. Sexual selection and the evolution of female choice. Evolution 36:1–12.
Kirkpatrick, M. and Ryan, M.J. 1991. The evolution of mating preferences and the paradox of the lek. Nature 350:33–38.
Kirkwood, T.B.L. and Rose, M.R. 1991. Evolution of senescence: late survival sacrificed for reproduction. Phil. Trans. Roy. Soc. Lond. B 332:15–24.
Kjellberg, F., Gouyon, P.-H., Ibrahim, M., Raymond, M. and Valdeyron, G. 1987. The stability of the symbiosis between dioecious figs and their pollinators: A study of *Ficus carica* L. and *Blastophaga psenes* L. Evolution 41:693–704.
Koenig, W.D. and Mumme, R.L. 1987. *Population Ecology of the Cooperatively Breeding Acorn Woodpecker*. Princeton University Press, Princeton.
Kondrashov, A.S. 1982. Selection against harmful mutations in large sexual and asexual populations. Genet. Res. 40:325–332.
Kondrashov, A.S. 1994. Sex and deleterious mutations. Nature 369:99–100.
Krebs, J.R. and Avery, M.I. 1985. Central place foraging in the European Bee-eater, *Merops apiaster*. J. Anim. Ecol. 54:459–472.
Krebs, J.R. and Davies, N.B. 1993. *An Introduction to Behavioural Ecology*, 3rd Ed. Blackwell, Oxford.
Krebs, J.R. and Kacelnik, A. 1991. Decision-making. In J.R. Krebs and N.B. Davies, eds., *Behavioural Ecology*, Blackwell, Oxford, pp. 105–136.

Lack, D. 1947. The significance of clutch size. Ibis 89:302–352.
Lack, D. 1954. *The Natural Regulation of Animal Numbers*. Oxford University Press, Oxford.
Lande, R. 1981. Models of speciation by sexual selection on polygenic traits. Proc. Nat. Acad. Sci. USA 78:3721–3725.
Lande, R. 1982. A quantitative genetic theory of life history evolution. Ecology 63:607–615.
Lande, R. 1988. Demographic models of the Northern Spotted Owl (*Strix occidentalis caurina*). Oecologia 75:601–607.
Law, R. and Hutson, V. 1992. Intracellular symbionts and the evolution of uniparental cytoplasmic inheritance. Proc. Roy. Soc. Lond. B 248:69–77.
Leslie, P.H. 1945. On the use of matrices in certain population mathematics. Biometrika 33:183–212.
Leslie, P.H. 1948. Some further remarks on the use of matrices in population mathematics. Biometrika 35:213–245.
Levins, R. 1968. *Evolution in Changing Environments*. Princeton University Press, Princeton.
Lewis, D. 1941. Male-sterility in a natural population of hermaphrodite plants. New Phytol. 40:56–63.
Lively, C.M. 1987. Evidence from a New Zealand snail for the maintenance of sex by parasitism. Nature 328:519–521.
Lively, C.M., Craddock, C. and Vrijenhoek, R.C. 1990. Red Queen hypothesis supported by parasitism in sexual and clonal fish. Nature 344:864–866.
Lloyd, D.G. 1974. Theoretical sex ratios of dioecious and gynodioecious angiosperms. Heredity 32:11–34.
Lloyd, D.G. 1975. The maintenance of gynodioecy and androdioecy in angiosperms. Genetica 45:325–339.
Lloyd, D.G. 1976. The transmission of genes via pollen and ovules in gynodioecious angiosperms. Theor. Pop. Biol. 9:299–316.
Lombardo, M.P. 1985. Mutual restraint in Tree Swallows: A test of the TIT FOR TAT model of reciprocity. Science 227:1363–1365.
Lotka, A.J. 1925. *Elements of Physical Biology*. Williams and Wilkins, Baltimore.

MacArthur, R.H. 1965. Ecological consequences of natural selection. In T.H. Waterman and H. Morowitz, eds., *Theoretical and Mathematical Biology*, Blaisdell, New York, pp. 388–397.
Macevicz, S. and Oster, G. 1976. Modeling social insect populations. II. Optimal reproductive strategies in annual eusocial insect colonies. Behav. Ecol. Sociobiol. 1:265–282.
McKelvey, K., Noon, B.R. and Lamberson, R.H. 1993. Conservation planning for species occupying fragmented landscapes: the case of the Northern Spotted Owl. In P.M. Kareiva, J.G. Kingsolver and R.B. Huey, eds., *Biotic Interactions and Global Change*, Sinauer, Sunderland, MA, pp. 424–450.
McLaren, I.A. 1966. Adaptive significance of large size and long life of the chaetognath *Sagitta elegans* in the Arctic. Ecology 47:852–855.
McNamara, J.M. and Houston, A.I. 1987. Partial preferences and foraging. Anim. Behav. 35:1084–1099.
Mangel, M. and Clark, C.W. 1988. *Dynamic Modeling in Behavioral Ecology*. Princeton University Press, Princeton.
Manning, J.T. and Thompson, D.J. 1984. Muller's ratchet accumulation of favourable mutations. Acta Biotheor. 33:219–225.

May, R.M. 1981a. Models for single populations. In R.M. May, ed., *Theoretical Ecology*, Blackwell, Oxford, pp. 5–29.
May, R.M. 1981b. Models for two interacting populations. In R.M. May, ed., *Theoretical Ecology*, Blackwell, Oxford, pp. 78–104.
May, R.M. and Oster, G.F. 1976. Bifurcations and dynamic complexity in simple ecological models. Amer. Natur. 110:573–599.
Maynard Smith, J. 1976. Sexual selection and the handicap principle. J. Theor. Biol. 57:239–242.
Maynard Smith, J. 1977. Parental investment: a prospective analysis. Anim. Behav. 25:1–9.
Maynard Smith, J. 1978a. Optimization theory in evolution. Ann. Rev. Ecol. Syst. 9:31–56.
Maynard Smith, J. 1978b. *The Evolution of Sex*. Cambridge University Press, Cambridge.
Maynard Smith, J. 1982. *Evolution and the Theory of Games*. Cambridge University Press, Cambridge.
Maynard Smith, J. 1989. *Evolutionary Genetics*. Oxford University Press, Oxford.
Maynard Smith, J. 1991. Theories of sexual selection. Trends Ecol. Evol. 6:146–151.
Maynard Smith, J. and Brown, R.L. 1986. Competition and body size. Theor. Pop. Biol. 30:166–179.
Maynard Smith, J. and Price, G.R. 1973. The logic of animal conflict. Nature 246:15–18.
Medawar, P.B. 1952. *An Unsolved Problem of Biology*. Lewis, London.
Metcalf, R.A. 1980. Sex ratio, parent–offspring conflict, and local competition for mates in the social wasps *Polistes metricus* and *Polistes variatus*. Amer. Natur. 116:642–654.
Metcalf, R.A. and Whitt, G.S. 1977a. Intra-nest relatedness in the social wasp *Polistes metricus*. A genetic analysis. Behav. Ecol. Sociobiol. 2:339–351.
Metcalf, R.A. and Whitt, G.S. 1977b. Relative inclusive fitness in the social wasp *Polistes metricus*. Behav. Ecol. Sociobiol. 2:353–360.
Michod, R.E. and Levin, B.R., eds. 1988. *The Evolution of Sex*. Sinauer, Sunderland, MA.
Milinski, M. 1979. An evolutionarily stable feeding strategy in sticklebacks. Z. Tierpsychol. 51:36–40.
Milinski, M. 1987. TIT FOR TAT in sticklebacks and the evolution of cooperation. Nature 325:433–435.
Milinski, M. 1993. Cooperation wins and stays. Nature 364:12–13.
Milinski, M. and Bakker, T.C.M. 1990. Female sticklebacks use male colorization in mate choice and hence avoid parasitized males. Nature 344:330–332.
Milinski, M. and Parker, G.A. 1991. Competition for resources. In J.R. Krebs and N.B. Davies, eds., *Behavioural Ecology*, Blackwell, Oxford, pp. 137–168.
Mirmirani, M. and Oster, G. 1978. Competition, kin selection, and evolutionary stable strategies. Theor. Pop. Biol. 13:304–339.
Møller, A.P. 1988. Female choice selects for male sexual tail ornaments in the monogamous swallow. Nature 332:640–642.
Møller, A.P. 1989. Viability costs of male tail ornaments in a swallow. Nature 339:132–135.
Møller, A.P. 1990. Effects of a haematophagous mite on the Barn Swallow (*Hirundo rustica*): a test of the Hamilton and Zuk hypothesis. Evolution 44:771–784.
Møller, A.P. 1994. *Sexual Selection and the Barn Swallow*. Oxford University Press, Oxford.

Monaghan, P. 1980. Dominance and dispersal between feeding sites in the herring gull (*Larus argentatus*). Anim. Behav. 28:521–527.
Morin, P. 1995. *Community Ecology*. Blackwell, Oxford.
Motro, U. 1982. Optimal rates of dispersal. II. Diploid populations. Theor. Pop. Biol. 21:412–429.
Motro, U. 1983. Optimal rates of dispersal. III. Parent–offspring conflict. Theor. Pop. Biol. 23:159–168.
Mukai, T. 1964. The genetic structure of natural populations of *Drosophila melanogaster*. I. Spontaneous mutation rate of polygenes controlling viability. Genetics 50:1–19.
Mukai, T., Chigusa, S.T., Mettler, L.E. and Crow, J.F. 1972. Mutation rate and dominance of genes affecting viability in *Drosophila melanogaster*. Genetics 72:335–355.
Muller, H.J. 1932. Some genetic aspects of sex. Amer. Natur. 66:118–138.
Muller, H.J. 1964. The relation of recombination to mutational advance. Mutat. Res. 1:2–9.

Nash, J.F. 1951. Non-cooperative games. Ann. Math. 54:286–295.
Nee, S. 1989. Antagonistic coevolution and the evolution of genotypic randomization. J. Theor. Biol. 140:499–518.
Nicholson, A.J. 1933. The balance of animal populations. J. Anim. Ecol. 2:132–178.
Nicholson, A.J. and Bailey, V.A. 1935. The balance of animal populations. Proc. Zool. Soc. Lond. 3:551–598.
Nilsson, P., Fagerström, T., Tuomi, J. andÅström, J. 1994. Does seed dormancy benefit the mother plant by reducing sib competition? Evol Ecol. 8:442–430.
Nonacs, P. 1986. Ant reproductive strategies and sex allocation theory. Quart. Rev. Biol. 61:1–21.
Nowak, M. and Sigmund, K. 1993. A strategy of win–stay, lose–shift that outperforms tit-for-tat in the Prisoner's Dilemma game. Nature 364:56–58
Nunney, L. 1989. The maintenance of sex by group selection. Evolution 43:245–257.

Oster, G.F. and Wilson, E.O. 1978. *Caste and Ecology in the Social Insects*. Princeton University Press, Princeton.

Packer, C. 1986. The ecology of sociality in felids. In D.I. Rubinstein and R.W. Wrangham, eds., *Ecological Aspects of Social Evolution*, Princeton University Press, Princeton, pp. 429–451.
Page, R.E. 1986. Sperm utilization in social insects. Ann. Rev. Entomol. 31:297–320.
Pamilo, P., Nei, M. and Li, W.-H. 1987. Accumulation of mutations in sexual and asexual populations. Genet. Res. 49:135–146.
Parker, G.A. 1970. The reproductive behaviour and the nature of sexual selection in *Scatophoga stercoraria* L. (Diptera: Scatophagidae). II. The fertilization rate and the spatial and temporal relationships of each sex around the site of mating and oviposition. J. Anim. Ecol. 39:205–228.
Parker, G.A. 1983. Arms races in evolution—an ESS to the opponent-independent costs game. J. Theor. Biol. 101:619–648.
Parker, G.A., Baker, R.R. and Smith, V.G.F. 1972. The origin and evolution of gamete dimorphism and the male–female phenomenon. J. Theor. Biol. 36: 529–533.
Parker, G.A. and Courtney, S.P. 1983. Seasonal incidence: adaptive variation in the timing of life history stages. J. Theor. Biol. 105:147–155.

Parker, G.A. and Hammerstein, P. 1985. Game theory and animal behaviour. In P.J. Greenwood, P.H. Harvey and M. Slatkin, eds., *Evolution: Essays in Honour of John Maynard Smith*, Cambridge University Press, Cambridge, pp. 73–94.

Parker, G.A. and Maynard Smith, J. 1987. The distribution of stay times in *Scatophaga*: reply to Curtsinger. Amer. Natur. 129:621–628.

Parker, G.A. and Sutherland, W.J. 1986. Ideal free distributions when individuals differ in competitive ability: phenotype-limited ideal free models. Anim. Behav. 34:1222–1242.

Partridge, L. and Barton, N.H. 1993. Optimality, mutation and the evolution of ageing. Nature 362:305–311.

Peck, J.R. 1994. A ruby in the rubbish: beneficial mutations and the evolution of sex. Genetics 137:597–606.

Perrins, C.M. and Moss, D. 1975. Reproductive rates in the Great Tit. J. Anim. Ecol. 44:695–706.

Pettifor, R.A., Perrins, C.M. and McCleery, R.H. 1988. Individual optimization of clutch size in Great Tits. Nature 336:160–162.

Pollard, E., Lakhani, K.H. and Rothery, P. 1987. The detection of density-dependence from a series of annual censuses. Ecology 68:2046–2055.

Pomiankowski, A.N. 1987. The costs of choice in sexual selection. J. Theor. Biol. 128:195–218.

Pomiankowski, A.N. 1988. The evolution of female mate preferences for male genetic quality. Oxford Surv. Evol. Biol. 5:136–184.

Poundstone, W. 1992. *Prisoner's Dilemma*. Doubleday, New York.

Price, M.V. and Waser, N.M. 1982. Population structure, frequency-dependent selection, and the maintenance of sexual reproduction. Evolution 36:35–43.

Price, T. and Liou, L. 1989. Selection on clutch size in birds. Amer. Natur. 134: 950–959.

Queller, D.C. 1992. A general model for kin selection. Evolution 46:376–380.

Queller, D.C. and Goodnight, K.F. 1989. Estimating relatedness using genetic markers. Evolution 43:258–275.

Redfield, R.J. 1994. Male mutation rates and the cost of sex for females. Nature 369:145–147.

Reid, W.V. 1987. The cost of reproduction in the glaucous-winged gull. Oecologia 74:458–467.

Rice, W.R. 1994. Degeneration of a nonrecombining chromosome. Science 263:230–232.

Roff, D.A. 1984. The evolution of life history parameters in teleosts. Can. J. Fish. Aquat. Sci. 41:989–1000.

Roff, D.A. 1992. *The Evolution of Life Histories*. Chapman and Hall, London.

Rose, M.R. 1991. *Evolutionary Biology of Ageing*. Oxford University Press, New York.

Rose, M.R. and Charlesworth, B. 1980. A test of evolutionary theories of senescence. Nature 287:141–142.

Roughgarden, J. 1979. *Theory of Population Genetics and Evolutionary Ecology: An Introduction*. Macmillan, New York.

Ryan, M.J., Fox, J.H., Wilczynski, W. and Rand, A.S. 1990. Sexual selection for sensory exploitation in the frog *Physalaemus pustulosus*. Nature 343:66–67.

Sandefur, J.T. 1990. *Discrete Dynamic Systems: Theory and Applications*. Oxford University Press, Oxford.

Schaffer, W.M. 1984. Stretching and folding in lynx fur returns: evidence for a strange attractor in nature. Amer. Nat. 124:798–820.
Schaffer, W.M. and Kot, M. 1986. Chaos in ecological systems: the coals that Newcastle forgot. Trends Ecol. Evol. 1:58–63.
Seger, J. 1983. Partial bivoltinism may cause alternating sex-ratio biases that favour eusociality. Nature 301:59–62.
Seger, J. 1985. Unifying genetic models for the evolution of female choice. Evolution 39:1185–1193.
Seger, J. 1988. Dynamics of some simple host–parasite models with more than two genotypes in each species. Phil. Trans. Roy. Soc. Lond. B 319:541–555.
Seger, J. 1991. Cooperation and conflict in social insects. In J.R. Krebs and N.B. Davies, eds., *Behavioural Ecology: An Evolutionary Approach*, 3rd Ed., Blackwell, Oxford, pp. 338–373.
Seger, J. and Brockmann, H.J. 1987. What is bet-hedging? Oxford Surv. Evol. Biol. 4:182–211.
Seger, J. and Hamilton, W.D. 1988. Parasites and sex. In R.E. Michod and B.R. Levin, eds., *The Evolution of Sex*, Sinauer, Sunderland, MA, pp. 176–193.
Selten, R. 1980. A note on evolutionarily stable strategies in asymmetric animal conflicts. J. Theor. Biol. 84:93–101.
Sherman, P.W., Jarvis, J.U.M. and Alexander, R.D., eds. 1990. *The Biology of the Naked Mole Rat*. Princeton University Press, Princeton.
Slagsvold, T., Lifjeld, J., Stenmark, G. and Briehagen, T. 1988. On the costs of searching for a mate in female Pied Flycatchers *Ficedula hypoleuca*. Anim. Behav. 36:433–442.
Stacey, P.B. and Koenig, W.D., eds. 1990. *Cooperative Breeding in Birds*. Cambridge University Press, Cambridge.
Stearns, S.C., ed. 1987. *The Evolution of Sex and Its Consequences*. Birkhäuser, Basel.
Stearns, S.C. 1992. *The Evolution of Life Histories*. Oxford University Press, New York.
Stephens, D.W. and Krebs, J.R. 1986. *Foraging Theory*. Princeton University Press, Princeton.
Stinson, C.H. 1979. On the selective advantage of fratricide in raptors. Evolution 33:1219–1225.
Strogatz, S.H. 1994. *Nonlinear Dynamics and Chaos*. Addison-Wesley, Reading, MA.
Strong, D.R. 1986. Density-vague population change. Trends Ecol. Evol. 1:39–42.
Sugihara, G. and May, R.M. 1990. Nonlinear forecasting as a way of distinguishing chaos from measurement error in time series. Nature 344:734–741.
Sundström, L. 1994. Sex ratio bias, relatedness asymmetry and queen mating frequency in ants. Nature 367:266–268.
Sutherland, W.J. 1983. Aggregation and the 'ideal free' distribution. J. Anim. Ecol. 52:821–828.
Sutherland, W.J. and Parker, G.A. 1985. Distribution of unequal competitors. In R.M. Sibly and R.H. Smith, eds., *Behavioural Ecology*, Blackwell, Oxford, pp. 255–274.
Swaddle, J.P. and Cuthill, I.C. 1994. Preference for symmetric males by female Zebra Finches. Nature 367:165–166.

Taylor, P.D. 1981. Sex ratio compensation in ant populations. Evolution 35:125–1251.
Taylor, P.D. 1988a. Inclusive fitness models with two sexes. Theor. Pop. Biol. 34:145–168.

Taylor, P.D. 1988b. An inclusive fitness model for dispersal of offspring. J. Theor. Biol. 130:363–378.
Taylor, P.D. 1989. Evolutionary stability in one-parameter models under weak selection. Theor. Pop. Biol. 36:125–143.
Taylor, P.D. 1992a. Altruism in viscous populations—an inclusive fitness model. Evol. Ecol. 6:352–356.
Taylor, P.D. 1992b. Inclusive fitness in a homogeneous environment. Proc. Roy. Soc. B 249:299–302.
Taylor, P.D. and Bulmer, M.G. 1980. Local mate competition and the sex ratio. J. Theor. Biol. 86:409–419.
Taylor, P.D. and Jonker, L.B. 1978. Evolutionarily stable strategies and game dynamics. Math. Biosci. 40:145–156.
Tong, H. and Smith, R.L., eds. 1992. *Royal Statistical Society Meeting on Chaos*. J. Roy. Statist. Soc. B 54:301–474.
Trivers, R.L. 1971. The evolution of reciprocal altruism. Quart. Rev. Biol. 46:35–57.
Trivers, R.L. and Hare, H. 1976. Haplodiploidy and the evolution of social insects. Science 191:249–263.
Trivers, R.L. and Willard, D.E. 1973. Natural selection of parental ability to vary the sex ratio of offspring. Science 179:90–92.
Tuljapurkar, S. 1989. An uncertain life: demography in random environments. Theor. Pop. Biol. 35:227–294.
Tuljapurkar, S. 1990. *Population Dynamics in Variable Environments*. Springer-Verlag, New York.
Turchin, P. 1990. Rarity of density dependence or population regulation with time lags? Nature 344:660–663.

Van Groenendael, J., de Kroon, H. and Caswell, H. 1988. Projection matrices in population biology. Trends Ecol. Evol. 3:264–269.
Volterra, V. 1931. *Leçons sur la Théorie Mathématique de la Lutte pour la Vie*. Gauthiers-Villars, Paris.
Von Bortkiewicz, L. 1898. *Das Gesetz der kleinen Zahlen*. Leipzig.

Wade, M.J. 1985. Soft selection, hard selection, kin selection, and group selection. Amer. Natur. 125:61–73.
Watson, P.J. and Thornhill, R. 1994. Fluctuating asymmetry and sexual selection. Trends Ecol. Evol. 9:21–25.
Werren, J.H. 1983. Sex ratio evolution under local mate competition in a parasitic wasp. Evolution 37:116–124.
Wilkinson, G.S. 1984. Reciprocal food sharing in the vampire bat. Nature 308:181–184.
Williams, G.C. 1957. Pleiotropy, natural selection and the evolution of senescence. Evolution 11:398–411.
Williams, G.C. 1966. *Adaptation and Natural Selection*. Princeton University Press, Princeton.
Williams, G.C. 1975. *Sex and Evolution*. Princeton University Press, Princeton.
Williams, G.C. 1985. A defence of reductionism in evolutionary biology. Oxford Surv. Evol. Biol. 2:1–27.
Williams, G.C. 1992. *Natural Selection: Domains, Levels, and Challenges*. Oxford University Press, New York.
Wilson, D.S., Pollock, G.B. and Dugatkin, L.A. 1992. Can altruism evolve in purely viscous populations? Evol. Ecol. 6:331–341.

Woolfenden, G.E. and Fitzpatrick, J.W. 1984. *The Florida Scrub Jay*. Princeton University Press, Princeton.

Yodzis, P. 1989. *Introduction to Theoretical Ecology*. Harper & Row, New York.

Yoshimura, J. and Clark, C.W. 1991. Individual adaptations in stochastic environments. Evol. Ecol. 5:173–192.

Yoshimura, J. and Clark, C.W., eds. 1993. *Adaptation in Stochastic Environments.* Lecture Notes in Biomathematics, Vol. 98. Springer-Verlag, Berlin.

Young, J.P.W. 1981. Sib competition can favor sex in two ways. J. Theor. Biol. 88:755–756.

Young, T.P. 1990. Evolution of semelparity in Mount Kenya lobelias. Evol. Ecol. 4:157–171.

Zahavi, A. 1975. Mate selection—a selection for a handicap. J. Theor. Biol. 53: 205–214.

Zahavi, A. 1977. The cost of honesty. (Further remarks on the handicap principle). J. Theor. Biol. 67:603–605.

Author Index

Subject Index